Subnational Hydropolitics

Subnational Hydropolitics

Conflict, Cooperation, and Institution-Building in Shared River Basins

SCOTT M. MOORE

OXFORD
UNIVERSITY PRESS

OXFORD
UNIVERSITY PRESS

Oxford University Press is a department of the University of Oxford. It furthers the University's objective of excellence in research, scholarship, and education by publishing worldwide. Oxford is a registered trade mark of Oxford University Press in the UK and certain other countries.

Published in the United States of America by Oxford University Press
198 Madison Avenue, New York, NY 10016, United States of America.

Library of Congress Cataloging-in-Publication Data
Names: Moore, Scott, 1985– author.
Title: Subnational hydropolitics : conflict, cooperation, and institution-building
in shared river basins / Scott M. Moore.
Description: New York, NY : Oxford University Press, 2018. |
Includes bibliographical references and index.
Identifiers: LCCN 2018000924 | ISBN 9780190864101 (hardback)
Subjects: LCSH: Water-supply—Case studies. | Water resources development—Case studies. |
Water rights—Case studies. | BISAC: POLITICAL SCIENCE / Public Policy /
Environmental Policy. | NATURE / Natural Resources. | NATURE / Ecology.
Classification: LCC HD1691.M66 2018 | DDC 333.91/62—dc23
LC record available at https://lccn.loc.gov/2018000924

9 8 7 6 5 4 3 2 1

Printed by Sheridan Books, Inc., United States of America

For Marisa, Sheri, and Louise

CONTENTS

PREFACE

Recent decades have seen growing strains placed on the water resources of arid and semiarid areas like the western United States, northern China, and the Middle East, as well as regions previously blessed with an abundance of water, including the southeastern United States, southern India, and even the United Kingdom. In 2015, a survey of global leaders conducted by the World Economic Forum identified water scarcity as the highest-impact risk for world development, while issues like pollution and flooding continue to grow more acute. Amid these challenges, scholars, policymakers, and the public at large have voiced increasing concern that conflict may arise between countries and social groups over shared water resources. These concerns are understandable: no other resource arouses quite so much passion as does water, and its management and governance inherently require navigating contentious ideas of fairness, equity, and access.

Yet of all the potential causes of conflict between peoples and nations, few are as misunderstood as water. Historically, instances of conflict between countries over water have been exceptionally rare. Far more common, and more economically and socially disruptive, are persistent political conflicts and disputes between water users at the subnational level, especially those that involve political jurisdictions like states, provinces, and prefectures. These conflicts, which in this book I call "subnational hydropolitics," have so far received little scholarly attention. This book therefore sets out to provide one of the first comprehensive accounts of the dynamics of conflict and cooperation in shared subnational river basins.

Such an understanding requires taking seriously both the institutional and ideational aspects of water. Basic institutional arrangements for decentralization and federalism shape relationships between water users in ways that have not previously been the subject of much research. At the same time, shared water resources are at the subnational level inextricably linked to ideational issues of identity and fairness that interact with, but are distinct from, institutions and institutionalized relationships. Only by probing both these dimensions of water can the politics of sharing it be completely understood. That is the primary purpose of this book.

The subnational level shows a different side of water conflict. It opens up an arena in which sectional tensions between water users, whether rooted in ethnicity, economics, or culture, intersect with a dense web of institutions and affiliations, including different layers of government, legal systems, and political parties. Bound by these institutions, subnational water users possess several inducements to cooperate—but also several avenues to compete for shared water resources. The ultimate message of this book is that water politics are shaped by both ideas and institutions, and that tackling the world's water resource challenges and preventing conflict over water requires reforming political as well as technical institutions for water resource management. So too is addressing the underlying tensions and grievances that are often magnified by shared water resources.

This book is indebted to several individuals and institutions without whom it would not have been possible. I owe much to the intellectual guidance of Nancy Bermeo, David Rueda, and Iain McLean of Oxford University; and Henry Lee of Harvard University, as well as to David Konisky of Indiana University, who encouraged me to write this book. I am also grateful to Ken Lieberthal of the Brookings Institution, Bill Antholis of the University of Virginia, Jeremy Schmidt of Carleton University, Katy Hansen of Duke University, and Tom Hale and Dustin Garrick of Oxford for providing feedback, advice, and encouragement on various elements of the book. Professor John Waterbury of New York University–Abu Dhabi was moreover generous enough to review the manuscript in its entirety. Jeremy Lewis, Anna Langley, Andrew Pachuta, and the rest of the editorial team at Oxford University Press, as well as Rajesh Kathamuthu at Newgen Knowledge Works, deserve my profound thanks for their support during the production process. Finally, I gratefully acknowledge the support of Merton College, Oxford; the Rhodes Trust; the Brookings Institution; the Belfer Center for Science and International Affairs and the Sustainability Science Program at Harvard University; and the Council on Foreign Relations International Affairs Fellowship program, all of which provided the time, sustenance, and good will which enabled me to write this book.

Scott M. Moore
Bryn Mawr, Pennsylvania
Fall 2017

Introduction

Subnational Hydropolitics

One of the lesser-known insurrections in American history occurred in 1934, when Arizona Governor B. B. Moeur declared martial law and deployed National Guardsmen armed with machine guns to prevent construction of Parker Dam, a project supported by both Congress and the Roosevelt administration. Arizona's troops ashore were accompanied by a specially assembled group of small boats, quickly termed the "Arizona Navy," that patrolled the waters of the Colorado River near the proposed dam's construction site. The threat posed by the dam appeared to be crystal clear to Moeur and his fellow rebels: if Parker Dam was to be completed, California, Arizona's downstream neighbor on the Colorado River, might secure in perpetuity preferential rights to the river's waters, leaving too little for Arizona to satisfy its own growing needs.

Moeur's rebellion is one of the more dramatic illustrations of conflict over water that occurs within countries instead of between them. Most writing and thinking about water conflict concerns the prospect of warfare between nation-states. But while the difficulties of securing cooperation on international transboundary rivers are relatively well known and understood, Moeur's rebellion highlights the distinctly different problem of preventing conflict on rivers shared by multiple subnational political jurisdictions, including states, provinces, prefectures, and governorates. Indeed, the problem of subnational cooperation is even more pervasive than that of international cooperation, for while many rivers are shared between countries, nearly all are shared between multiple subnational units. At the same time, even as scholars and policymakers devote growing attention to improving cooperation between countries that share common water resources, many waterways remain mired in protracted, acrimonious disputes between lower-level jurisdictions. This state of contention, which I call "subnational hydropolitics," is often thought of as an isolated phenomenon—the result of unique historical tensions between the states of the Colorado or Murray-Darling River basin, for example. But it is in fact a systemic challenge for waterways across the globe, with common sources of conflict—as well as common solutions.

This book describes common patterns of subnational conflict and cooperation over water in four countries, the United States, India, China, and France, each chosen to represent a different form of subnational water conflict. In doing so, it attempts to redress the fact that while entire library sections are devoted to understanding international water conflict, few studies have addressed the distinctive issue of subnational jurisdictional conflict and cooperation over water. This inattention represents a serious lacuna: across the four countries examined in this book, the problem of securing cooperation between subnational jurisdictions over shared water resources rivals, and sometimes exceeds, that between countries. Apart from its potential to foster domestic political instability or even violence, interjurisdictional water conflict imposes considerable economic and environmental costs, impeding economic development in shared river basins and hindering attempts to protect riparian ecosystems. Moreover, subnational water conflict is a problem that is likely to only grow more acute as a result of population growth, climate change, and other pressures in the world's major river basins.

Fortunately, the cases examined in this book reveal common features that can help scholars and policymakers better understand and prevent conflict at the subnational as well as international levels. The dynamics of conflict and cooperation between subnational jurisdictions are shaped by the interaction of three factors: decentralization, sectional identity, and political opportunity structure, which enables nongovernmental organizations and interests to participate in the political process. Decentralization largely explains why subnational water conflicts arise, while sectional identity explains why they persist over time, often despite multiple attempts to resolve them. Political opportunity structure, finally, explains how subnational hydropolitics can become more or less conflictual over time. This theory of conflict and cooperation over water departs from existing ones in suggesting that ideational factors like sectional identity are just as important as institutional ones. Water is highly emotive; it reflects and refracts tensions between ethnic, religious, and other social groups that share it. Where institutions reinforce sectional divisions between social groups, whether over race, language, or ethnicity, shared water resources can both reflect and magnify intergroup conflicts and disputes. Conversely, where institutions promote the inclusion of third parties that bridge divides between water users, cooperation can flourish. The theory of subnational hydropolitics developed in this book helps to explain why subnational water conflict occurs—but also that it is far from inevitable. The real political salience of water lies not in its capacity to incite warfare or violence but rather in its ability to mobilize key actors either to promote collective action or to channel sectional discord. Shared water resources are a stage on which these deeply political dramas unfold.

This understanding of water conflict and cooperation within countries carries three broad implications for the study of environmental and water politics. First, it shows that the difference between international and subnational hydropolitics is one of degree rather than kind. While the existence of shared institutions like political parties, laws, and judiciaries provides more potential avenues to prevent interjurisdictional conflict, it does not do so automatically. On the contrary, in many countries sectional tensions are amplified by institutional arrangements for decentralization. Where this is the case, interjurisdictional collective action problems within countries can be just as complex as those of international transboundary river basins. Second, this new understanding of hydropolitics suggests that ideational factors are just as important as institutional ones in causing water conflict. Shared natural resources can be a powerful locus for sectional identities and grievances, often with grave consequences for conflict and cooperation. Third, and perhaps of broadest relevance, the cases examined in this book indicate that the cooperative management of shared natural resources requires local stakeholder participation—but not too much. Interjurisdictional collective action works best within institutional frameworks that balance national and subnational interests, as well as those of resource users and third parties. Despite the focus of much scholarly attention on localized collective action, the cases examined in this book show that subnational leaders are often the instigators of resource conflicts, while national political elites are crucial champions of cooperation.

THE PUZZLE OF WATER CONFLICT: MORE PENS THAN SWORDS

Lying at the heart of the study of water conflict is a fundamental puzzle: the voluminous literature on the subject has so far struggled to articulate or identify a clear relationship between water and warfare. In the early 1990s, a senior World Bank official famously predicted that "the wars of the next century will be fought over water," while in 2014 Indian strategist Brahma Chellaney made a splash by claiming that the continent's shared river basins would become "Asia's New Battleground" (Kramer et al. 2013; Chellaney 2014). But it has not quite turned out that way. The world has so far avoided interstate warfare over water in the twenty-first century. On the other hand, it has witnessed many more localized conflicts over water, including between members of different tribes, agriculturalists and urban-dwellers, and states and provinces. This book focuses on this last form of subnational water conflict, namely political disputes that occur between first-tier subnational administrative jurisdictions over shared water resources. It does so because although these interjurisdictional conflicts represent only one form of water conflict, they hold up a mirror to the wider question of what causes it. While subnational jurisdictions are subject to a single sovereign government, the dynamics of water conflict and cooperation between them closely resemble those of nation-states operating in the absence of higher authority. Explaining how and why subnational water conflicts arise and persist therefore promises to help scholars and policymakers better understand water conflict in general—as well as how it can be prevented.[1]

The significance of subnational water conflict is heightened by the fact that it appears to be considerably more common than the international variety. Despite the conventional wisdom that sharing water poses a high risk of war between countries, the most comprehensive effort to quantify water-related conflict, the International Water Event Database, which includes some 6,400 water-related events from 1948 to 2008, identified no cases of outright warfare and fewer than thirty instances of interstate violence of any kind (De Stefano et al. 2010). Other large-scale empirical assessments bolster this conclusion by finding only a weak association between shared water resources and measures of both violent and nonviolent international conflict (Toset, Gleditsch, and Hegre 2000; Gleditsch et al. 2006; Brochmann and Gleditsch 2012). Qualitative case study assessments have moreover cast doubt on the role of water in conflicts like the 1967 Six Day War, often alleged to have been triggered by an attempt to divert the waters of the Jordan River (Shaheen 2000). Instead of conflict, thousands of cooperative agreements exist to govern international transboundary rivers, even in river basins subject to acute water scarcity (Wolf 1995; Stinnett and Tir 2009; Tir and Stinnett 2012). Furthermore, where conflict does arise between countries over shared water resources, parties often settle differences by agreeing to cooperate in other areas (Uitto and Duda 2002; Delli Priscoli and Wolf 2009; Susskind and Islam 2012). Contrary to popular belief, therefore, international water politics appear to be defined by both the rarity of conflict and the prevalence of cooperation.

In contrast, subnational water conflict appears to be more widespread. The available evidence suggests that if water conflict is surprisingly rare between countries, it is puzzlingly common within them. In the late 1990s, a widely cited study of water conflict observed that "while no 'water wars' have occurred . . . what we seem to be finding, in fact, is that geographic scale and intensity of conflict are inversely related" (Yoffe and Wolf 1999, 201). Subsequent research has bolstered this conclusion by documenting extensive water-related conflict between social groups, communities, and regions at the subnational level (Bernauer, Bohmelt, and Koubi 2012; Bernauer, Bohmelt, Buhaug, et al. 2012). Recent work in political economy has similarly demonstrated linkages between "water shocks" like droughts and floods and various forms of conflict. Dry years appear to drive a spike in land invasions in rural Brazil, for example, while in India communal riots become more commonplace following floods, and property-related violence increases during periods of below-average rainfall (Hidalgo et al. 2010; Hsiang and Burke 2014; Sarsons 2015). Yet even in these cases, the precise causal link between water and conflict remains devilishly difficult to determine. It is not for nothing that the relationship between water and conflict has been referred to as a "conceptual minefield" (Dimitrov 2002).

The very puzzle embodied in the fact that subnational water conflict appears to be more common nonetheless presents an opportunity: while the extensive literature on the prospect of water wars has so far struggled to identify direct pathways from water to warfare, the causes and consequences of subnational water conflict represent more uncharted theoretical and empirical terrain. The purpose of this book is accordingly to describe and explain the causes of conflict and cooperation between subnational administrative units in a limited number of carefully selected shared river basins. Based on this case study analysis, I suggest that persistent interjurisdictional conflicts occur when subnational politicians in decentralized political systems link water to preexisting sectional tensions, especially those concerning ethnicity, language, or geographic identity. In combination, these sectional identities and political institutions can create interjurisdictional collective action dilemmas just as acute as those which afflict international transboundary river basins. Conversely, durable interjurisdictional cooperation occurs when subnational officials are induced by third parties like nongovernmental organizations to create collective action institutions, including watershed associations and river basin management organizations, that restrain the temptation to manipulate sectional tensions over water. Countries at risk of subnational water conflict must therefore create interjurisdictional institutions that transcend rather than channel sectional discord.

Given the relative paucity of academic work on the subject, the objective of this book is to propose an initial theory of subnational hydropolitics that can be further tested and refined against additional case studies. The analysis of subnational water conflict presented here should accordingly be viewed as initial rather than definitive. But, as I elaborate in the remainder of this book, this theory of subnational hydropolitics highlights the previously underappreciated problem of securing cooperation between political jurisdictions below the nation-state level. It also points the way toward a mature conceptual understanding of the politics of water, perhaps the world's most critical natural resource, at both national and subnational levels of political organization. In this book, I use the term *subnational hydropolitics* to reflect the belief that the use of shared water resources is inherently shaped by politics at the subnational level as well as the international level. But I do so in the equally strong belief that the study of water politics must move beyond the determinism embodied in traditional hydropolitical theory. Water conflict is not simply the product of one jurisdiction's geographic position up- or downstream of another. The most relevant geographies are sectional rather than riparian. Upstream and downstream riparian relations, whether cooperative or conflictual, are shaped by a complex constellation of subnational political actors, all with their own conception of their interests in a shared river basin. Moreover, sectionalism interacts with two other aspects of political power, namely decentralization and political opportunity structures, which creates the space for cooperation as well as conflict over shared water resources. Accordingly, the study of subnational hydropolitics attempts to explain how political contests create not only competition but also collaboration (Gerlak, Heikkila, and Lubell 2008).

This book poses several primary lessons for understanding and preventing water conflict. First, the underlying cause of the cases of water conflict examined in this book is not physical geography or water scarcity, as is often alleged, but rather sectional identity. As we shall see in Chapter 1, while water shortages, pollution, flooding, and other issues can provide the spark for water conflict, deeper tensions between social groups are the fuel. As this chapter describes in detail, the traditional focus of hydropolitical theory on riparian geography, and whether a jurisdiction lies up- or downstream of another, is largely misplaced. Riparian geographies are given life by political opportunists, especially subnational leaders who seize on underlying tensions between sectional groups to build cross-party political support within their jurisdictions. Second, the structure of decentralization plays a much more important role in water resource management than is commonly assumed, and largely determines institutional capacity for interjurisdictional and intersectoral coordination. While many different types of institutions bear on conflict and cooperation, including water rights systems, laws, and policies, decentralization itself shapes these different arrangements and therefore acts as a kind of macrovariable that influences other factors. Indeed, as Chapter 2 describes, decentralization is a crucial enabling condition of subnational water conflict. Without a significant degree of decentralization, there is little incentive or ability for subnational leaders to engage in water conflict.

Third, finally, and perhaps in greatest contrast to the prevailing scholarly consensus, it is not the quantity of institutional arrangements that matters for cooperation over water, but rather their quality. Collaborative, participatory, and adaptive interjurisdictional institutions do not develop easily, and only rarely on their own. Instead, durable interjurisdictional collective action requires the involvement of third parties, especially nongovernmental organizations, to induce cooperation between subnational jurisdictions. To be clear, interjurisdictional water cooperation need not *always* stem from third-party involvement. Individual subnational politicians may occasionally agree to cooperate with their counterparts on an ad hoc basis. But the key issue is whether this cooperation outlasts individual actors and proves to be durable over time. As Chapter 3 further describes, third-party involvement is almost always necessary to develop robust institutions that sustain such cooperation over the long term. Before presenting the analysis that gives rise to these conclusions, the remainder of this Introduction lays out and defines the key concepts of decentralization, sectionalism, and political opportunity structure, and then progresses to a description of methods and the structure of the book.

THE PERVASIVE PROBLEM OF SUBNATIONAL COOPERATION OVER WATER

The point of departure for this book is the realization that despite the overwhelming concentration of both scholars and policymakers on the international level, the problem of interjurisdictional conflict and cooperation over shared water resources is even more pronounced at the subnational level. While most river basins are shared between multiple countries, the realities of spatial scale mean that nearly all are shared between multiple subnational administrative jurisdictions. These jurisdictions include municipalities, counties, indigenous areas, water districts, and other political units. But the problem of interjurisdictional cooperation over water is illustrated most vividly by the extent to which most river basins are shared between multiple "first-tier" jurisdictions like states and provinces. A basic overlay of subnational and river basin boundaries across the globe, depicted in Figure I.1, shows that only two major river basins, the Kuskokwim in Alaska and the Skeena in British Columbia, lie entirely within a single, first-tier subnational jurisdiction.[2] By far the majority of the world's other river basins are shared between multiple subnational political jurisdictions, including many that possess considerable autonomy and powers distinct from that of their parent nation-state. According to another data set, 319 rivers, including half of the world's international rivers, lie at least partly within a federal country (Garrick et al. 2013). These basins are subject to subnational, as well as international, dynamics of conflict and cooperation.

Interjurisdictional water conflict differs in several important respects from other forms of conflict, including those that occur between countries or communal groups. First, and most fundamentally, interjurisdictional water conflicts only

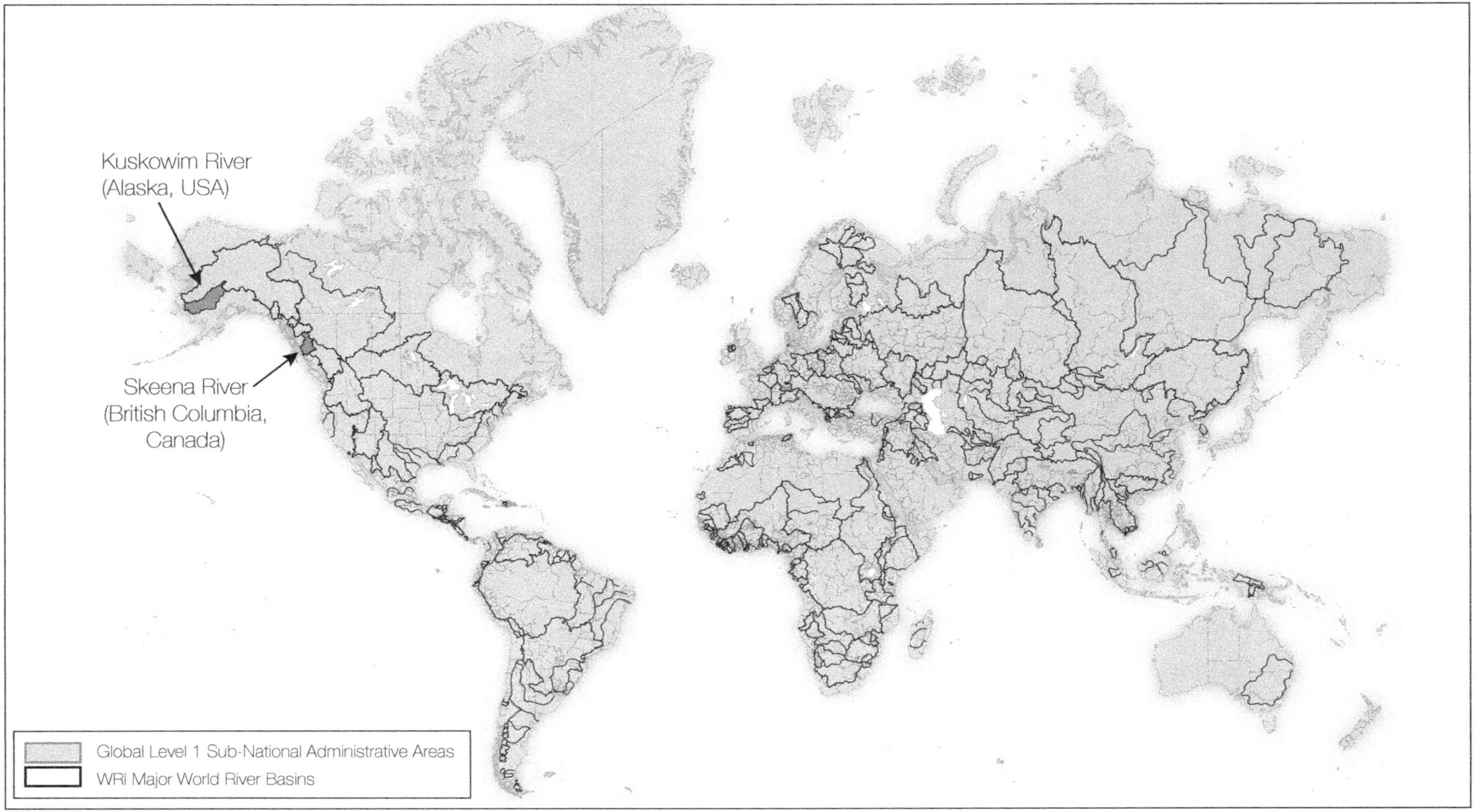

Figure I.1 Global Discontinuities Between River System and Administrative Boundaries

occasionally involve violence. Unlike nation-states, which can usually call upon diplomatic services and armed forces, subnational jurisdictions rely on less weighty means to pursue disputes, especially legislative and judicial maneuvers. However, as we shall see in Chapter 3, this competition is often sustained by underlying sectional tensions, which hinder interjurisdictional cooperation. Second, unlike forms of intercommunal conflict, interjurisdictional water conflict takes place between subnational political entities that enjoy some measure of autonomy from the central government, as well as a defined and discrete territory. As we shall see in Chapter 2, these "type I" territorial jurisdictions play a crucial role in water politics because they often serve as the basis for political representation and the implementation of nationwide policies. Indeed, as Chapter 3 details, the representative functions of subnational jurisdictions render them a potent platform from which communal tensions over water or other shared resources can be amplified. Third, unlike nation-states, subnational political actors are bound by a web of formal institutional relationships to a single sovereign government, as well as to each other. Subnational entities can therefore employ a broad range of strategies to pursue their individual interests, including lobbying, legal action—and outright confrontation. These distinctive features make interjurisdictional water conflict in many ways the most complex and conceptually interesting form of subnational water conflict.

Unfortunately, no consistent cross-national data exist on the occurrence of interjurisdictional water conflicts, so it is difficult to say how prevalent they are with respect to other forms of international or subnational water conflict. However, the available evidence shows that interjurisdictional water conflicts are prevalent in many countries of major political and economic importance, and often impose considerable costs. During a few months of 2016 alone, India's Karnataka state suffered an estimated US$3.75 million in damages following a spate of violence related to a long-running dispute over the Cauvery River (Seetharaman, Irfan, and Aravind 2016). Interjurisdictional water conflicts are often nonviolent, however, and take the form of lawsuits, competition for legislative influence and public support, and various forms of protest. As we shall see in Chapters 4 and 5, respectively, notable examples of long-running interjurisdictional conflicts over water include the Colorado River basin in the United States, which has endured a "long and bitter feud" between the seven states that share its waters (Getches 1984–1985), as well as almost all of India's major rivers, so much so that a senior water resource official has warned that interstate "hydro-politics is threatening the very fabric of federalism" (Menon 2003). Similar cases of protracted interjurisdictional conflict have been described in countries as diverse as Pakistan, Australia, and Nigeria (Garrick et al. 2014). The existence of cases in such a wide range of countries suggests that, at a minimum, interjurisdictional water conflict represents a more prevalent form of water politics than has previously been recognized by most scholars and policymakers.

Perhaps just as important, the problem of achieving collective action among subnational political jurisdictions has grown more acute in light of a persistent global trend toward political and economic decentralization. Since the 1970s, many formerly centralized states such as Denmark, Indonesia, and Uganda have devolved powers in most policy areas to local governments (Hooghe and Marks 2003). Even countries like France whose political cultures remain highly-centralized have devolved considerable fiscal and administrative powers to local levels. At the same time, many already decentralized countries have transferred additional powers to subnational levels, so much so that "Few spaces around the world have remained untouched" by this global trend (Rodriguez-Pose and Gill 2003, 343). This widespread process of decentralization has invested subnational political units, and particularly first-tier jurisdictions like states and provinces, with a far greater range of responsibilities, often accompanied by correspondingly increased access to financial, technical, and administrative resources. This combination of autonomy and capacity serves, as we shall see, to provide the means for subnational jurisdictions to engage in water conflict. It is no exaggeration to say that, in some countries, the capacity of subnational jurisdictions to compete for water rivals that of their parent nation-states.

Apart from their empirical prevalence, interjurisdictional hydropolitics also present three important conceptual issues for the broader study of conflict and cooperation over water. First, the prevalence of interjurisdictional conflict challenges the common presumption that resources shared between countries are more difficult to govern effectively. Scott Barrett neatly summarizes this view: "Resources that lie entirely within a nation's territorial borders can be effectively managed; [while] shared resources are prone to overuse" (Barrett 2003, 33). The apparent commonality of interjurisdictional water conflict suggests that, on the contrary, the existence of a single sovereignty is no guarantee of cooperation over shared natural resources. A recent statement by the chief minister of the Indian state of Goa in response to a dispute with Karnataka, its upstream neighbor on the Mahadayi River, illustrates the extent to which subnational water conflict rivals that which occurs between nation-states: "In the age of globalization, even two countries cannot be inimical to each other. It is not right . . . Karnataka and Goa are two states in the same country." Yet with this very statement, Goa's chief minister proceeded to halt bus service to its neighbor to protest Karnataka's attempt to divert water from the Mahadayi, the subnational equivalent of sealing the border (IANS 2016).

The second, closely related theoretical issue raised by interjurisdictional water conflict is whether the existence of shared institutions alone is sufficient to achieve cooperation between resource users. The idea that common institutions facilitate cooperation is fundamental to the natural resource management and environmental governance literature (Ostrom 2005). But interjurisdictional water conflict suggests that the existence of shared institutions may be merely necessary, but not sufficient, to prevent conflict. Unlike its international equivalent, subnational conflict over water occurs between actors bound by a web of institutions and relationships, including political parties and social networks, all of which might be expected to compel or promote cooperation, on the one hand, and sanction conflict, on the other. The waters of the Colorado River in the United States and the Krishna River in southern India are both apportioned by interstate agreements, for example, but have been the site of long-running interjurisdictional disputes that have defied multiple attempts at resolution by the central government. The failure of existing national institutions to prevent such conflict cries out for explanation.

A third, equally intriguing question arises from the fact that, as the examples of the United States and India suggest, interjurisdictional water conflict appears considerably more prevalent in some countries than others, and even within single river basins the intensity of conflict can shift markedly over time. In river basins like the Colorado and Murray-Darling, the overall level of cooperation between subnational jurisdictions has improved over time. A final issue arises from the fact that, as these examples further suggest, interstate river basin disputes feature prominently in the politics of large federal countries like the United States, India, and Australia, while they are less readily discernible in more centralized countries like China, South Africa, and France. These different forms of decentralization constitute one of the most fundamental ways in which political systems vary from one another, but they are rarely the subject of attention from scholars of environmental politics (Meadowcroft 2016). However, as we shall see in Chapters 1 and 2, decentralization plays a crucial role in subnational hydropolitics. A common feature of the case studies explored in this book is the dominant role of these jurisdictions, despite and often in opposition to the attempts of central states to exert greater influence. Indeed, subnational states often frustrate the efforts of central governments to promote reform, not vice versa as is usually assumed.

A coherent theory of subnational hydropolitics must address each of these issues. It must explain first why subnational cooperation can in some cases be just as difficult to achieve as international cooperation; second, how conflict arises and persists despite the existence of shared institutions, many explicitly designed to prevent it; and third, why the intensity of subnational conflict over water varies across countries and in some cases changes over time. My answer to these questions relies on the interaction of ideational and institutional factors. As we shall see in Chapter 1, the economic and ideational importance of water makes it a powerful agent of political mobilization, and can under some circumstances make interjurisdictional water conflicts even more intractable than international ones. At the same time, institutional arrangements for decentralization can impede interjurisdictional collective action by haphazardly dividing responsibilities between levels and branches of government. Consequently, as Chapter 2 describes, many of the mechanisms that might be expected to help prevent interjurisdictional conflict, like partisan affiliation or fiscal dependence on the central government, in practice often fail to contain conflict over water. In some basins, however, third-party actors like national political leaders or environmental groups are able to build alliances to induce subnational jurisdictions to build collaborative, participatory, and adaptive interjurisdictional institutions, thereby resolving and preventing water conflicts. But as Chapter 3 explains, the ability of third parties to form coalitions and influence decisions varies widely across countries, and their influence often shifts over time. In combination, these factors explain not only why subnational units fight over water just as countries sometimes do but also how they do so, and why in some cases more than others.

METHODS, DEFINITIONS, AND CASE SELECTION

In the previous section, I noted that subnational water conflict raises three important conceptual questions, namely why it appears to be just as difficult to resolve as at the international level, why existing institutional arrangements sometimes fail to prevent or resolve it, and why it appears to be more common in some countries than others. My methodological approach in this book attempts to address each of these questions through the analysis of several carefully selected case studies. This approach is more suitable than quantitative analysis because, given the paucity of available data and theory, the objective of this book lies more in theory-building than theory-testing. Detailed case study analysis provides a means to weigh the many factors that bear on subnational hydropolitics and to be sensitive to the geographic, historical, social, and other contexts that frame conflict, cooperation, and institution-building in individual shared river basins (Batterbury and Ferdando 2006). This approach nonetheless carries some significant limitations. It is extremely challenging to account for the many differences between countries. However, the historical case study approach helps to reduce the risk of identifying spurious causal factors by enabling them to be assessed across time (King, Keohane, and Verba 1994).[3] Indeed, although this book is fundamentally a cross-national comparative study, I stress as much as possible comparisons between cases within the same country, as well as different historical periods within individual basin case studies.

Before describing the selection of these case studies, it is necessary to better specify what both conflict and cooperation mean in the context of shared water resources. I define *conflict* as a persistent state of competition over shared water resources, as manifested by legislative maneuvering, legal disputes, and rhetorical rivalry, all of which undermine attempts to manage shared water resources at an interjurisdictional scale. In keeping with the prevailing literature, I define effective *cooperation* in managing shared water resources in terms of three criteria: collaboration, participation, and adaptability. *Collaboration* refers to the practice of making decisions based on shared values and norms of reciprocity (Koontz et al. 2004; Lubell 2004; Sabatier et al. 2005). *Participation*, meanwhile, refers to the inclusion of relevant stakeholders in decision-making. Both practices, the consensus runs, are essential to adapting to changing environmental and social objectives over time (Blatter and Ingram 2001; Ribot 2002; Weber 2003). Decades of research have similarly established that these objectives are best promoted through collective action institutions, which range from formal organizations to informal social norms, and serve to constrain the actions of individuals or groups (Ostrom 2005; Pahl-Wostl et al. 2012; Marshall 2013). To be effective, collective action institutions must also possess sufficient capacity to negotiate putative conflicts between water users and reach management decisions in the context of complexity, ecosystem feedbacks, and pervasive uncertainty (Garrick 2015).

My approach in this book is to examine the development of these collective action institutions at a particular spatial scale, that of the interjurisdictional river basin. Accordingly, I further define *cooperation* in terms of the ability of collective action institutions to operate at a scale sufficient to solve both interjurisdictional and intersectoral coordination problems. To do so, institutions must be able to navigate potential conflicts between the various political jurisdictions that share a river basin as well as various stakeholder groups, including agricultural, industrial, and domestic water users; nongovernmental organizations; and representatives of different functional bureaucracies, including those concerned with water pollution flood control as well as irrigation and water supply. There is no ideal form or structure to accomplish these aims, and river basin collective action institutions can take the form of singular entities like the Tennessee Valley Authority or polycentric networks like linked sets of watershed associations (Teclaff 1996; Imperial 2005; Lankford and Hepworth 2010). Irrespective of their form, the important thing is for collective action institutions to prevent most conflicts in shared river basins even under changing circumstances, and to be able to resolve any that nonetheless arise.

But while many previous observers have pointed to the spatial collective action challenges inherent to interjurisdictional river basins, I emphasize a different underlying factor: sectionalism and sectional identity. Apart from spanning multiple political jurisdictions, river basins are also usually populated by multiple ethnic, religious, linguistic, and cultural groups, as well as different economic water-user groups. These distinct groups are typically concentrated in different geographic regions, creating marked spatial divisions or cleavages. As a general term, *sectionalism* applies to any situation where geographically concentrated social groups are united by a common identity distinct from that of neighboring regions (Gimpel and Schnuknecht 2004). In the case of water, sectionalism therefore implies that water users are motivated primarily by shared regional identity, though this often coincides with shared economic interests and cultural values. Most water conflicts feature some sort of sectional divide, whether between farmers and urban-dwellers, different linguistic groups, or citizens of one state or province and their neighbors. My emphasis on sectionalism is accompanied by an important caveat: it plays a more significant role in some countries than others. As further described in Chapter 3, federal systems like the United States and India give much greater voice to sectional groups and interests than do unitary states, and consequently these countries tend to feature more contentious subnational hydropolitics.

In order to account for this variation, I select individual river basin case studies in four countries: the United States, India, China, and France. These four countries capture a range of outcomes, including some where subnational water conflict is common and others where it is rare, as depicted in Table I.1. Within each country, I select specific case studies of institutional development within individual river basins. These individual cases represent a range of specific institutional outcomes, including sustained interjurisdictional cooperation as well as protracted conflict. Several cases evince periods of both conflict and cooperation, further multiplying variation in outcomes. In the United States, the archetypal federal system, I examine the long-running but now largely resolved interstate dispute in the Colorado River basin, and contrast this with examples of more durable institutionalized collective action, such as the Delaware River Basin Commission. In India, a much more centralized federal system, I examine the similarly protracted interstate dispute over the Krishna River and contrast this with the case of the Damodar Valley Corporation, one of India's only true regional river basin governance institutions. In China, which centralizes political decision-making but decentralizes policy implementation, I examine the Yellow River Conservancy Commission, a uniquely powerful bureaucratic entity that has nonetheless enjoyed only mixed success in resolving conflicts between subnational actors. Finally, in France, I describe the Water Agency system, a network of regional river basin institutions that has proven highly successful in preventing conflict in a centralized and vertically integrated state.

The analysis of individual case studies attempts to identify specific common factors that lead to cooperative and noncooperative outcomes across countries. The case of the United States features a wide diversity of institutional forms for river basin governance, as well as a range of outcomes in the intensity of conflict between subnational jurisdictions. Accordingly, the U.S. case facilitates assessment of how institutional diversity affects the propensity for such conflict. In addition, because the Colorado is an international river, the U.S. case also sheds light on the significance of international versus subnational conflict dynamics. The Indian case, meanwhile, presents a particular puzzle in that, unlike the United States, its constitution specifically enumerates powers concerning river basin governance at both the central and subnational levels. These powers have nonetheless been sparingly exercised, even as subnational conflict has intensified. Moreover, because its Constitution enables the central government to redraw state boundaries, the case of India enables assessment of how the reorientation of political geography within a river basin influences conflict and cooperation. Several new states were created during the course of the Krishna basin case study, for example, illustrating how geographic position influences their behavior.

The case of China, on the other hand, presents an opportunity to isolate the role of civil society and popular participation in shaping water conflict and cooperation. While China features several regional river basin governance institutions, its authoritarian system severely constrains nongovernmental participation. China's distinctive combination of political centralization and fiscal decentralization also enables an assessment of how political control and single-party rule influence the behavior of subnational jurisdictions even under conditions of economic and practical autonomy. Finally, the French case illustrates the role of technocratic elite decision-making and of a highly integrated, centralized political system and culture. In France, water resource management has historically been dominated by engineering bureaucracies, and its political institutions and customs closely link officials at central and local levels. At the same time, however, France is one of the few countries to feature strong governmental structures at the regional scale, providing an opportunity to assess their implications for conflict and cooperation in shared river basins. But while each case study adds leverage to the motivating question of why interjurisdictional water conflict arises and persists, the U.S. and Indian country cases serve a particularly important "theory-building" role in my analysis because they feature a higher level of interjurisdictional conflict.

This research design borrows heavily from the comparative historical analysis tradition within comparative politics and shares its concern for how fundamental political institutions, especially the allocation of political power, shape outcomes over time (Skocpol 1979; Thelen and Steinmo 1992; Amenta 2009). In particular, this design employs detailed, multidecadal historical case studies to describe, analyze, and explain how competing demands on shared waterways have been accommodated by collective action institutions in interjurisdictional river basins. This approach emphasizes the development and function of institutions over time, and also strives to consider the multiple social, economic, and environmental factors that affect them. Case studies are constructed using a wide range of evidence, including public and legislative records, archival material, official reports, biographies, and newspapers. This evidence was gathered during periods of fieldwork in Beijing, New Delhi, and other parts of both China and India. This sample of countries is of course limited, and the conclusions to be gleaned from it must accordingly be qualified. Even so, the implications of the sample itself are potentially wide-ranging, for they question the overall understanding of water conflict that has hitherto guided both scholars and policymakers, especially the assumption that international rather than subnational conflict poses the greater challenge. In the next section, I sketch the structure of the book and preview the case studies that give rise to these tentative conclusions.

STRUCTURE AND PLAN OF THE BOOK

The remainder of this book is organized into two main parts. The first part consists of three theoretical chapters which lay out a conceptual framework for understanding interjurisdictional conflict and cooperation over water, and set the stage for the case studies that make up the second, empirical portion of the book. The three conceptual chapters describe in turn the three elements of the theory of subnational hydropolitics presented above, namely decentralization, sectional identity, and political opportunity structure. The first chapter, "Deconstructing the State," attempts to reorient hydropolitical theory away from its traditional fixation with both international relations and upstream–downstream riparian geographies toward a focus on sectional geographies and divides. This chapter also explains why states face inherent constraints on managing water within their boundaries, and presents an initial survey of subnational hydropolitics around the globe. Chapter 2 explains how, contrary to much existing theory, certain aspects of decentralization can impede cooperation and collective action among subnational jurisdictions. These gaps and weaknesses explain why, in many cases, existing institutions fail to prevent or resolve interjurisdictional water conflict. Chapter 3 examines the primary determinants of conflict and cooperation over water, namely sectional identity and political opportunity structure. It explains how water can become a locus for sectional tensions and how subnational officials often gain political advantage from competing with neighboring jurisdictions for access to and control of shared water resources. This salience explains why interjurisdictional water conflict can sometimes rival its international equivalent in intensity and duration. Chapter 3 also describes how, in contrast, third parties create a constituency for building strong interjurisdictional institutions that reduce the risk of conflict. Crucially, the ability of third parties to influence water resource management varies across countries depending on the political opportunity structure. This variation helps to explain why water conflicts arise and persist between some subnational jurisdictions, and why cooperation emerges in others.

Readers less interested in these conceptual issues may wish to proceed directly to the case studies that comprise the second section of the book. This second section contains four chapters, each addressing one of the four countries selected in the methods section above. Chapters 4 and 5, which describe cases in the United States and India, respectively, are longer and more detailed than Chapters 6 and 7, which pertain to China and France, respectively. This is because, as depicted in Table I.1, the United States and India serve a theory-building function and require more elaborate and in-depth treatment. Because they feature different levels of development and institutional capacity but a similarly high level of interjurisdictional water conflict, the United States and India present an opportunity to isolate common factors that explain why interjurisdictional conflicts emerge. To provide additional leverage in isolating causal factors, I also examine secondary case studies in each country which represent more cooperative outcomes. I then further contrast the U.S. and Indian cases against those of China and France in Chapters 6 and 7, respectively. Because these two cases serve a theory-testing function, I examine them in comparatively less detail. Readers primarily interested in a particular country may wish to proceed directly to the relevant chapter.

Each case study follows a uniform structure. First, the applicable political actors in each case are described, along with their basic roles in water resource management. Second, institutional arrangements for both decentralization and water resource management are outlined, with emphasis on how the structure of central–local relations influences civil society participation and interjurisdictional collective action. These features of each country case are summarized in a modified form of Table I.2. The table summarizes both the institutional landscape in each country case as well as sectional cleavages and third-party political opportunity structures. Together, these features characterize the dynamics of water conflict and cooperation in each country case. Third, an introduction is given to the individual case study or studies examined in each chapter, including a summary of the case and, where relevant, a synopsis of

Table I.1 Case Study Selection: Political System Type and Conflict Outcomes

Outcome/Political System Type	Federal	Unitary
Cooperative	Delaware River Basin Commission (USA) Damodar Valley Corporation (India)	River basin agencies (France)
Conflictual	Colorado River basin (USA) Krishna River basin (India)	Yellow River Conservancy Commission (China)

Table I.2 Features of Country Case Studies

Case Study Feature	Explanation
Driving water resource management issues	Challenges in each river basin that governance institutions must address
Institutional collective action problems	Institutional barriers to addressing these challenges
Integration mechanisms	Potential solutions to these institutional barriers
Sectional cleavages	Noninstitutional barriers to collective action
Political opportunity structure	Degree of access for third-party involvement in interjurisdictional collective action

major trends and patterns evident in the case. The case study narrative follows this introduction. Because of their duration and complexity, the Colorado and Krishna case study narratives are divided into several historical time periods. Finally, each chapter concludes with a summary relating its conclusions to the three primary factors that I propose shape subnational hydropolitics, namely decentralization, sectional identity, and political opportunity structure. In the process, I also assess the performance of river basin governance institutions in each country based on the general criterion of adaptive efficiency, as indicated by their legitimacy in the eyes of resource users, inclusion of civil society, and environmental performance (Imperial and Yandle 2005).

Two of the case studies suggest a direct, negative relationship between the strength of river basin collective action institutions and the intensity of conflict: India features both weak river basin governance institutions and high levels of subnational political conflict over water, while France features strong, collaborative, and participatory institutions along with low levels of conflict. However, the remaining cases betray a less direct relationship. China features strong river basin governance institutions but also exhibits a surprisingly high level of conflict between subnational jurisdictions. The United States, meanwhile, features a unique diversity of both strong and weak river basin governance institutions, as well as both cooperative and noncooperative outcomes. Together, these cases illustrate the fact that institutions alone are not sufficient to generate positive water resource management outcomes. Instead, collective action institutions must balance national and subnational interests and include resource users as well as third parties.

This observation also previews one of the ultimate conclusions of this book, namely that third party–induced cooperation is essential to sustaining collaborative, participatory, and adaptive management of water resources shared between multiple political jurisdictions. Moreover, the cases examined in this book suggest that nongovernmental advocacy organizations, in alliance with national political elites, are most effective at inducing this kind of cooperation. The French river basin agencies and the Delaware River Basin Commission are both the product of nongovernmental advocacy and have been largely successful in addressing complex management challenges such as water quality. In contrast, the Yellow River basin and Damodar Valley Corporation cases both represent examples of interjurisdictional collective action imposed by central governments; while they have been successful in addressing some water resource management challenges, they have proven ineffective in resolving complex and diffuse issues like non-point-source pollution. Finally, the Colorado and Krishna River basin cases were shaped primarily by sectional interests and subnational elites, with the result that interjurisdictional collective action has largely failed to address pressing management issues. These cases show that while it can be just as difficult to secure cooperation between subnational jurisdictions as between nation-states, third-party participation greatly increases the chances of success.

This book concludes with a chapter focused on the implications of these findings. The governance of shared water resources in general, and of shared river basins in particular, is a matter of increasing concern in light of trends in population, economic growth, and climate change. Accordingly, in the Conclusions and Policy Recommendations, I summarize the scholarly implications of this study and detail the lessons it yields for water resource management and policy in the United States, China, and India. I devote special attention to the necessity for both civil society and central government involvement to prevent and resolve conflict over water. Many credible third parties are capable of acting as bridging organizations, but in practice, civil society organizations, and national political leaders, are most likely to effectively induce cooperation between subnational water users.

Finally, the book also speaks to pressing policy debates regarding pollution and scarcity, particularly in China and India. Based on the case study evidence examined in this book, I advise policymakers to focus on creating robust interjurisdictional institutions to resolve conflicts and prevent disputes between different water users, as well as to facilitate bargaining and dialogue between subnational jurisdictions. These institutions must be carefully designed to facilitate effective third-party participation, without being unwieldly or overly politicized. Such institutions are underdeveloped in the United States and nearly absent in China and India. Developing these institutions therefore represents a critical political challenge in the twenty-first century. But even readers primarily interested in policy will want to persist through the case studies that follow, for they illustrate this book's central message that subnational water conflict stems from a combination of institutional and ideational factors.

NOTES

1. This section is adapted in part from "Rethinking Conflict over Water" by Scott M. Moore, *Oxford Research Encyclopedia of Environmental Science*, 2016, reproduced by permission of Oxford University Press, http://environmentalscience.oxfordre.com/view/10.1093/acrefore/9780199389414.001.0001/acrefore-9780199389414-e-104.
2. This result was obtained by visualizing river basin and administrative area shapefiles in a Geographic Information System (ArcMap 10), and using the spatial join command to append river basin data to the administrative area to which it is spatially most closely related. River basin data were obtained from shapefiles produced by the World Resources Institute and provided by the UN Food and Agriculture Organization, *WRI Major Watersheds of the World Delineation*, July 11, 2006, available at http://www.fao.org/geonetwork/srv/en/metadata.show?id=30914&currTab=simple (accessed April 2, 2011), while major "level-1" subnational administrative boundary data were obtained from the Global Administrative Areas Project, *Global Administrative Areas*, August 29, 2009, http://www.gadm.org/ (accessed April 2, 2011). Importantly, although the Kuskowim and Skeena are contained within a single first-tier jurisdiction, both nonetheless span other subnational political units, including boroughs and indigenous political unit boundaries.
3. A common pitfall of cross-national, comparative case study analysis is to identify spurious common factors across disparate cases. When comparing different countries, there is a considerable temptation to seize on any available commonality to explain the phenomenon of interest, while neglecting other potential causes of variation. Related methodological issues, such as pervasive endogeneity in candidate explanatory variables, are especially pronounced in comparative studies. For this reason, confidence in candidate causal factors can be enhanced by multiplying within-case variation, such as across different historical periods.

1

Deconstructing the State

Subnational Conflict and Cooperation over Water

The concept of hydropolitics outlined in the Introduction defines conflict and cooperation over water as part of a set of interactions that occur at both international and subnational levels. On this definition, conflicts between states, provinces, or municipalities differ in degree from those that occur between countries but not necessarily in kind. Indeed, as this chapter details, there are many more similarities than differences between international and subnational hydropolitics. But reconceptualizing hydropolitics as a continuum rather than a dichotomy between international and subnational levels requires considerable elaboration, which this chapter attempts in three sections. The first seeks to better define subnational hydropolitics by comparing and contrasting it with its international equivalent, noting in particular a long-term trend toward greater institutionalized cooperation at the international, but not necessarily the subnational, level. The second explains this divergence by exploring the many constraints that central governments face in attempting to manage water resources, especially when political power is decentralized. As this section demonstrates, contrary to popular belief, shared water resources can be difficult to manage even when they are contained within the same country. Finally, the third builds on this conceptual foundation by adding an empirical survey of subnational hydropolitics in ten countries, in the process sketching its three basic dimensions: interjurisdictional, intergovernmental, and state–society relationships. In reconceptualizing hydropolitics along these lines, this chapter attempts to address the first of the key questions raised in the Introduction, namely why interjurisdictional collective action in shared river basins can be equally or even more difficult at the subnational than the international level. The answer has much to do with the often unappreciated limits of the state in managing shared water resources—and the likewise unrecognized role that subnational political elites can play in hindering its involvement. But states are not without their advantages: as Chapter 3 details, they can offer third parties opportunities to support cooperation instead of conflict.

RECONCEPTUALIZING HYDROPOLITICS: CONFLICT AND COOPERATION OVER WATER AT DIFFERENT LEVELS

The first and perhaps most basic issue to be tackled in reconceptualizing hydropolitics is that, as described in the Introduction, its traditional focus on the prospect of interstate warfare has not been borne out by events. Instances of violence and conflict between states over water have turned out to be exceptionally rare. Indeed, instead of conflict, many international transboundary river basins have followed a long-term trend towards more institutionalized cooperation, thanks to the involvement of non-state and multi-lateral actors. The misplaced focus of traditional hydropolitical theory on the sovereign nation-state is further indicated by the fact that, in contrast to this long-term trend toward greater institutionalization, no comparable global trend has occurred in water resource management within states. All of these empirical realities therefore point to the need to reconceptualize hydropolitics not as a contest between nation-states over water but rather as a spectrum of both cooperative and noncooperative relationships between both national and subnational political units that share common water resources (Zeitoun and Mirumachi 2008).

In its original definition, *hydropolitics* referred to the prospect for conflict between riparian countries, especially when one lies upstream of another. At the international level, the absence of a single, overarching sovereign authority means that countries can at least in principle exploit asymmetries in political, economic, and military power to control shared waterways. The seminal work in the hydropolitics literature, John Waterbury's *Hydropolitics in the Nile Valley*, describes how Egypt's political, military, and economic dominance relative to its upstream neighbors allowed it to effectively control the Nile's water resources, despite its disadvantaged position as the downstream riparian (Waterbury 1979). Similar dynamics, in which countries exploit power asymmetries to control shared water resources, have been widely described elsewhere, especially in the Middle East (Zeitoun and Warner 2006; Zeitoun and Allan 2008). Several cross-national studies have likewise concluded that geographic contiguity, and especially an upstream–downstream orientation, is associated with a higher probability of conflict between riparian countries (Toset, Gleditsch, and Hegre 2000; Furlong, Gleditsch, and Hegre 2006; Brochmann and Gleditsch 2012). The traditional conception of hydropolitics therefore seeks to explain international riparian relations in terms of the mutual interaction of power and geography, which tends to make them both tense and unstable.

In practice, however, geography is not always destiny when it comes to shared water resources, and most situations are not zero-sum even when countries lie up- and downstream of one another. An upstream country may agree to sell power generated from a hydroelectric dam to its downstream neighbor at concessionary rates, for example, thereby providing benefits to both countries. Indeed, despite the costs entailed in securing cooperation between sovereign nation-states, the existence of shared water resources appears to be more likely to induce cooperation than to provoke conflict. Thousands of cooperative agreements exist to govern international transboundary rivers, even in river basins subject to acute water scarcity (Wolf 1995; Stinnett and Tir 2009; Tir and Stinnett 2012). Moreover, where conflict does arise between countries over shared water resources, parties often settle differences by agreeing to cooperate in other areas (Uitto and Duda 2002; Delli Priscoli and Wolf 2009; Susskind and Islam 2012). Even in the paradigmatic case of the Nile River basin, important periods of cooperation have existed between the upstream and downstream riparian states, belying the notion that conflict between them is inevitable (Waterbury 2002; Tvedt 2009). Most recent studies in the hydropolitics tradition have been careful to temper any implication of geographic determinism, stressing that numerous confounding variables exist that mediate relationships between geography and the possibility of international conflict over water (Oates 1999; Gleditsch et al. 2006; Matthews and St. Germain 2007).

In recognition of the interaction between political geography and institutions, most recent scholarship in the hydropolitics tradition has begun to focus on nation-states not as unitary actors but rather as a collection of political actors and institutions (Sneddon and Fox 2006; Alam, Dione, and Jeffrey 2009). While the state remains a crucial actor and one that serves as the locus for many subsidiary institutions like water management laws and bureaucracies, it is no longer seen as a purely coherent, isolated entity locked in conflict or cooperation with its neighbors. In a related development, several scholars have begun to describe "hydropolitical complexes" that encompass a diverse range of actors at these different levels. Within these constructs, international tensions and rivalries are seen as reflected in competition over shared water resources (Schultz 1995; Turton, Ashton, and Cloete 2003). These complexes, and the conflict dynamics that occur within them, have also been identified in certain subnational regions (Musemwa 2006; Turton and Ohlsson 2000). A few researchers have moreover proposed tentative linkages between water-related conflict and disputes at the subnational level and tensions in international transboundary basins, particularly in South Asia (Giordano, Giordano, and Wolf 2002; Wirsing 2007). This trend in hydropolitical theory has the advantage of highlighting subnational actors and politics, while upholding the state's significance as a means of enforcing some greater level of coherence in policymaking than is the case among sovereign nation-states.

Even so, the transition toward a greater focus on subnational hydropolitics remains incomplete: a few studies have observed that conflict appears to be more common between subnational jurisdictions oriented upstream and downstream of one another, just as is often thought to be the case for countries (Gyawali 1999; Postel and Wolf 2001). But, as we shall see further in Chapter 3, these upstream–downstream dynamics are not the result of geographic determinism. Instead, upstream–downstream relationships embody deeper sectional divides that separate water users by ethnicity, culture, language, economic status, and many other categories, which in turn shape their preferences with respect to shared water resources. Often, these divergent preferences stoke tensions over the use of shared water resources, which, given the fundamental importance of water to every economic sector and its emotional resonance, often proves a potent tool for political mobilization that subnational politicians are eager to exploit. Extant hydropolitical literature is therefore incomplete; while it helpfully refocuses attention on localized politics, tensions, and rivalries, instead of the high geopolitics typical of early work in the hydropolitics tradition, it does not fully address the role of underlying sectional tensions in driving water conflict.

The importance of refocusing attention on the subnational level, and the sectional tensions that characterize it, is further suggested by the existence of a long-term trend, difficult to reconcile with traditional hydropolitical theory, toward stronger, more institutionalized cooperation on international transboundary rivers. Prior to about 1945, natural resources like water were viewed as the property of the nation-state in which they happened to be deposited (Schrijver 2008). Particularly in Europe, water resources were subject to a historical process of "territorialization," which molded both populations and natural resources to the boundaries of fixed, sovereign states (Vaccaro 2008). By the mid-nineteenth century, for example, the Franco–Spanish Border Treaties stipulated that "All standing and flowing waters, whether they are in the private or public domain, are subject to the sovereignty of the State in which they are located, and therefore to that State's legislation" (United Nations Food and Agriculture Organization 1998, 253). This territorialization tended to create high costs to collective action between the countries that shared a given waterway. The treaties in effect divided transboundary waters between the two countries, providing that "Flowing waters change jurisdiction at the moment when they pass from one country to the other" (United Nations Food and Agriculture Organization 1998, 253). The principle of territorial sovereignty continued to pose profound conceptual difficulties for the cooperative management of international transboundary rivers. As one international legal scholar summarized the problem in the late 1950s, "The contradictory character of flowing water raises complicated questions in international law. It belongs by definition to the domestic jurisdiction of the territorial sovereign. But if the territorial sovereign acts to the full extent of his powers, he interferes with the legitimate expectations of the lower riparian" (Berber 1959). Under this traditional interpretation, states have few inherent incentives to cooperate over shared water resources, particularly when no state possesses significant military, diplomatic, or economic leverage over another.

Since the end of the Second World War, however, a variety of nonstate and multilateral organizations have helped to promote norms and institutions that have lowered these barriers to cooperative institution-building. In particular, international judicial bodies have formalized two general principles, reasonable use and consultation in the allocation of water resources between nations, that have underpinned a growing number of cooperative water agreements. A landmark moment in the development of international water law occurred in 1957 in the form of the "Lake Lanoux" decision of the International Court of Arbitration. The decision, which concerned the modification of flow on a river shared between France and Spain, replaced the principle of territorial sovereignty with a new obligation between upstream and downstream states. This novel obligation created a norm of reciprocity between the countries that share a river under which, "according to the rules of good faith, the upstream State is under the obligation to take into consideration the various interests involved, to seek to give them every satisfaction compatible with the pursuit of its own interests, and to show that in this regard it is genuinely concerned to reconcile the interests of the other riparian State with its own" (*France v. Spain, Lake Lanoux Arbitration* 1957). The Lake Lanoux decision thereby created an institutional constraint on the power asymmetry between upstream and downstream riparians that had long frustrated international cooperation. By addressing this asymmetry, the new norm of cooperation has helped to lower the costs of collective action on international transboundary rivers.

The influence of this new norm was illustrated by the 1971 Ramsar Convention on international wetlands, which signaled a departure from the sovereign-resource model by including a novel provision obliging states to cooperatively manage transboundary wetlands. The codification of international water law was advanced most dramatically, however, by the 1997 Convention on the Law of the Non-Navigable Uses of International Watercourses. The convention codified at least four important new principles into international law. First, states are limited to equitable and reasonable, rather than unlimited, utilization of transboundary water resources. Second, under the convention state parties assume an obligation to refrain from activities which substantially harm downstream states. Third, the convention establishes a mandate that nations inform neighboring countries of planned activities, such as dam construction, that may affect other countries. Fourth and finally, the convention provides that all countries should consider environmental impacts and the need to protect ecosystems in their respective national water resource policies, representing an almost complete break with the tradition of nation-state sovereignty over natural resources. The convention entered force in 2014 upon ratification by the requisite number of thirty-five state parties, marking a major success for the cooperative approach to transboundary river management (McCaffrey 1997).

In parallel, multilateral organizations and advocacy networks have promoted similar principles of collaboration, participation, and equity in water resource decision-making. In its 2000 Millennium Declaration, the United Nations General Assembly pledged "To stop the unsustainable exploitation of water resources by developing water management strategies at the regional, national, and local levels, which promote both equitable access and adequate supplies" (2000, 4). Similarly, a vocal transnational advocacy network has emerged around the "right to water" that emphasizes the principles of participation and equity (McCaffrey 1992–1993; Gleick 1998; Salman and McInerney-Lankford 2004). The influence of the collaborative–participatory model is further evidenced in several widely accepted water resource management frameworks. The 1992 Dublin Statement on Water and Sustainable Development, widely considered the "Magna Carta" of water resources management, calls for implementing several core principles including holistic management, participatory governance, gender equity, and efficient and equitable use (International Conference on Water and the Environment 1992; Salman and Bradlow 2006). The associated paradigm of integrated water resources management includes a focus on the equitable allocation of water resources and the consultation of diverse stakeholders (Chene 2009). These principles have become firmly established in international water governance, if not always fully implemented.

At the subnational level, however, no comparable degree of convergence has occurred. Below the nation-state level, water resource policies remain highly diverse and shaped less by internationally agreed best practice than by historical contingency (World Bank 2004). Why has the relative success of nonstate and multilateral organizations in institutionalizing cooperation in international transboundary basins not transferred to the substate level? Despite the intuition that the existence of a single central government would help to simplify management of shared natural resources, this is often not the case for water. The next section explores in greater detail the constraints that central states face in managing water resources, even those that fall entirely within their borders.

LIMITS OF THE STATE: CONSTRAINTS ON CENTRAL GOVERNMENT INVOLVEMENT IN WATER RESOURCE MANAGEMENT

The first and most fundamental question posed by subnational hydropolitics is why they should occur at all, given the many tools that sovereign central governments possess to prevent the presumably unfavorable occurrence of interjurisdictional conflict within their borders. The answer to this question lies in the many constraints that central states face in managing water resources. First, certain inherent characteristics of water make it difficult to apply traditional regulatory devices, including property rights and price rationing, to manage it. Second, the large water infrastructure works, such as reservoirs and canals, necessary to support an industrial economy typically require central state financing, leading states to create specialized hydroengineering bureaucracies. These technocratic entities are generally ill-suited to addressing the complex management issues that arise in interjurisdictional river basins. Third and finally, electoral incentives often discourage national political leaders from becoming too deeply involved in contentious interjurisdictional issues. Each of these constraints applies to some countries more than others, and they are likely to be greater for more decentralized systems, as the next section explains in more detail. But in combination, these three constraints explain why central states are often unable to prevent the occurrence of subnational hydropolitics, and sometimes to resolve them once they arise.

The historically dominant role of the state stems in large part from the difficulties of relying on market forces to allocate and arbitrate conflicts between water users. In principle, at least, markets offer a much more compelling mechanism for governing water resources than does the state. Most economists agree that water, like other resources, is most effectively and efficiently allocated by means of marginal cost pricing or by efficient assignment of property rights (Zarour and Isaac 1993; Fisher and Huber-Lee 2006). On this understanding, there is little need for state involvement in allocating and supplying water and no economic reason for conflict to occur between water users. Even under conditions of acute scarcity, the theory goes, those who value it most highly either pay more for the privilege of using it or compensate others for use of the common resource. Trade in water-intensive goods and commodities between countries, which has come to be known as "virtual water," can further compensate for localized water scarcity (Conca 2012). Other sources of conflict between resource users, such as pollution and harms arising from the construction of dams, can likewise be addressed through transfer payments from those who benefit from externalities to those harmed by them. Although even adherents of the economic view acknowledge numerous qualifications to these arguments, a pure economic analysis questions whether conflict over water should occur at all (Easter, Rosegrant, and Dinar 1999; Chong and Sunding 2006; Debaere et al. 2014).

In practice, however, the role of central states in allocating and managing water has persisted despite the appeal of the market. The dominant role of the state reflects several features of water that make it difficult to apply traditional economic rationing devices such as pricing and property rights. The fugitive nature of water as a transient resource, constantly cycling between surface, subsurface, and atmospheric deposition, has frustrated attempts to assign property rights to it since at least the time of the Romans (Geltzler 2004). Furthermore, the complex connections between surface and subsurface water make it difficult to monitor water use by individual resource users and by extension costly to enforce water rights (Feitelson 2006). The pricing of water to reflect its scarcity value, meanwhile, immediately raises normative objections. Raising prices on water is typically highly regressive and places disproportionate burdens on the poor and those without ready access to water supplies. As a practical matter, such objections make politicians around the world extremely reluctant to expose important constituencies, especially the urban middle class and agriculturalists dependent on irrigation, to full marginal cost pricing of water (Dinar, Rosegrant, and Meinzen-Dick 1997; Lichtenthaler 2000; Boberg 2005). As a result of these constraints, water resources have long "lain at the limit of the sphere of applicability of the market as a social institution for allocating resources" (Bakker 2003, 33).

At the same time, water sits uneasily within the bureaucratic structure of modern states. As has often been noted, the flowing and fungible nature of water stands sharply at odds with the rationality of the modern nation-state, which often seeks to simplify and control complex, boundary-defying natural ecosystems (Scott 1998). The involvement of the state in managing and supplying water is nonetheless a feature of nearly every industrialized society and has emerged at similar points in the process of economic development (Garrick 2015). Where both population density and levels of economic development are low, as in areas dominated by subsistence agriculture, herding, or hunting and gathering, there is little need for either technology or institutions to mediate direct access to communal water supplies. In this "local subsistence mode," authority and control over water resources usually remain centered in communities, and conflicts and disputes are primarily handled through informal institutions (Worster 1985).

However, at higher levels of economic development, the institutional and technological complexity of water management increases expotentially, requiring state involvement. Growing populations demand higher water withdrawals for more intensive forms of agriculture, while new economic uses like power generation, coal processing, and steel production result in higher levels of both water use and water pollution, thereby reducing the availability of water for drinking and recreation. At the societal level, rising incomes and changing social norms generates higher water demand for hygienic and aesthetic uses, even as demand increases for water-intensive products and commodities. This increased water demand was often regional and centered on areas dominated by irrigated agriculture or urban centers. Because water withdrawals often outstripped local water sources, however, technological means were applied to increase the supply of available water. Historically, the most common such technology involved interbasin water transfer via canal, reservoir, and aqueduct.

The high cost of supplying these growing water demands naturally favored the involvement of central states. Because water is exceptionally heavy, and therefore expensive to transport, per unit volume, supplying water to critical constituencies like irrigated farmland and population centers could in most cases be accomplished only by mobilizing the financial resources of the central state (Bakker 2010). Accordingly, by the late nineteenth century, water supply had become widely recognized as a core responsibility of the modern state (Bakker 2003).

But states often proved ill-equipped to address the problem of competing water uses, especially those associated with the growth of cities and industry. As one historian has remarked, "In all modern urban societies, water management quickly became a national priority. Urbanely conceived and controlled water management programs, however, seldom take into account the needs of rural populations" (Vaccaro 2008, 227).

In the face of these challenges, most states adopted a technobureaucratic approach to water resource management, which emphasizes elite decision-making, technical solutions, and a tendency toward static rather than dynamic planning. In particular, states turned to specialized bureaucracies to fulfill their newfound responsibilities to provide and manage water for a growing number of uses. These bureaucracies, such as the U.S. Army Corps of Engineers and the French Corps des ponts et chaussées, played a critical role in the early phases of economic development in their respective countries. Their efforts to control flooding, improve waterborne navigation, and supply water for growing urban areas transformed entire regions and ranked as one of the most important aspects of state-building in most industrialized countries, including Germany, France, and the United States (Blackbourn 2011; Pritchard 2011; McCool 2012). Like other bureaucratic organizations, however, these agencies sought to maximize political support for the functions they perform. One of the most important means of doing so was to develop specialized, technocratic knowledge and capabilities that distinguished each bureaucratic organization from its rivals and aided in its appeals to accumulate resources, authority, and autonomy (Carpenter 2001). These forms of bureaucratic strategic behavior are well documented in the literature on American political development, and indeed the U.S. Army Corps of Engineers represented a prime example of such behavior during the first half of the twentieth century (Downs 1967; Maass 1974; Pisani 2002). These features of the technobureaucratic approach meant that it impeded rather than promoted the ability of central states to effectively manage water resources within their borders.

The prevalence of rent-seeking behavior among hydroengineering bureaucracies has also helped to sow the seeds of conflict between water users within countries. Despite their influence, in many cases hydroengineering bureaucracies found it expedient to form political alliances with other bureaucracies and political officials to construct large-scale water infrastructure projects. This was especially true in more decentralized countries like the United States, where the division of power between central and local levels required bureaucratic officials to seek the support of subnational officials, and vice versa. In the United States, for example, "Most states have never been able to finance the great water resource projects undertaken within their limits. . . . In case after case, states mobilized . . . to seek federal assistance" (Elazar 1984, 39). These alliances served the interests of all parties: agencies gained support and created demand for their technical expertise, while political leaders gained access to the financial and technical resources needed to overcome localized water scarcity. However, these alliances rarely benefited all water-user groups in a given river basin or watershed since the construction of large dams, canals, and reservoirs inevitably served some regions at the expense of others (Faith 1980; Rucker and Fishback 1983; Ostrom, Schroeder, and Wynne 1993). Such exclusionary alliances often attempted to silence opposition from these losers by "shortcutting both a general discussion of goals" to be achieved by specific projects, as well as preventing "a consideration of alternative means" that might avoid some adverse impacts on project losers (Ingram 1971, 105).

One particular inadequacy of the technobureaucratic approach in the face of these challenges bears emphasis, namely that of information asymmetry. Because water use is highly localized, and monitoring it correspondingly difficult and expensive, central states face an inherent disadvantage to individual water users in gaining access to water data. Hydroengineering bureaucracies often develop a detailed knowledge of natural hydrology but much less clarity on the scope and scale of human water uses, particularly where groundwater is involved. While recent advances in remote sensing have to some extent ameliorated these issues, it remains a substantial challenge for most central governments to develop an adequate understanding of how much water is both used and available in a given basin (Moore and Fisher 2012). At the same time, information asymmetries often create tensions between water users and undermine collaborative and participatory decision-making (Organization for Economic Cooperation and Development 2015). By the late twentieth century, these strains had become marked in many regions by the combination of population growth, rising incomes, and changes in regional water availability as a result of climate change (Vorosmarty et al. 2000; Arnell 2004; Alcamo, Florke, and Marker 2007; Gleick and Palaniappan 2010). Inadequate provision of information both by and to central states therefore plays a significant role in fomenting subnational water conflict.

But even as the institutional structure of the modern state leaves it ill-equipped to prevent water conflicts, it also provides scarce political incentive to resolve them. In many countries, political representation is based around subnational jurisdictions so that electoral districts coincide with state or provincial boundaries, for example. This structure means that national legislative representatives effectively represent particular subnational jurisdictions, which therefore gain a direct voice in government (Tsebelis 2002). At the same time, representatives of individual jurisdictions can gain the "cheap support" of local constituencies within these units, rather than having to support agendas common to the nation as a whole. This electoral structure makes national political leaders reluctant to become embroiled in issues like interjurisdictional river disputes because involvement is likely to force them to choose winners and losers and cost them the support of representatives of disfavored jurisdictions (Macey 1990; Gibson and Calvo 2000; Samuels and Snyder 2001). This disincentive is strongest where malapportionment of electoral seats gives less populous subnational jurisdictions equal voice or voting rights to their more populous neighbors or where subnational jurisdictions are directly represented in legislative upper houses. Both are common features of federal political systems; direct upper-house representation existed in the United States prior to the Seventeenth Amendment, for example, and persists in Germany and many other federal countries (Swenden 2010).

The existence of these electoral incentives obviates the use of other mechanisms that central states might employ to compel the cooperation of subnational jurisdictions in shared river basins. In principle, for example, central governments can either offer or withhold financing to encourage particular behaviors on the part of subnational officials. In practice, however, national political leaders often use such fiscal transfers to gain support at the state level. While there is little cross-national study on the subject, evidence from India in particular suggests that fiscal transfers are shaped by electoral politics at the state level, rather than the other way around (Arulampalam et al. 2009). At the same time, however, it is clear that the strength of electoral disincentives to negotiating interjurisdictional water resource disputes varies across countries. In particular, as the next section details, decentralization makes it considerably more difficult for states to solve certain coordination and collective action problems inherent to shared water resources. The extent of decentralization, which differs greatly in countries like Canada and China, helps to explain why the management of interjurisdictional water issues is difficult for all central states but for some more so than others.

DIMENSIONS OF SUBNATIONAL HYDROPOLITICS: INTERJURISDICTIONAL, INTERGOVERNMENTAL, AND STATE–SOCIETY RELATIONSHIPS

Subnational hydropolitics, unlike the international variety, occur within a single sovereign jurisdiction. And while the dynamics of interjurisdictional conflict and cooperation over water are fundamentally similar at both levels, the existence of a single sovereign government does carry several important implications. As this section details, within a single nation-state political actors and water users are bound by a variety of formal and informal relationships that link them at central and local levels, across jurisdictional boundaries, and across social and economic sectors. These relationships vary considerably across countries, with important consequences for the nature of subnational hydropolitics. While some countries decentralize political power and provide few formal mechanisms to encourage interjurisdictional cooperation, others feature favorable political opportunity structures that encourage the participation of third parties and cooperation in shared river basins. The density of these institutional arrangements distinguishes the subnational from the international level. Third parties seeking to promote institution-building have more tools and avenues to pursue them at the domestic than the international level. The significance of this distinction, however, is lessened by the sheer diversity of institutional arrangements across countries. Rather than establishing a dichotomy between subnational and international hydropolitics, they are more usefully viewed as a continuum.

As noted in the Introduction, a major challenge to describing and analyzing subnational water politics is the lack of comparable cross-national data. In this section, I attempt to discern the basic features of subnational hydropolitics by using what is to my knowledge the most complete data set on subnational water politics to date, namely papers covering ten countries across the developed and developing world prepared for a workshop on Water Security and Federal Rivers held at Oxford University in April 2012. All of these case studies describe water politics in federal or decentralized political systems; as noted in the Introduction and further described in Chapter 2, a meaningful degree of decentralization is effectively a prerequisite for interjurisdictional hydropolitics to occur. The cases vary substantially across political system type, level of development, and ethnic, religious, and linguistic diversity, among other factors. However, the water politics of all countries can be usefully analyzed across three primary dimensions: interjurisdictional relationships between province-level political units; intergovernmental relationships between central and local levels of government, as well as between bureaucratic entities at the same level; and finally, state–society relationships between governmental and nongovernmental actors. Although subnational water conflict occurs along each of these dimensions, interjurisdictional conflict tends to exert the broadest political, economic, and environmental impacts.

When analyzed along these principal dimensions, the ten cases can be distinctly, although not dichotomously, divided into two categories based on whether they employ a federal or unitary political system. While it is difficult to precisely characterize federal systems as opposed to some forms of decentralization, hydropolitics in federal systems display two common features rooted in intergovernmental and interjurisdictional institutions typical among federations (Stepan 1999). First, federal systems are more apt to feature powerful judiciaries, especially among the many federations that employ the English common law legal system. Lawsuits are especially common in federal systems, many of which feature a number of interstate water-sharing disputes (Richards and Singh 2002; Zimmerman 2011). Second, federations typically codify the division of powers and responsibilities between central and subnational levels of government but also are marked by continual debates concerning the proper role of the federal versus local government.

Subnational hydropolitics in federal systems therefore tend to be characterized by debates over water rights and the division of powers between the state and the federal government. In contrast, unitary states tend to rely less on litigation to address intergovernmental or interjurisdictional disputes, in part because unitary states tend to feature civil law systems. Moreover, unitary states tend to bifurcate water politics to a greater extent between elements of the central state, especially between different bureaucratic departments, than on an interjurisdictional basis. What really separates subnational hydropolitics in federal as opposed to unitary systems, however, are their relative scale and complexity and the greater prominence of interjurisdictional conflict. Accordingly, subnational political jurisdictions like states and provinces play a greater role in the hydropolitics of decentralized and federal political systems.

In large part, this increased role stems from the space that decentralized systems offer to subnational leaders to pursue independent preferences, including competition with their neighbors over water. First, unlike centralized systems, decentralized government often produces a decentralized political party structure, which can increase the independence of local officials who are not expected to adhere to a tightly defined national platform (Riker 1987; Robertson 2012). As Alfred Montero writes, "Governors and mayors have interests emanating from the offices they hold, and not just the parties they represent" (2005, 64). Second, decentralization often increases the fiscal independence of subnational leaders. While cross-national data are sparse, many studies of the effect of fiscal transfers on subnational behavior tend to stress how subnational interests shape fiscal transfers, rather than vice versa (Beland and Lecours 2006; Arulampalam et al. 2009). As fiscal federalism scholar Jonathan Rodden observes, subnational officials "follow a political, not an economic, logic" (2006, 273). A dramatic example of the extent to which subnational politicians can insulate themselves from central states and national political priorities is the existence of "authoritarian enclaves" found throughout Latin America. By cultivating a network of patronage and clientelist relationships, local leaders are able to effectively insulate their jurisdictions from nationwide political reforms (Gibson 2005, 2010; Falletti 2005). Where decentralized settings enable subnational politicians to gain such freedom of action, they are free to pursue a variety of behaviors, including conflict as well as cooperation with their neighbors.

These features of water politics in decentralized and federal systems are well illustrated by the case of Pakistan. Pakistan's hydropolitics, both international and subnational, revolve around water-sharing agreements for partitioning the Indus River system and the extensive irrigation system that it supports. Interjurisdictional water disputes began to emerge following the re-creation of the provinces of Punjab and Sindh in 1970 and prevented the construction of planned storage and hydropower dams. However, the determined involvement of the federal government led to the conclusion of the 1991 Water Apportionment Accord that has had some success in meeting water requirements for cotton production in both Punjab and Sindh. Even so, within months of the accord's conclusion, a fundamental disagreement emerged between the two provinces over whether or not upstream Punjab could lay claim to a portion of the Indus' flow based on historic rather than current water use; and this disagreement continues to animate interprovincial hydropolitics. In downstream Sindh, this tension has taken on a sectional tone, with "sloganeering and street protests" reflecting a "fusion of hydropolitics with identity politics of Sindhi nationalists." Even in the midst of these sectional tensions, the federal government has since conclusion of the accord been largely unwilling to engage in resolving the dispute. At the same time, the credibility of water resource agencies, including the Indus River System Authority which is supposed to ensure operation of the Indus River irrigation system consistent with the accord, has been weakened by mismanagement (Ahmad, Aziz, and Khan 2014). Pakistan's experience therefore highlights both the difficulties central governments often face in addressing interjurisdictional water conflict and the danger of mapping such conflict onto preexisting sectional tensions.

The complex interplay between interjurisdictional and intergovernmental relationships in federal systems is further illustrated by the case of Australia in the form of a long-running interstate dispute in the Murray-Darling River basin. Indeed, relations between the riparian states of the Murray-Darling played a key role in Australia's constitutional development. Before consenting to federation, the upper and middle riparian states, New South Wales and Victoria, respectively, demanded the insertion of Section 100 to the Constitution, which provides that "the Commonwealth shall not . . . abridge the rights of the State . . . to reasonable use of the waters of the rivers." During the early years of federation, this provision proved sufficient to constrain central involvement in the basin; and in 1914, the two states, joined by the downstream riparian, South Victoria, concluded a water-sharing agreement which promised to provide sufficient water for irrigation in each jurisdiction (Hammer 2010). However, competition soon emerged between the states as each sought to develop water infrastructure projects that primarily benefited each individual jurisdiction (Connell 2007). By the 1980s, it had become clear that a greater federal role was essential to effective basin governance. In 1987, a new Commonwealth–state agreement subjected state water use to review by a national commission; and in 1995, a cap was introduced on overall water use in the basin, followed by creation of a world-leading water rights system and finally, in 2007, passage of the Water Act and subsequent amendments, which created a robust institutional architecture for basin governance. Under the act, an independent entity, the Murray-Darling Basin Authority, was created to provide impartial advice and formulate basin master plans to guide water use, and special provisions were included to ensure water availability for environmental use. While the Australian example is a largely positive one, effective basin governance required the extensive involvement of the federal government in order to contain interjurisdictional conflict in the Murray-Darling basin and promote more institutionalized cooperation instead.

The influence of decentralization on water politics is further illustrated by variation among federal systems. In more centralized federations, interjurisdictional issues tend to play a more muted role than in their more decentralized counterparts, while intergovernmental tensions are more acute. In the case of Mexico, water resource management has been marked by central–local tensions as governments have shifted budgetary allocations between state and federal governments. In recent decades, the decentralization of expenditure responsibilities has broadly enhanced the role of state and municipal governments relative to the center (Barrios 2014). In Spain, which has become progressively more decentralized, growing water scarcity has strained relationships between central and subnational governments, as well as between these levels and that of the European Union. A planned long-distance water transfer in the Ebro River basin, for example, provoked the states slated to receive water to pass laws encouraging the project, while those planned to export water passed countervailing legislation prohibiting such transfers (Albaic, Calvo, and Esteban 2014). Despite the absence of any comparable water scarcity issues, a similar dynamic has played out in western Canada (Saunders and Wenig 2007). On the other hand, in Brazil's highly centralized federal system, the federal government plays a strong role in supporting local river basin institutions that are to a large extent fiscally and administrative independent of the state governments that share most of Brazil's major river basins (Coelho et al. 2014). The contrast between the more cooperative intergovernmental relationships characteristic of more centralized federal countries is also evident with respect to water data. While in Pakistan interprovincial management is continually hindered by the reluctance of provinces to share hydrologic information, in Brazil the close involvement of the highly capable National Water Agency provides crucial support to the formulation of individual river basin management plans (Coelho et al. 2014).

The distinction between federal and unitary systems remains significant with respect to the third dimension of hydropolitics, namely state–society relationships. This relationship varies widely across countries; in China, for example, the role of civil society is tightly constrained. However, in some centralized federal systems, the close integration of bureaucratic and nongovernmental organizations at the river basin level has helped to facilitate cooperative management. In Brazil's Doce River basin, for example, a nongovernmental organization, the Bioatlântica Institute (IBIO), has been given the responsibility for exercising administrative functions, including the preparation of a basin plan and management of water charges levied on users within the basin. As such, IBIO works closely with both the National Water Agency and state governments, aiding coordination and reducing the prospect of major disputes between them (Coelho et al. 2014). Spain's strong tradition of water-user participation in water management has likewise helped to bolster implementation and enforcement of water policy. In central Spain, for example, where water use is concentrated in irrigation associations with a long history of collective management, water scarcity pressures have been reduced because of better water use regulation, especially of groundwater (Albaic, Calvo, and Esteban 2014). As these examples suggest, while relationships between state and nonstate actors are important to both federal and centralized systems, in general they tend to be more highly institutionalized in more centralized systems, while federal systems tend to be characterized more by sectional politics, including long-running interjurisdictional disputes.

However, it is equally clear that the distinction between federal and centralized systems is neither dichotomous nor deterministic. In Spain, for example, widely regarded as a quasi-federal state, interstate water disputes have proven to be in some ways just as contentious as in more federal systems, and have been the source of street protests and become major issues in state elections. Indeed, an assessment of Spain's subnational hydropolitics could just as easily apply to the paradigmatic case of the Colorado River basin: "Water conflicts among states are politically driven to gain electoral support, at the cost of threatening to break down federal basins" (Albaic, Calvo, and Esteban 2014, 144). Moreover, the experience of South Africa, a decentralized, though not federal, political system, suggests that even in decentralized countries there is nothing inevitable about the existence of interjurisdictional water conflict, provided that institutions are carefully designed. Crucially, South Africa's approach has been to circumvent province-level water politics altogether by carefully allocating functions between central, local, and catchment authorities. During the formulation of South Africa's current constitution, a special standing committee was formed to consider the division of powers concerning water management, which concluded that "it is important to ensure that the allocation of powers and functions to second tier water management agencies should be separate from the attribution of powers and functions to regional government." This approach has by all accounts been successful in avoiding the interjurisdictional water conflict common to other decentralized systems, while providing the necessary flexibility to adapt to water management challenges over time (Muller 2014, 189).

This brief survey of cases suggests that the distinction between federal and unitary political systems is meaningful, though not determinative, in explaining water politics across countries. However, it also underlines the fact that contentious interjurisdictional hydropolitics often coincide with international cooperation. In the case of Spain, intense controversy between the autonomous communities of Castilla-La Mancha, Murcia, and Valencia has not prevented Madrid from concluding a successful agreement with Lisbon guaranteeing Portugal quarterly and weekly flows from the shared Tajo River (Albaic, Calvo, and Esteban 2014). Perhaps more than anything else, these cases instead illustrate the core message of the Introduction, namely that subnational hydropolitics are often more complex and contentious than the international variety. The three primary dimensions of subnational water politics, interjurisdictional, intergovernmental, and state–society relationships, describe a wide range of interactions between central, local, and nonstate political actors. But they also highlight the distinctive role played by subnational political actors, who are often, and contrary to popular belief, those most opposed to water policy reforms and the establishment of cooperative institutions in shared river basins. Subnational elites can pursue a range of strategies to achieve their own independent objectives, including "undermining civil society, excluding alternative sources of knowledge and supporting and legitimizing business-as-usual decision making" (Matthews and Schmidt 2014, 38). Central governments accordingly face marked, and often unappreciated, challenges in managing water within their borders—challenges that nonstate actors can often help to overcome by catalyzing and supporting collective action between water users. In order to be effective, such cooperation must often address underlying sectional tensions between subnational jurisdictions. Even within the universe of subnational hydropolitics, these complex interjurisdictional dynamics are often the most interesting and important, as the following chapters detail.

2

Constrained Collective Action

Decentralization, Autonomy, and Institutionalized Cooperation in Shared River Basins

As noted in the Introduction, it is frequently assumed that the costs of collective action between political jurisdictions are lower when they are part of the same country. A web of shared institutions and relationships, the thinking goes, helps to lower these costs relative to the international level, where cooperation is likely to be rarer and costlier. But, as this chapter explains, where political power is extensively decentralized, considerable constraints are placed on interjurisdictional collective action. In particular, because it distributes power between different levels of government, decentralization exacerbates the interjurisdictional and intersectoral coordination problems that are inherent to water resource management. This chapter explains how existing institutions often fail to prevent and resolve interjurisdictional water conflicts. It proceeds in three sections. The first section sets the stage for this discussion by exploring in greater detail the relationship between shared institutions and conflict potential, particularly in relation to other proposed sources of water conflict like geography and scarcity. The second section explains why decentralization creates specific barriers to interjurisdictional collective action, especially at the river basin scale. In particular, bureaucratic fragmentation, electoral incentives, and information asymmetries often create disincentives to establish institutional structures for river basin management. The third section, finally, explains why, as a result, collective action in shared river basins is often ad hoc and confined to relatively simple issues like point-source pollution control, rather than more complex and contentious issues like allocation. In combination, these challenges explain why existing institutional mechanisms so often fail to prevent interjurisdictional water conflicts from arising and to resolve them once they begin.

THE CAUSES OF WATER CONFLICT: DETERMINISM VERSUS INSTITUTIONALISM

This primary purpose of this chapter is to explore the institutional dimensions of subnational hydropolitics and in particular the role of decentralization. Doing so requires first understanding how institutions influence conflict and cooperation over shared water resources, especially in contrast to factors like geography and scarcity. This section accordingly discusses the two primary theoretical traditions concerning the causes of water conflict. The first, which I call *deterministic,* holds that water conflict occurs when particular exogenous conditions apply, such as when one party sits upstream of another or when water is sufficiently scarce that it cannot satisfy the demands of all water users. The second tradition, which I call *institutionalist,* proposes in contrast that conflict results primarily from failures or deficiencies in the design and functioning of shared institutions, including norms, rules, and governance authorities. Each of these traditions maps onto several primary literatures, including environmental security, political ecology, and collective action. In this section, I critique the causes of water conflict proposed by both of these theoretical traditions, emphasizing in the process the importance of institutional arrangements for shaping subnational hydropolitics. These critiques are summarized in Table 2.1, which suggests that while existing literatures contribute key pieces of the puzzle of subnational hydropolitics, none fully account for the distinctive dynamics of water conflict and cooperation at the subnational level. Accordingly, the remainder of this chapter describes why collective action in interjurisdictional river basins is often constrained and how this sets the stage for conflicts to arise.

Both the deterministic and institutionalist traditions propose that water conflict is multicausal. However, the two are also distinguished by the deterministic belief that water conflict arises in large part from exogenous factors, especially physical water scarcity. The proposed link between water scarcity and conflict is long-standing, but it has found its broadest influence in the relatively recent literature on environmental security. This literature has developed since the end of the Cold War, amid growing attention to the question of whether renewable resources like water and timber might be the cause of conflict in the twenty-first century, much as minerals and metals had been in the twentieth. The essential claim of the environmental security literature is that a causal pathway exists between physical scarcity of critical, nonrenewable resources like water and violence. Two contemporaneous accounts serve to illustrate this proposed relationship: Peter Gleick's prediction that "water and water-supply systems are increasingly likely to be both objectives of military action and instruments of war as human populations grow, as improving standards of living increase the demand for fresh water, and as global climactic changes make water supply and demand more problematic and uncertain" (Gleick 1993, 79) and Frederick Frey's assertion that "Water has four primary characteristics of political importance: extreme importance, scarcity, mal-distribution, and being shared"

Table 2.1 Proposed Causal Factors in Water Conflict

Theoretical School	Contributing Literatures	Proposed Causes of Water Conflict	Critique
Deterministic	Environmental security	Physical water scarcity + weak political and social institutions	Scarcity is a subjective and constructed concept and heavily mediated by institutional factors.
	Political ecology	Unequal resource distributions + exclusionary institutions	Distributive politics are important, but subnational hydropolitics are primarily influenced by arrangements for decentralization.
Institutionalist	Collective action	Lack of enabling conditions like trust and social capital + lack of robust institutions to support collective action	Collective action is extremely useful for describing conflict and cooperation dynamics at local levels but less so at the interjurisdictional level.

(Frey 1993, 55). As both formulations make clear, the underlying logic is that conflict between social groups or political jurisdictions is likely to result when too little water exists in a given place to satisfy the needs of all water users.

However, the existence of such a direct link is conceptually problematic. A physical shortage of water in fact means little by itself because many water users are able to adapt with the aid of technology and other resources. Given sufficient wealth and capacity, localized water scarcity can often be ameliorated through long-distance water transfers, for example (Otero et al. 2011). More meaningful conceptions of water shortage include economic scarcity, in which water users lack the financial means to obtain adequate supplies of water, and technical scarcity, where water users lack access to the necessary technology to obtain adequate supplies of water. These measures of water scarcity are dependent not on the physical supply of water but rather on the resources and institutions available to a given set of water users (Ohlsson 2000; Turton and Ohlsson 2000; Yohe and Tol 2002). Unsurprisingly, countries vary widely on these measures of economic and technical water scarcity, and just as this variation suggests, virtually all studies that find a relationship between physical water scarcity and conflict potential do so only when accounting for factors like state capacity and regime type as well as physical water scarcity (Raleigh and Urdal 2007; Gizelis and Wooden 2010; Bernauer and Bohmelt 2014). More recent work in the environmental security literature accordingly attempts to chart such "pathways" linking water scarcity to conflict through such intervening institutional variables (Kahl 2006; Bernauer, Bohmelt and Koubi 2012; Bohmelt et al. 2014). The pathways approach recognizes that the relationship between scarcity and conflict is not direct but is rather mediated by institutional capacity.

Environmental security owes much to a second literature, the Marxist-inspired subfield of political ecology, which emphasizes the role of political power in shaping access to resources. In some ways, political ecology incorporates Karl Wittfogel's influential, though now discredited, theory that autocracy arose in some regions because of the need to forcibly mobilize resources to maintain irrigation systems (Wittfogel 1957). Political ecology holds that asymmetric power structures enable elites to deprive certain social groups of access to critical resources such as water (Swyngedouw 1997; Mollinga 2008; Bakker 2010). Accordingly, the environmental security literature outlines processes by which water resources can be "captured" by particular groups or political actors, thereby stoking grievances among disenfranchised water users. Fundamental to the process of capture is the use of political power and, often, the co-opting of state institutions to secure control over water resources. According to one influential description of the capture process, "powerful groups recognize that a key resource is becoming scarce, and use superior institutional resources to grab it" (Homer-Dixon 2000, 15). In its emphasis on intergroup dynamics and power asymmetries, the political ecology literature adds an important political element to the scarcity and institutional factors emphasized in environmental security. However, the political ecology literature's focus on intergroup relationships does not directly address the primary form of conflict explored in this book, namely interjurisdictional water conflicts. Moreover, the political ecology literature, along with others within the deterministic tradition, does not satisfactorily explain why, as in several of the cases described in Chapters 4–7, both conflict and cooperation can arise even under difficult circumstances, such as weak states and acute water scarcity.

The literature on collective action, on the other hand, offers just such an account. The collective action literature has produced a detailed prescriptive theory of how institutional arrangements at multiple levels of organization together sustain robust, cooperative exploitation of shared resources without the aid either of state regulation or of the assignment of private property rights. In a seminal article, Elinor Ostrom and her collaborators demonstrated that cooperation over water can emerge in what is seemingly the most difficult of circumstances, namely a rural Nepalese irrigation system which depends on both upstream and downstream communities to work together (Ostrom and Gardner 1993). Subsequent work in the collective action tradition has outlined how collective action can produce sustainable and cooperative resource management at several "levels of action." At the lowest level, known as the "operational" level, informal norms and practices structure day-to-day interactions between resource users, while at higher levels, known as the "collective action" and "constitutional" levels, respectively, formal rules and procedures facilitate collective action (Imperial 1999; Ostrom 2011; Blomquist and deLeon 2011). At the next highest level, the collective action level, rules must exist which clearly define resource users, establish compliance mechanisms, and maintain social trust (Ostrom 1994a, 1994b, 2000). Indeed, this trust is the bedrock of successful collective action.

At higher levels of political and social organization, it becomes more difficult to establish and maintain such trust. As a result, collective action theory specifies that institutions be "nested" at different levels of governance, such that collective action can be undertaken even at the level of an interjurisdictional river basin, where individual social trust is insufficient to sustain cooperation (Marshall 2013). Nesting requires that political authorities like nations, states, and provinces create the necessary space for lower-level collective action to take place, for example, by empowering community decision-making (Andersson, Gibson, and Lehoucq 2006; Andersson and Ostrom 2008; Hardy and Koontz 2009). Effective nesting also requires "cross-scale" institutions like watershed associations or water policymaker networks to link actors at different scales of governance (Berkes 2002, 2006; Adger, Brown, and Tompkins 2005). Collective action theory holds that this form of "polycentric" governance enables water users to cooperatively govern shared water resources, even under changing circumstances (Schneider et al. 2003; Oakerson and Parks 2011; Pahl-Wostl et al. 2012). However, the theory of polycentric governance does not directly address the dynamics of conflict and cooperation between political jurisdictions.

Another branch of the collective action literature, known as "institutional collective action" (ICA), partly redresses this lacuna by explaining cases of self-organizing cooperation between neighboring municipalities in the United States (Ostrom, Tiebout, and Warren 1961). In particular, the ICA literature has identified several factors that encourage collaboration between neighboring subnational political jurisdictions. These factors include a prior history of collaboration, repeated interaction, and higher social and human capital (Lubell et al. 2002; Leach and Sabatier 2005; Feiock and Scholz 2010). The ICA literature finds that neighboring subnational jurisdictions that possess these characteristics, including strong personal relationships between mayors and shared perceptions of common interests, can establish institutions that have proven capable of solving collective action problems related to crime and economic development, for example (Post 2004; Bickers, Post, and Stein 2010). A related literature on collaborative and grassroots ecosystem management concludes that similar factors enable networked forms of cooperation between political units that share water bodies such as estuaries (Weber 2003; Sabatier et al. 2005; Berardo and Scholz 2010). By explaining the emergence or absence of cooperation between water users in terms of institutional design and function, these institutionally centered literatures present a compelling theory of water conflict and cooperation.

Nonetheless, as applied to the central problem in this book, that of interjurisdictional conflict in shared river basins, this extant theory suffers from several shortcomings. First, the factors that ICA identifies as being conducive to cooperation appear never to have been the subject of comparative analysis, so the applicability of its conclusions to countries and contexts beyond the United States remains unsubstantiated. Second, while collective action theory does acknowledge that diversity complicates cooperation over shared resources, it does not fully account for the power asymmetries highlighted in the political ecology literature, especially underlying sectional tensions and grievances between water users. These are important lacunae because, as we shall see in Chapter 3, such fault lines frequently prove to be the primary source of interjurisdictional water conflict. Third and finally, collective action theory has only recently and tentatively begun examining the role of political system structure in shaping institutional development (Rahman, Hickey, and Sarker 2012; Ratner et al. 2013; Persha and Andersson 2014). Decentralization is nonetheless a critical factor in this development: as we shall see in the remainder of this chapter, decentralization exacerbates many of the collective action challenges inherent to managing shared water resources. Indeed, as noted in the Introduction, one of the major implications of the case studies explored in Chapters 4–7 is that collective action theory should take more seriously how the interaction between sectional identity politics and decentralization affects collective action potential between resource users.

THE PROBLEM OF SCALE: DECENTRALIZATION AND COLLECTIVE ACTION POTENTIAL IN INTERJURISDICTIONAL RIVER BASINS

The central problem of hydropolitics is that water resources, including rivers, lakes, and subsurface aquifers, almost always occupy a different spatial scale from that of the people who use them. While water bodies such as lakes, aquifers, and river basins are usually hydrologically linked, human water uses are fragmented across different economic sectors, socioeconomic groups, and political jurisdictions. This fundamental problem of scale in turn imposes three distinct coordination problems for the management of shared water resources, which Richard Feiock (2013) calls institutional collective action dilemmas. The first of these problems, functional dilemmas, arise when responsibility for various aspects of water resource management, such as land use planning, pollution control, and navigation, is divided between multiple governmental authorities, who often work at cross purposes. Vertical dilemmas, in contrast, arise when responsibility for various aspects of water resource management is divided between central and local levels of government. Horizontal dilemmas, finally, arise because political jurisdictions are often improperly sized to efficiently capture externalities like transboundary pollution flows or migratory fish populations; this applies equally to nation-states that share a common waterway. Each of these dilemmas is exacerbated by a common feature of political organization, decentralization, though not evenly. As this section describes, decentralization imposes special challenges for interjurisdictional cooperation—the fundamental issue in subnational hydropolitics.

In general terms, decentralization can be thought of as the transfer of some combination of political decision-making, fiscal resources, and administrative responsibility from central to subnational levels of government (Treisman 2007). Nearly all countries feature some form of decentralization because national governments, no matter how centralized, cannot hope to fulfill all responsibilities of governing large territories as well as to accommodate local differences. However, the form and extent of decentralization vary widely across countries. While federal countries like India combine all three forms of decentralization, more centralized states such as China devolve some fiscal and administrative powers but retain those of decision-making and policy formulation (Riker 1964; Lijphart 1999; Treisman 2007). Given this variation, there is little consensus among scholars concerning whether or how decentralization influences conflict and cooperation over shared water resources. To the extent that decentralization features explicitly in the collective action and related literatures, it is generally viewed as a necessary but not sufficient condition for self-organizing collective action institutions to emerge. Such institutions are unlikely to emerge in the presence of a hegemonic central state (Ostrom 1990, 2010; Ostrom, Walker, and Gardner 1992). Decentralization and institutional collective action therefore often coincide. Two prominent collective action scholars argue, for example, that "Polycentric structures of overlapping organizations—networks, federal systems, and other multiple-organization arrangements" nurture the creation of flexible institutions that are more adaptable than centralized or bureaucratic management structures and, therefore, better able to deal with the challenges of managing entire watersheds (Schlager and Blomquist 2008, 17).

Other research traditions tend toward a less sanguine view of the effect of decentralization on the governance of shared natural resources. Scholars concerned with the fiscal aspects of decentralization and federalism, for example, point out that numerous economic incentives exist for subnational jurisdictions to compete for tax revenue, employment-generating business activities, and natural resource rents (Filippov, Ordeshook, and Shvetsova 2004; Cai and Treisman 2005; Volden 2005). Many, though by no means all, scholars extend the idea of such competition to suggest that decentralization can induce a "race to the bottom" as each subnational unit attempts to maximize its own share of economic resources, usually creating negative environmental and social externalities as a result (Desai, Freinkman, and Goldberg 2005; Gervasoni 2010; Sonin 2010). Several empirical studies of environmental outcomes appear to support this interpretation. Increased subnational competition appears to explain a large fraction of global deforestation (Burgess et al. 2012), for example; and neighboring jurisdictions are less likely to control pollution on waterways which flow into downstream jurisdictions, thereby passing the cost of compliance on to their neighbors (Lipscomb and Mobarak 2011). In sum, while these accounts diverge, they suggest that different types of decentralization carry significant consequences for the complex collective action problems inherent to water resource management.

Indeed, as this chapter details, the complex nature of water resource management sits uneasily within decentralized political systems. The dean of the study of fiscal federalism, Wallace Oates, observed, for instance, that America's highly decentralized federal system is "quite poorly designed to deal with the provision of certain important public goods, notably environmental resources," and that "a much more rational map would probably entail some fairly sizeable regional governments that extend over watersheds" (Oates 1999, 1131). A frequently cited essay, "The Dangers of Decentralization," meanwhile, cites France's highly successful system of regional river basin governance entities, known as "water agencies," as a chief example of how even highly centralized political systems can effectively solve collective action problems (Prud'homme 1995, 214). As these diverging accounts suggest, decentralization helps explain the institutional aspects of subnational hydropolitics, especially why it occurs despite the existence of a sovereign central government theoretically capable of containing it.

Under decentralization, both functional and vertical dilemmas are exacerbated by the division of responsibility for water resource management among multiple governmental units at multiple levels. Water quality, for example, is often regulated by both provincial- and national-level environmental protection agencies. Although in some cases this shared responsibility can provide redundancy and flexibility, in others the pursuit of similar policy objectives by different levels of government exercising similar powers creates a risk of duplication, inefficiency, or even intergovernmental competition (Scharpf 1988; Besley and Coate 2003; Volden 2005). These combined functional–vertical collective action dilemmas tend to be especially marked for federal countries. One of the defining characteristics of federal systems is a constitutional limitation on the power of central governments relative to subnational levels and a general attempt to confine exclusive central powers to high-spillover issue areas and the provision of public goods (Weingast 1995). As a result, federal constitutions typically grant the central level only limited powers over water resource management (United Nations Food and Agriculture Organization 2010). In the United States, for example, the lack of a coherent federal water resource policy has led states to adopt widely varying water rights doctrines, frustrating judicial attempts to settle interstate river disputes (Hundley 1975). The combination of vertical and functional institutional collective action dilemmas means that for decentralized political systems in general, and federal systems in particular, water resource issues are often placed among those "matters where the central government has no clear power [and] the states have, but because of territorial limitations cannot proceed singly" (Watts 1970, 79).

The problem of fragmented authority and responsibility for handling various aspects of water resource management is exacerbated by the diffusion of power to nonstate as well as local authorities. Beginning in the 1970s, and in parallel to the movement to promote political decentralization, efforts began to improve water resource management by reducing the role of the state in delivering and managing water. The number and role of functional, quasi-governmental bodies like utilities, water districts, and water-user associations have increased dramatically as central states sought alternatives to the perceived wastefulness, inefficiency, and elitism of command-and-control approaches (Vorosmarty et al. 2000; Mullin 2009; Bakker 2010). In many cases, states have combined devolution with the privatization of water supply and distribution systems (Bakker 2010). This effort has exposed many more water users and water supply utilities to market forces, but the results of privatization efforts have been highly uneven across countries (World Water Assessment Programme 2009). Although in some cases privatization has markedly improved the performance of water supply systems, in others it has simply stoked the belief that the state had "exercised . . . power on behalf of narrow, special interests" (Ingram, Whiteley, and Perry 2008, 17). Indeed, instead of correcting for the failures of central states in encouraging a collaborative and participatory approach to water resource management, in several cases water sector privatization has led to widespread civil unrest and political instability centered on concerns that the commoditization of water effectively reduced access to a critical resource (Olivera 2004). These nonstate dimensions of decentralization have further strained the ability of water resource managers to coordinate across different sectors and agencies. As economist Charles Howe has written, "It seems unlikely that . . . states, and all the special districts that currently have a say in water planning and management will surrender their prerogatives to unified river basin initiatives" (2005, 29). Whatever its benefits, privatization has certainly exacerbated certain coordination problems inherent to water resource management.

As this observation suggests, even these challenges of intersectoral and intergovernmental coordination are eclipsed by the third and most pronounced institutional collective action dilemma under decentralization, namely horizontal dilemmas. The problem of scale inherent to hydropolitics is most pronounced within river basins, which are topographical features that effectively represent different paths for any given drop of water that originates on a land surface to a terminal body of water, usually the ocean. The problem of scale is so acute within shared river basins because the management of shared river water is subject to high-spillover effects, meaning that it is difficult to confine either the costs or benefits of collective action to a single political jurisdiction within the basin (Young 2002; Bulkeley 2005; Imperial and Yandle 2005). Negative economic externalities like water pollution are transmitted from upstream to downstream regions of the river basin, while public goods like free navigation and flood control benefit all parties along a river (Dorfman 1974). Where their boundaries do not match those of a river basin or other water resource, any group of political units that shares a river basin must devise some form of collective effort to address common water resource management issues, including for the provision of aids to navigation, the control of pollution, the allocation of water, and the construction of large water infrastructure projects.

This basic collective action problem is also structured by the spatial orientation of political units relative to the water resources that they share. Although in practice shared waterways often flow between jurisdictions in some places and form the boundary between them in others, there are two basic configurations for political units which share a given river basin, upstream–downstream and co-riparian (see Figure 2.1). This overlay of political and physical geography creates a spatial incentive structure for conflict and cooperation. Because the total quantity of water in a basin is finite for most intents and purposes, basin water resources are generally rivalrous (Waterbury 2002). However, where jurisdictional boundaries are such that units are co-riparians, as in Geometry I, water uses such as hydropower are for practical purposes nonexcludable, providing a ceteris paribus incentive for cooperation if the actors involved are relatively few. Where jurisdictions lie up- and downstream of one another, the inexorable flow of river waters downstream means that while actions taken by an upstream actor almost inevitably affect those downstream, the reverse is not necessarily true. In Geometry II, this upstream–downstream orientation creates a fundamental asymmetry in power dynamics: the most important water uses, such as industrial effluent release and non-return-flow irrigation, are both subtractable and excludable, providing little inherent motivation for upstream

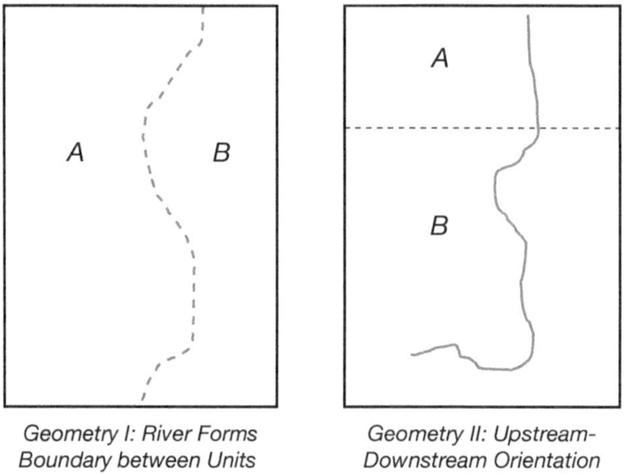

Figure 2.1 River Basin Political Geometries

units to cooperate with their downstream neighbors and a considerable incentive to simply pass on the costs of pollution, water consumption, and other negative externalities (LeMarquand 1976, 887; Shumeli 1999; Brochmann and Gleditsch 2012). This basic spatial mismatch often makes it more difficult to convince neighboring jurisdictions to cooperate regarding these shared river basin management issues.

Decentralization exacerbates such horizontal cooperation problems by imbuing subnational jurisdictions with considerable powers, including some more traditionally ascribed to nation-states. This is poignantly illustrated by their growing tendency to exercise functions traditionally ascribed to sovereign governments, such as diplomacy and foreign trade (Soldatos 1990; Paul 2005; Amen et al. 2011). In the early 1990s, scholars began to observe that state and provincial governments had come to play a noticeable role in foreign affairs. One early observer stated, for example, that "noncentral governments are progressively reaching beyond national borders to enhance their influence and economic well-being" (Fry 1993, 23). Subsequently, a variety of "para-diplomatic" activities have been observed that essentially replicate the functions traditionally ascribed to nation-states within the international system (Kincaid 2003). The foreign activities of American states, for example, have expanded to include a wide array of educational and economic activities, and it is now commonplace for senior state government officials to interact widely with their foreign counterparts (McMillan 2012). Even in Japan's highly centralized system, subnational units have begun to conduct substantial foreign economic relations and disburse forms of development aid to their foreign counterparts (Jain 2005). Less noted in the extant literature is the fact that, along with these powers and roles, subnational jurisdictions sometimes behave like independent nation-states—and produce similar challenges for institutional collective action in shared river basins.

In particular, the growing role and power of subnational jurisdictions under decentralization have made it more difficult to employ some of the institutional methods mechanisms, which Richard Feiock calls "integration mechanisms," that can help overcome collective action dilemmas. These mechanisms vary chiefly in their degree of institutionalization and range from informal networks that can create opportunities for cooperation on specific interjurisdictional challenges to centralized regional authorities that undertake a specific governance task, such as river basin planning and allocation, on behalf of multiple, overlapping subnational authorities (see Table 2.2). Importantly, as Feiock observes, these integration mechanisms can also be distinguished by the extent to which they preserve or reduce the autonomy of the existing political jurisdictions that establish them (2013, 401). Integration mechanisms that least threaten subnational autonomy are less costly and difficult to establish and, therefore, more common. However, they are also weaker, less comprehensive, and less capable of addressing the most complex governance challenges (Marshall 2008). More specifically, Table 2.2 shows that, to a certain extent, a trade-off exists between the strength of institutional collective action and the likelihood that it will occur. All other things being equal, and in the absence of external, third-party pressure or inducement, subnational leaders are more likely to pursue weaker and less institutionalized forms of cooperation over shared water resources.

In the absence of such institutions, subnational leaders are more likely to engage in competitive strategies to pursue their interests. In many cases, decentralized political systems provide avenues for such competition, with judicial lawsuits being perhaps the best example. A common feature of many federal systems, the use of judicial action provides a means for subnational units to secure additional water rights at the expense of their neighbors through simple judicial award, rather than

Table 2.2 Types of Integration Mechanisms and Degrees of Institutionalization

Integration Mechanism	Degree of Institutionalization
Informal networks	Essentially none; relies on repeated informal interaction between individuals
Contracts and mandated agreements	Low; creates mutually agreed responsibilities but no formal institutional structure
Working groups, partnerships, and constructed networks	Low; creates some institutional decision-making structures and obligations for participants but few constraints on individual freedom of action
Multiplex self-organizing systems and councils of governments	Moderate; creates joint decision-making structures, obligations, and some pooling of authority but preserves autonomy in most policy areas
Centralized regional authorities	High; creates separate decision-making structures and lines of authority that may constrain freedom of action and usurp the powers of existing jurisdictional units

Adapted from Feiock (2013).

collective action or political alliance-building. Because judicial decisions do not require political bargaining, judicial strategies are especially appealing to subnational states that lack large legislative delegations or that seek to address fundamental asymmetries in political power. Unsurprisingly, water-sharing disputes are often explicitly sectional, pitting upstream and downstream riparians against one another. In one recent example, Florida's governor justified a lawsuit against the neighboring upstream state of Georgia as an attempt "to stop Georgia's unmitigated consumption of water" (Editorial Board 2013). Similarly, a senior politician in the southern Indian state of Karnataka recently urged his fellow leaders to "explore the legal route of approaching the [Indian] Supreme Court" because the state had "been awarded less water share than what it should have rightfully got" from its upstream neighbor (Staff Correspondent 2013).

The concept of integration mechanisms therefore helps to bring the problem of cooperatively managing shared river basins into sharper focus. These mechanisms serve to illustrate that while the problem of scale in shared river basins vexes all countries, those that feature extensive decentralization are likely to face more acute collective action dilemmas, particularly of the horizontal kind. However, the dynamics of conflict and cooperation in shared river basins are not simply a static game between state actors. Third parties, including nongovernmental actors, can help implement stronger rather than weaker integration mechanisms in response to river basin collective action problems. For the purposes of this study, third parties are entities other than economic water users or subnational governments who nonetheless possess an interest in the governance of a given, shared water resource (Weinthal 2002). While less institutionalized integration mechanisms can and do arise organically, more institutionalized integration mechanisms are likely to arise only with encouragement, inducement, or resources provided by a third party, such as a higher-level government (Feiock 2013). But the development of collaborative and participatory institutions is a complex process. As Andrea Gerlak, Tanya Heikkila, and Mark Lubell observe, a key uncertainty in the literature on collaborative water resource management institutions is "how individual collaborative programs interact with other institutional arrangements, such as how states govern water resources generally" (2008, 424). In the next section, I describe how, under decentralization, these arrangements often lead to weak forms of cooperation in shared river basins.

THE DILEMMA OF AUTONOMY: INSTITUTIONAL PATCHWORKS AND INTERJURISDICTIONAL COOPERATION UNDER DECENTRALIZATION

Much of the existing literature on water resource management relies on the idea of self-organizing collective action—that the users of a shared resource will usually perceive it to be in their interest to cooperate and, when empowered to do so, will create appropriate institutions to facilitate cooperation. But as the preceding section has suggested, decentralization tends to exacerbate the collective action dilemmas inherent to cooperatively managing interjurisdictional river basins. This is especially true of horizontal collective action—the very type of interjurisdictional cooperation that is most crucial when multiple subnational units share a river basin. But while decentralization does make it more difficult for neighboring subnational jurisdictions to cooperate, it does not do so fatally. Most real-world examples of interjurisdictional relationships are characterized by a mixture of conflict and cooperation, even in highly decentralized political systems (Feiock and Scholz 2010). As noted in the Introduction, cases of intense and protracted interjurisdictional water conflict also often coincide with extensive cooperation in other regions and policy areas. This section explains this apparent contradiction in the form of institutional patchworks, wherein interjurisdictional collective action does take place but cooperation usually takes the form of weak agreements that are themselves unable to prevent water conflict from occurring. The concept of institutional patchworks highlights a key feature of subnational hydropolitics: it's not the existence of shared institutions that matters for preventing water conflict but rather their strength and flexibility.

Underlying the question of what kind of institutions prevent water conflict is an even more basic one: why would subnational leaders choose not to cooperate with their neighbors over water, or any other issue of common concern? The answer, as I elaborate in this section, has to do with the fact that cooperation sometimes comes at the expense of autonomy, which subnational officials prize because it enables them to pursue independent interests, including of course their own. Subnational autonomy is often tenuous: unlike nation-states in the international system, states, provinces, counties, and prefectures are subject to a single, sovereign central government. While nearly all sovereign nation-states divide the responsibilities and powers of government between central and local levels, each of which has some capacity for independent action, in many cases power is delegated rather than devolved (Duchacek 1970). Local governments in France and Denmark possess considerable practical autonomy, for example, but the center retains veto power and can coerce subnational officials through political sanction, financial inducement, or the threat of recentralizing delegated powers.[1] Even in federal systems where subnational powers are constitutionally protected, practical autonomy waxes and wanes over time, and subnational leaders must often be on guard against overreach by the federal government (Wheare 1963; Braun 2000). To be clear, in most cases subnational officials are likely to find it in their interest to cooperate with their neighbors. But the precarious nature of subnational political power can also discourage subnational leaders from engaging in institutionalized collective action with their neighbors.

The general value of autonomy stems from the fact that it allows subnational officials to pursue their own interests and objectives rather than those of the center. Such preference divergence is common: virtually all countries suffer from some form of principal-agent problem between central and local levels. Local officials, no matter how tightly controlled by the center, inevitably develop preferences, interests, and objectives that differ from those of the national capital simply by sitting outside of it. China, for instance, despite being a hierarchical, authoritarian country, is characterized by pervasive noncompliance with central directives at the local level, where officials frequently disregard them when they are too expensive or difficult to follow (Lieberthal 1992; Mertha 2005). While the extent of this preference divergence differs among countries depending on how much power is granted to central versus local levels of government, it creates a common dilemma for subnational leaders across countries: how to promote their own interests and objectives while avoiding sanction and preserving access to resources from the central government. Such autonomy helps to bolster the legitimacy of subnational leaders to govern and to enhance their ability to build political alliances and pursue policy priorities.

Preserving this sphere of autonomy can lead subnational officials to oppose highly institutionalized interjurisdictional collective action and to prefer looser, weaker forms instead. Institutional collective action entails agreeing on methods of coordination, the distribution of costs and benefits among actors, and preventing free-riding and defection. In the case of shared water resources, each of these issues is complicated by pervasive information asymmetries and the difficulties of adequately pricing water. Each of these requirements of institutional collective action therefore carries a distinct risk of failure, such that subnational officials risk surrendering their jurisdiction's authority or autonomy without gaining any permanent benefits in return (Feiock 2013). Instead of entering into agreements that may obligate them to follow certain procedures or cause them to foreclose options like pursuing interjurisdictional lawsuits, subnational officials are likely to prefer to join interjurisdictional bodies whose primary functions are communication and consultation rather than decision-making. Joseph Zimmerman notes, for example, that in the United States, "The modus operandi of most states does not encourage extensive interstate joint ventures because states, as semiautonomous entities, naturally are reluctant to engage in such ventures due to the loss of exclusive control accompanying them" (2011, 201).

At the same time, autonomy-seeking subnational leaders are likely to approach institutionalized cooperation with their riparian neighbors strategically, choosing to cooperate in some cases and not in others. As a seminal study of U.S. river basin commissions observed, for example, "Most states appear to join commissions for defensive purposes. . . . They seek defenses against one another as well as against federal action" (Derthick 1974, 151). As this observation suggests, conflict and cooperation between riparian jurisdictions are not mutually exclusive. While the jurisdictions that share a river basin may form a joint body to lobby for central government financing for regional flood control, they may also repeatedly challenge each other's use of a shared river in court. In the interest of preserving their autonomy vis-à-vis the center as well as against neighboring jurisdictions, subnational states are likely to prefer weaker, less institutionalized arrangements that preserve their freedom of action. A typical result of decentralization is therefore what I call an institutional patchwork, in which some strong interjurisdictional institutions exist but weaker forms of cooperation predominate.

An important implication of the dilemma of autonomy is that subnational officials concerned with preserving it tend to strongly resist the creation of new institutions and bodies that dilute or duplicate powers existing given to their jurisdictions. Where powerful special-purpose jurisdictions like the Tennessee Valley Authority were created over the vociferous opposition of state leaders, for example, subnational officials devoted considerable resources to ensuring that the power of such entities was severely circumscribed (Chandler 1984). As a result, such centralized regional authorities are comparatively rare in decentralized countries. Special-purpose functional jurisdictions like water districts and river basin management committees, on the other hand, appear to be more common. While little consistent cross-national data are available, they do appear to bear out the proposition suggested by the cases examined in this book, that subnational states in decentralized political systems tend to engage in weaker and less institutionalized forms of cooperation. Local governments that share estuaries, for instance, are more likely to cooperate through working groups, partnerships, and constructed networks than formal institutions like commissions or centralized regional authorities (Schneider et al. 2003; Berardo and Scholz 2010; Floress, Prokopy, and Broussard 2011). In the United States, water-sharing agreements rather than commissions represent the most common form of institutional collective action in shared river basins (Schlager and Heikkila 2009).

Institutional patchworks in which weaker cooperative institutions coincide with stronger, more centralized authorities do possess important advantages. They are by definition more diverse than monocultures and can foster innovation and adaptation to emerging environmental challenges (Pagano 2007; Clarke 2007; Schlager and Blomquist 2008). Functional jurisdictions, in particular, possess a number of advantages; unlike states and provinces, they are relatively inexpensive to create and modify (Hooghe and Marks 2010). Because they are intended to exercise a specific set of functions, they can provide "services to specific areas without following the jurisdictional lines of an existing city or county," thereby presenting a useful response to "problem[s] delineated by natural features, such as a watershed" (Mullin 2009, 6). There are indeed many examples of successful, special-purpose water jurisdictions: Nebraska's Natural Resource Districts, for example, are organized around watershed lines and have helped to implement innovative markets for groundwater, a notoriously complex water resource management problem.[2]

However, the problem with institutional patchworks in shared river basins is that they set the conditions for conflict to emerge as well as cooperation. In some basins, shared institutions are strong enough to prevent conflict between riparian jurisdictions, while in others they are either too weak or lacking altogether. Weak agreements and institutions, though numerous, are often incapable of addressing large-scale collective action dilemmas or of preventing conflict between water users as water allocation requirements change over time (Garrick 2015). In particular, these institutions typically lack features that raise the cost of compliance, such as monitoring, enforcement, and dispute resolution mechanisms, making them easier to forge but less effective in practice (Mandarano, Featherstone, and Paulsen 2008; Zimmerman 2012; Schlager, Heikkila, and Case 2012). Moreover, these agreements typically include few provisions related to participation and often limit decision-making to state water resource managers. As John Thorson has observed, "No compacts have truly succeeded in eliminating externalities because the regional governments that they establish do not coordinate well with pre-existing governments which create the compacts" (1994, 140). Thus, where subnational states do cooperate, they usually do so in the form of weak institutions that are unlikely to contain or prevent conflict as water resource management challenges increase. Patchworks of weak institutions instead set the stage for subnational hydropolitics to unfold, as the following chapter explains.

NOTES

1. The threat of future regulation is often one of the most effective instruments central governments possess to promote reform over the objections of local governments. However, like all threats, it must be credible in order to be effective.
2. I am indebted to Richael Young of Mammoth Water Trading for the example.

3

Dynamics of Hydropolitics

Sustaining Conflict and Catalyzing Cooperation over Water

As noted in the Introduction, it is far from clear why subnational jurisdictions like states and provinces would choose to fight over water, especially for prolonged periods of time. Enmeshed in deep political, social, and economic relationships, these jurisdictions and the politicians who lead them would seem to have much more to gain from cooperation than conflict. In this chapter, I explain the emergence and persistence of subnational conflict and cooperation over water in terms of two oppositional factors, sectional identity and political opportunity structure. As I describe further in this chapter, persistent interjurisdictional water conflict usually results from the conflation of shared water resources and sectional identity. In particular, subnational political leaders often seek to gain political support from geographically concentrated ethnic, linguistic, or other groups by alleging that neighboring groups are effectively stealing water at their constituents' expense. Because of the unique economic importance and emotional resonance of water, such allegations are highly effective in mobilizing political support, and successive subnational politicians can find it expedient to engage in confrontational rather than cooperative behavior. Using water as a tool for political mobilization can also help subnational politicians achieve other objectives, including enhancing their autonomy, distracting from other political challenges, and outflanking opposition parties. On the other hand, where the political opportunity structure in a given country enables third parties to form alliances with national political elites, these alliances create a constituency for strong interjurisdictional institutions that can prevent and resolve conflicts over shared water resources.

This chapter proceeds in three sections. First, it explains how the conflation of sectional identity and shared water resources sustains interjurisdictional conflict over long periods of time. The desire of subnational politicians and geographically concentrated water user groups to seek broader support for their objectives leads them to link water-sharing issues to sectional identity. The second section describes how, provided they enjoy access to the political process, third parties can help to induce collective action in shared river basins. The third and final section integrates these processes to present a framework for how subnational hydropolitics unfold and how cooperation can emerge over time. The role of third parties provides the third and final piece of the puzzle posed in the Introduction, explaining why some countries feature high levels of interjurisdictional water conflict and others none, as well as why in some river basins interjurisdictional conflict has slowly been transformed into cooperation. Third parties and their elite allies are essential to counteracting the influence of subnational political leaders, who are often driven to use shared water resources as a tool for political mobilization. If the value of autonomy explains why subnational officials sometimes choose conflict over cooperation, its alter ego, sectional identity, explains why interjurisdictional water conflict can not only arise but also persist over long periods of time.

SUSTAINING CONFLICT: SECTIONAL IDENTITY AND POLITICAL MOBILIZATION OVER WATER

The previous chapter explained why, although cooperation is usually in the interest of subnational leaders, their concern for autonomy often causes them to oppose the creation of strong interjurisdictional institutions to manage shared river basins. In the absence of these institutions, minor disputes can morph into sustained interjurisdictional water conflict. Given the existence of so many incentives to engage in cooperation, one would expect even those subnational leaders who choose to engage in conflict to do so for only limited periods and that few of their successors would perpetuate these conflicts. While in the short run subnational officials might wish to compete for water for purely economic reasons, for example, these are unlikely to lead to persistent conflict. In this section, I describe how the power of water to mobilize political actors on the basis of shared sectional identity explains why interjurisdictional water conflicts are so hard to stop once they start. While many observers stress the normative resonance of water—the fact that debates over its allocation, quality, and distribution ignite particular passions—fewer touch upon its political salience. Given the unique importance of water to all economic sectors and the fact that water uses are closely tied to individual localities, shared water resources prove to be a potent locus for political mobilization. Subnational political leaders and water-user groups often find it convenient to adopt shared waterways as a kind of political foil, galvanizing support based on alleged inequities between neighboring groups and regions. This linkage between water and sectional identity can sustain subnational water conflict over long periods of time and often despite multiple attempts to resolve it.

The degree to which identity politics shape water management has been only partly recognized in the existing literature. Scholars concerned with the anthropological dimensions of water have long emphasized that cultural values of water vary widely between groups and across regions (Ingram, Whiteley, and Perry 2008). At the same time, scholars of collective action have long acknowledged that resources distributed between different ethnic, linguistic, or other social groups are more difficult to manage cooperatively. This kind of heterogeneity among resource users may inhibit creation of the trust that underpins collective management of shared resources and is likely to make it more difficult to reach consensus on contentious issues like assigning water usage rights (Ostrom et al. 1999; Feiock 2013). But the relationship between sectionalism and shared water resources often goes much deeper.

In northern Yemen, for example, water use was historically governed by an elaborate system of rules that determined the right to divert runoff from the annual spring rains. For centuries, this system effectively allocated water between the several tribes who shared the region's scarce water resources. The system had no provision for regulating groundwater use, however, and following the advent of pumping in the 1960s, subsurface water resources were seen as "fair game" open to unlimited use. Water users split between various tribes, who were often in conflict both with each other as well as with the central state, were unable to agree on new rules to govern the use of this novel resource; and groundwater mining has continued mostly unimpeded ever since (Lichtenthaler 2003). To a very large extent, Yemen's continuing water crisis, one of the world's most severe, is a reflection of its tribal divisions. Indeed, around the globe, many water-user conflicts map in one way or another onto deeper tensions between groups united by different identities.

Divisions between social groups, whether differentiated by ethnicity, language, or other characteristics, often coincide with differences in water use, translating existing tensions into the sphere of shared water resources. These value-based sectional conflicts can occur between social groups or between regions, and sometimes both (Boelens, Getches, and Gil 2010). A common example of social group–based sectional conflict is disputes between water users who value recreational, religious, or environmental uses of water and those who favor agricultural and industrial uses (Moore, Mulville, and Weinberg 1996). This is the case in California's San Joaquin Valley, for example, where a largely rural agricultural region is challenging federal regulations which limit withdrawals for irrigation in order to protect endangered species—rules that, the San Joaquin farmers argue, reflect a misguided set of urban-elite priorities (Wines and Medina 2015). In many places, sectional water conflict is far more acute. Studies of local water conflict in regions as diverse as Darfur, northeastern India, and the Andes, for example, suggest that group territorial identity and perceptions of fairness underpin long-running feuds and violent disputes over common water supplies (Assal 2006; Chowdhury and Kipgen 2013; Boelens 2015).

Similar grievances can also mobilize sectional groups against central or local governments. In a tragic recent example, violence flared in India's Karnataka state when farmers protested the state government's inability to prevent a Supreme Court order prohibiting Karnataka from diverting water from the Mahadayi River, which is shared with the neighboring states of Goa and Maharashtra. A village leader's allegation that "We have been demanding water from the government, but what we received in return is verbal abuses, boots and laathis [clubs]" caused a minor political crisis, with state governing and opposition party leaders trading accusations over the others' ability to fight for Karnataka's interests. The state legislature soon became consumed by accusations of the opposing party having sold out Karnataka's interests in inter-state water-sharing issues (Jamwal 2016).

This conflation of water and sectional identity is amplified by institutional arrangements that codify sectional cleavages through subnational boundaries. Most countries recognize sectional divisions in the territorial makeup of government. Boundaries between states and provinces, for example, often reflect political and cultural regions that predate the formation of the country itself (Muthien and Khosa 1995; Larkins 2010). This is especially true of federal political systems, which typically delineate political units like states and provinces around ethnoreligious, linguistic, or cultural cleavages (Sharansky 1970; Elazar 1984, 1987). Formal recognition of these sectional cleavages in the design of most federal systems reflects the fact that federations are often formed through bargains between preexisting political units which each possess a strong sense of identity and which demand autonomy and recognition in return for uniting in a single country (Rector 2009). Precisely because they are drawn around these preexisting geographic identities, subnational jurisdictions can become focal points for sectional conflicts and rivalries (Gibbons 1987; Stewart 1990; Bowman 2007). As Erik Wibbels observes, "Whereas unitary systems encourage the formulation of national platforms and policies, federal systems explicitly give voice to geographically concentrated interests" (2006, 173). India's postindependence leaders, for instance, drew state boundaries along ethnolinguistic lines, believing that doing so would simplify administration (Singh 2007). However, the codification of ethnolinguistic cleavages has instead promoted the growth of state-based political parties that appeal to sectional, rather than national, identity (Mawdsley 2002).

Within individual subnational jurisdictions, officials face incentives to treat water as a distributive good to seek the support of sectional groups. In most countries, water is allocated by political rather than economic means, making it a valuable instrument of political patronage. Most urban water prices around the world remain very low, and in many countries and regions agricultural water is virtually free (Rogers, de Silva, and Bhatia 2002; Zetland and Gasson 2013; Cooper, Crase, and Pawsey 2014). In the absence of market-based allocation of water, it becomes essentially a distributive good, and water resource management policies come to reflect the interests of individual political constituencies within counties (Dinar, Rosegrant, and Meinzen-Dick 1997). Shaped by political interests, water resource management decisions often create distinct sets of winners and losers, such as those favored with subsidized water or those who suffer as a result of dam construction (Bennett 1974; Hart 1974; Ingram, Whiteley, and Perry 2008). Under political allocation, providing access to water become a useful means for politicians to deliver benefits to favored constituents. The political value of water as a distributive good creates a certain incentive for subnational politicians to maximize control over water shared with neighboring jurisdictions, in order to increase the quantity of goods that can be distributed to one's own constituents. Control over water therefore represents a form of political power that subnational political leaders are reluctant to surrender.

The conflation of sectional tensions and shared water resources can create considerable pressure on subnational leaders to compete rather than cooperate with neighboring jurisdictions. This dynamic is well illustrated by the Cauvery (Kaveri) River dispute, which has pitted Karnataka against the neighboring state of Tamil Nadu for decades. In the early 1990s, "regional jingoism" over the dispute in Karnataka led to attacks on Tamil schools, while a decade later a flare-up attracted the support of film stars in both states for their respective linguistic groups, and Karnataka resorted to blocking all Tamil television channels. Matters have scarcely improved: as a recent news report records, "the Cauvery water dispute is a perennial tinder box waiting to be lit by activists and political parties looking to appease their vote banks." This prediction proved prescient as both Tamil Nadu's major parties revised their election manifestos to include a pledge to force Karnataka to release water to its downstream neighbor (Murali 2016). In Karnataka, meanwhile, a powerful farmers' organizer led protests in which Tamil Nadu's chief minister was burned in effigy and Karnataka's politicians were warned that they would "face strong agitation if water was released to Tamil Nadu" (Press Trust of India 2016). The extent of the countervailing pressure placed on the Tamil Nadu government, meanwhile, was demonstrated by a six-party meeting called at the behest of the state's leading farming association in August 2016. At the meeting, the association's powerful president excoriated the Tamil Nadu government for not acting quickly to secure the state's allocation of water, charging that "The Tamil Nadu government went to the Supreme Court only after we held protests . . . demanding release of Cauvery water" (*The Hindu* 2016a).

But while sectionalism can sometimes pressure state politicians to adopt a confrontational policy with respect to shared natural resources, it can also allow them to manipulate the emotive politics of water to gain support. In 2016, the long-serving chief minister of Odisha sensed an opportunity to distract from political controversy by manufacturing a dispute with the upstream state of Chhattisgarh over the shared Mahanadi River. The Mahanadi had not previously been a source of significant political debate in Odisha, a relatively water-rich state. Chief Minister Naveen Patnaik therefore originally found little reason to oppose Chhattisgarh's plans to develop water infrastructure on its portion of the river, especially since Odisha's own booming industrial sector was expected to benefit. However, Patnaik's stance changed once his government became imperiled by a corruption scandal. Patnaik had previously tended to cast New Delhi as a scapegoat in times of political difficulty. As a newspaper account put it, "One of his successful political strategies is to blame the central government when the going gets tough at home." However, this time, the opposition Congress Party began to criticize Patnaik for failing to stand up to Chhattisgarh's development plans for the Mahanadi, charging that his government was either "inefficient or hand-in-glove with the industrial houses expected to get benefits from the water projects in Chhattisgarh." Stung by the criticism,

Patnaik calculated that there was more to be gained by engaging in conflict with Chhattisgarh than cooperation. He soon began to publicly chastise neighboring Chhattisgarh for stealing Odisha's water (Swain 2016).

The linkage between sectional grievances and shared water resources is easy to make but difficult to break. As a news report noted, the chief minister's sudden, repeated, and strident attacks on Chhattisgarh made it "awkward to Patnaik to scale down his rhetoric on the water row" (Swain 2016). Meanwhile, Odisha Congress Party workers initiated a statewide work stoppage and blocked highways with burning tires to protest Chhattisgarh's planned dam construction on the Mahanadi, creating further pressure on Patnaik from the opposition (NewsX Bureau 2016). As the Odisha example highlights, the charge that state politicians are failing to protect their constituents' access to water is especially potent and one they are consequently desperate to avoid. Given the effectiveness of water issues at galvanizing supporters, the political calculus for subnational politicians would seem to weigh in favor of taking an aggressive stance, lest they be outflanked by political opponents. This logic certainly seems to apply in the case of the Cauvery, where an editorial in *The Hindu* proclaimed that "The Cauvery water dispute is turning out to be less about water and irrigation and more about linguistic chauvinism and regional identity. . . . No party or state government appears to believe it can afford to be seen as taking even so much as a conciliatory step toward defusing the crisis" (2016c). This potential for political mobilization helps to explain why subnational leaders persistently allow sectional tensions and grievances to shape their approach to shared water resources, often favoring competition over cooperation.

As these examples illustrate, subnational states in decentralized political systems compete for water much as countries can but do so through political and judicial, rather than diplomatic or military, means (Bakker 1999; Sneddon and Fox 2006; Harris and Alatout 2010). But when sectionalism is mapped onto shared water resources, this competition can be every bit as harmful to the cause of interjurisdictional cooperation. The testimony of a Nevada congressman in opposition to federal efforts to enhance interstate cooperation along the Colorado River illustrates this point well: "I think the states should guard most zealously the water that belongs to them. . . . The water is our natural resource" (U.S. House Committee on Irrigation and Reclamation 1928, 51). But if the conflation of water and sectional identity explains why interjurisdictional water conflicts persist, what about cooperation? In the Introduction, I emphasized that the intensity of interjurisdictional conflict can wax and wane over time. Periods of cooperation are distinguished by alliances between third parties, who lobby for more collaborative and participatory approaches to governing shared river basins, and national elites, who can help translate this advocacy into action. These alliances can overcome sectional interests by lobbying for the creation of collaborative, participatory institutions, in essence creating an alternative political constituency for cooperation instead of competition. As the next section describes, third-party participation is therefore a critical determinant of the degree of conflict and cooperation in shared subnational river basins.

INDUCEMENTS TO COLLECTIVE ACTION: POLITICAL OPPORTUNITY STRUCTURE AND THIRD-PARTY PARTICIPATION

The previous section explored how sustained interjurisdictional water conflicts are rooted in sectional and identity politics. The conflation of these tensions with shared water resources makes it even more difficult to establish strong, shared institutions to effectively manage river basin water resources. This section explores how institutionalized cooperation can nonetheless emerge with the aid of third parties like multilateral institutions and nongovernmental organizations (NGOs). These third parties create a crucial constituency for cooperation. The involvement of third parties takes several forms, including the formation of transnational advocacy networks, direct lobbying of governments, and the use of "side payments" to persuade otherwise reluctant national governments to engage in cooperation over shared water resources (Waterbury 2002; Weinthal 2002; Bobekova 2015). The involvement of third parties in attempts to strengthen cooperation in international transboundary river basins is well documented, but their impact at the subnational level has received much less attention. Nonetheless, accounts suggest that domestic third parties, such as environmental NGOs, can also promote international cooperation by shaping national political discourse in favor of collaboration (Zawahri and Hensengerth 2012). Whatever its source, the crucial feature of third-party involvement in managing shared water resources is that they possess a shared belief in the value of creating strong collective action institutions and of collaborative, participatory approaches to decision-making (Myint 2012). To be sure, this belief is often rooted in self-interest: without institutions in which they possess a voice, third parties would be unable to participate in decision-making. But this very self-interest makes third parties reliable advocates of the kind of institution-building that promotes cooperation in shared river basins.

Before proceeding, it is worth elaborating on the definition of third parties. The term *third party* arises from the fact that in many cases entities exist who are concerned with how water is managed and used in a given basin but are not themselves direct water users. Although they are not necessarily neutral, third parties by definition do not have a direct role to play in deciding issues like how much water is allocated to different states in a shared river basin or in how best to control pollution that flows downstream from one province to another. On the other hand, third parties often have a significant stake in the outcome of such decisions. By this definition, there are many possible third-party actors. For example, multinational corporations may be concerned about whether interjurisdictional legal battles over water might impact them, even if the corporations have no concrete stake in the outcome of litigation itself. Other potential third-party actors include multilateral and international organizations, civic associations, and academic institutions. However, not all third parties are created equal. Those with greater resources, prestige, or interest in a given river basin are more likely to enjoy influence.[1] Moreover, third parties must often form coalitions that include political elites in order to gain access to and influence in the political process. Brazil's highly successful participatory model to basin governance, for example, "depends on the ability of reform-oriented networks" composed of technocrats and NGOs to band together to "resist and diffuse policy opposition" (Lemos and de Oliveira 2004).

In the case studies explored in this book, two types of third-party actors emerge as especially important: national governments and NGOs. National governments are not usually directly party to interjurisdictional water conflicts, but they have a clear interest in the outcome: such conflicts are often politically embarrassing and can undermine economic growth in affected regions. The interest of NGOs in such conflicts, meanwhile, arises because of their issue-based orientation (Mertha and Lowry 2006). The importance of environmental NGOs in advocating for national environmental policy reforms has long been noted, for example (Harrison 1996; Weaver and Rockman 1996; Scott 2000). In the context of shared river basins, environmental NGOs usually seek to end interjurisdictional conflict because it threatens attempts to protect riparian ecosystems as a whole. Both central governmental and NGO third-party actors can play a key role in promoting cooperation over conflict in shared river basins. However, both possess individual constraints: national governments are often hesitant to take the lead in resolving interjurisdictional disputes because they often depend on subnational politicians for electoral support. NGO third parties, meanwhile, often lack the influence to directly influence policymaking on their own. Both national government and NGO third parties are therefore most effective when they form coalitions, as described further in the final section of this chapter.

The importance of these third-party coalitions lies in their ability to act as bridging organizations in river basins shared between multiple water users residing in multiple political jurisdictions. Bridging organizations, also known as "boundary organizations," play a critical role in environmental governance by transferring knowledge and expertise and arbitrating interests across central, regional, and local levels of governance (Brown 1991; Cash et al. 2006; Berkes 2009). Thus equipped, bridging organizations perform three important functions in managing shared river basins. First, bridging organizations link actors and decision-making at both central and local levels of government. Second, they help to circumvent the opposition of subnational politicians to robust interjurisdictional collective action. Third, bridging organizations lobby for the creation of collaborative and participatory water resource management institutions in which they enjoy both voice and representation. This last function also highlights the fact that third parties provide the most powerful, and often the only, constituency for robust interjurisdictional collective action, but they do so out of profound self-interest. In the case of shared river basins, though, this self-interest serves a crucial purpose: creating strong institutions that reduce the risk of conflict.

Strong institutions that are collaborative, participatory, and adaptive serve to reduce the risk of conflict by mitigating its proximate causes. In game-theoretic terms, such institutions alter the payoff structures of the users of a shared water resource, reducing the incentive to free-ride and pass costs on to other users.[2] They do so by providing multiple opportunities for interaction, thereby increasing the cost of "defecting" by, for example, allowing unimpeded groundwater abstraction from a shared aquifer or alternatively building a large storage dam to deprive a neighboring jurisdiction of dependable downstream water flow. By the same token, institutions also reduce the transaction costs entailed in cooperation. They provide a forum and mechanism for decision-making, information-sharing, and consensus-building, which aids in addressing complex or contentious issues like intersectoral water allocation, environmental flow restoration, and climate change adaptation. Such institutions ultimately serve the interests of all actors but especially third parties. By lobbying for the creation of strong collaborative institutions, groups concerned with water resource management can gain both voice and influence into decision-making, allowing them to pursue agendas like environmental protection, improved water quality, or changes to planned infrastructure projects. Third parties thereby create a natural constituency for cooperative institution-building in shared river basins.

However, the ability of third parties to effectively lobby for cooperative management of shared water resources is highly contingent on their ability to participate in the political system and to forge alliances with political elites. More specifically, the ability of nongovernmental actors to play a role in political decision-making, social mobilization, and other activities depends on the prevailing political, social, and economic environment in a given country. Social movement scholars refer to the degree of such inclusion as the "political opportunity structure" (Kitschelt 1986; Kriesi 1995; Van der Heijden 1997). *Political opportunity structure* refers to a wide range of factors, including the legal basis for nongovernmental participation in political processes, the degree of repression or recognition employed under different forms of government, and the level of political consciousness and mobilization in a given society (Benford and Snow 2000). Social movement scholars also distinguish between states that actively seek nongovernmental participation in decision-making, as is the case for many European countries, and those like the United States, which accept such participation but do little to actively encourage it (Dryzek et al. 2003). The definition of political opportunity structure is open to some debate among sociologists, but as a concept it is useful in distinguishing variation in the degree of openness and acceptability of nongovernmental participation across countries (Tarrow 1988; Meyer 2004).

Political opportunity structure also shapes the ability of third parties to forge alliances with elites, including political leaders, at central and local levels of government. This capacity depends in turn largely on the allocation of political power within a given country, which ranges from highly centralized, authoritarian regimes, in which decision-making is tightly controlled by a small elite, to highly decentralized federal systems, in which political power is constitutionally allocated between actors at national and subnational levels of government. In general, the more decentralized the political system, the more points of access it may be expected to offer to third parties, especially to NGOs (Kitschelt 1986; Kriesi et al. 1992; Marks and McAdam 1996). Several accounts suggest that the dispersion of policymaking powers across levels of government makes it more difficult for elites to resist grassroots pressure to enact reforms, thereby increasing the chances that third parties will succeed in influencing policy at one level or another. If third parties fail to gain access to decision-makers at the local level, for example, they can also choose to pursue their cause at the provincial or national level (Cheikbossian 2004, 2008). Whatever the causal mechanisms, multilevel governance critically shapes the course of environmental policy reforms. In both the United States and France, for example, conservation organizations effectively exploited points of access at both local and national levels to gain support for national water quality legislation in the early twentieth century, for example (Paavola 2006; Bouleau 2009).

The importance of environmental movements and organizations in achieving sustainable resource management has long been noted by scholars. Third-party NGOs have been especially important in catalyzing opposition to dams and other environmentally destructive, water-related infrastructure projects in countries as diverse as Brazil, the United States, and India (Keck and Sikkink 1998; Wood 2007; McCool 2012). However, these organizations can also create a constituency for cooperative, integrated approaches to managing shared river basins. In the midst of a bitter interstate dispute over the Mahanadi River basin in eastern India, for example, nongovernmental Riverkeeper chapters from the two states formed a "Mahanadi Peace Initiative" dedicated to "connect people of both the states to the Mother River" (Orissa Diary 2016). The case of Brazil moreover illustrates how civil society groups and organizations perform several critical functions in lobbying for the creation of collaborative, participatory approaches to governing shared water resources. NGOs proved crucial in several respects: first, overcoming bureaucratic inertia to proposed participatory reforms; second, channeling resources to agencies charged with implementing reforms; and third, formulating principles and objectives for new governance institutions that were created at the river basin scale. Fourth, and even more importantly, civil society groups also bestowed legitimacy on water policy reforms and institutions through their participation and support. Because the reforms were intended in large part to coordinate a diverse set of public and private sector actors in pursuit of common goals, this legitimacy proved to be an essential element of institutional success and performance (Abers and Keck 2006, 2009, 2013).

The significance of the Brazilian example is further heightened by the fact that NGOs were not merely *accepted* as participants in water resource decision-making but were actively *encouraged* to play such a role. Indeed, the government viewed nongovernmental actors as critical sources of legitimacy as well as expertise, advice, and insight in creating new water resource management institutions (Abers and Keck 2006, 2009, 2013). As the Brazilian case suggests, governments can either encourage or inhibit the involvement of third-party NGOs in the political process in a number of ways, including in the definition of policy issues as either technical or political, designing participation rules in decision-making structures, and offering resources to civil society in the form of grants, information, or capacity-building (Koontz et al. 2004). The Brazilian example moreover suggests how formal state resources and nongovernmental participation can complement one another: the state provides the financial and technical resources that NGOs often lack, while NGOs provide the legitimacy and convening power that technobureaucratic organizations inherently find it difficult to provide. The case of Brazil thus suggests that when third parties are accorded sufficient influence in policymaking and made an equal partner to the state, they can greatly aid in the development of cooperative and participatory decision-making institutions. Third-party involvement can also bestow legitimacy upon these institutions in the eyes of political actors at different levels—including states and provinces that might otherwise oppose interjurisdictional collective action. In the following section, I present a model to describe how these dynamics of hydropolitics play out within countries, including the case studies described in Chapters 4–7.

FROM CONFLICT TO COOPERATION: SECTIONALISM, AUTONOMY, AND THIRD-PARTY INSTITUTION-BUILDING

The preceding two sections of this chapter have examined how both interjurisdictional conflict and cooperation can emerge in shared subnational river basins. The two key factors that shape cooperative versus conflictual outcomes, sectionalism and political opportunity structure, vary widely across countries, helping to explain why subnational hydropolitics are more common in some nations than others. Yet one question raised in the Introduction remains outstanding: how interjurisdictional river basins characterized by conflict can become more cooperative over time. In this section, I propose a framework to explain how shared river basins can not only evince both cooperative and noncooperative outcomes but also transform outcomes over time. This framework, which is based principally on the case studies described in Chapters 4–7, explains the three-part puzzle laid out in the Introduction: why interjurisdictional water conflict arises despite the existence of central authority, why such conflict can be just as difficult to resolve as international water conflict, and why the intensity of interjurisdictional water conflict appears to vary so greatly both across countries and across time. As this section details, the fundamental answer to all these questions relates to the interaction of three actors influenced by the three key factors identified in Chapters 1–3: subnational officials who have a large measure of autonomy and seek to preserve it, an expression of shared sectional identity among water users who live within the various subnational jurisdictions which share a common water resource, and a political opportunity structure that allows third-party actors to ally with national political elites to build cooperative institutions for river basin management.

The first and most fundamental element of this framework is the existence of powerful, first-tier administrative jurisdictions. Decentralization enhances the role of many subnational actors in water resource management relative to that of the central government. But within this constellation of subnational political actors, first-tier administrative jurisdictions like states and provinces tend to play a pivotal role for two reasons. First, in contrast to special-purpose jurisdictions like school or water districts, first-tier political jurisdictions possess both a defined territory and a general set of responsibilities for many governmental functions within their boundaries. In sharp contrast to irrigation districts, water-user associations, or other water governance institutions, subnational jurisdictions are also characterized by permanence and durability. Because states and provinces tend to exercise a wide range of governmental functions, and often embody long-standing regional divisions, they tend both to "make reform costly and unusual" and "to outlive the conditions that brought them into being" (Hooghe and Marks 2010, 19). Given their distinctive stature, subnational jurisdictions tend to play the single most important role in all matters affecting shared water resources.

The distinctive spatial scale of first-tier jurisdictions moreover gives rise to a second important characteristic, namely their role as implementation agents for central states and preference-aggregators for other subnational political actors. States and provinces occupy a unique meso-level of governance, being small enough to address localized issues and concerns while still being large enough to efficiently exercise critical functions like economic development, law enforcement, and tax collection. For this reason, even relatively centralized governments often rely on first-tier jurisdictions to implement policies and reforms. By the same token, decentralization reforms often result in an accumulation of power and responsibility at the provincial level (Keating 1998; Cammisa 1995; Mertha 2005). Particularly where decentralization establishes or strengthens subnational legislative bodies, these subnational states assume a correspondingly important role as interest-brokers for political constituencies both within their jurisdictions as well as between higher and lower levels of government (Tarrow 1977). Indeed, an important feature of first-tier jurisdictions is that subsidiary tiers are subordinate to them so that villages or towns report to districts and districts to provinces, for example (Sharansky 1970). Subnational states like states, provinces, and prefectures therefore represent the dominant political actor below the central government.

Where subnational political leaders possess sufficient autonomy, the jurisdictions they govern can behave much like independent nation-states, to the point of frustrating the efforts of central governments to improve cooperation over shared water resources. Even as new norms of cooperation were being developed on Europe's international transboundary waterways during the 1970s, for example, German state governments made common cause with industrialists to oppose federal water pollution regulations, thereby undermining the central government's attempts to cooperate with downstream countries along the shared Rhine River (LeMarquand 1976). As the case studies described in Chapters 4–7 illustrate, similar central–local conflicts have hindered institutional collective action in shared river basins. In these cases, the behavior of first-tier jurisdictions like states and provinces resembles that of independent nation-states. However, interjurisdictional water conflict is neither a direct nor an inevitable result of decentralization but rather the interaction of three factors, namely autonomy and sectional identity, decentralization and interjurisdictional collective action, and third-party involvement and political opportunity structure.

The framework, depicted in Figure 3.1, illustrates how both conflict and cooperation can emerge between political jurisdictions which share a common water resource. In the figure, the three key factors identified in Chapters 1–3, namely decentralization, sectional identity, and political opportunity structure, are depicted at the top. Notably, as discussed in the Introduction, decentralization acts as a kind of macrovariable, influencing both the intensity of sectional identity as well as political opportunity. The four central political actors identified in Chapter 1 are listed in italics in the middle portion of the figure and grouped into two clusters. In the first, toward the left-hand side of the figure, subnational political leaders ally with geographically concentrated water-user groups, like irrigation associations, to mobilize

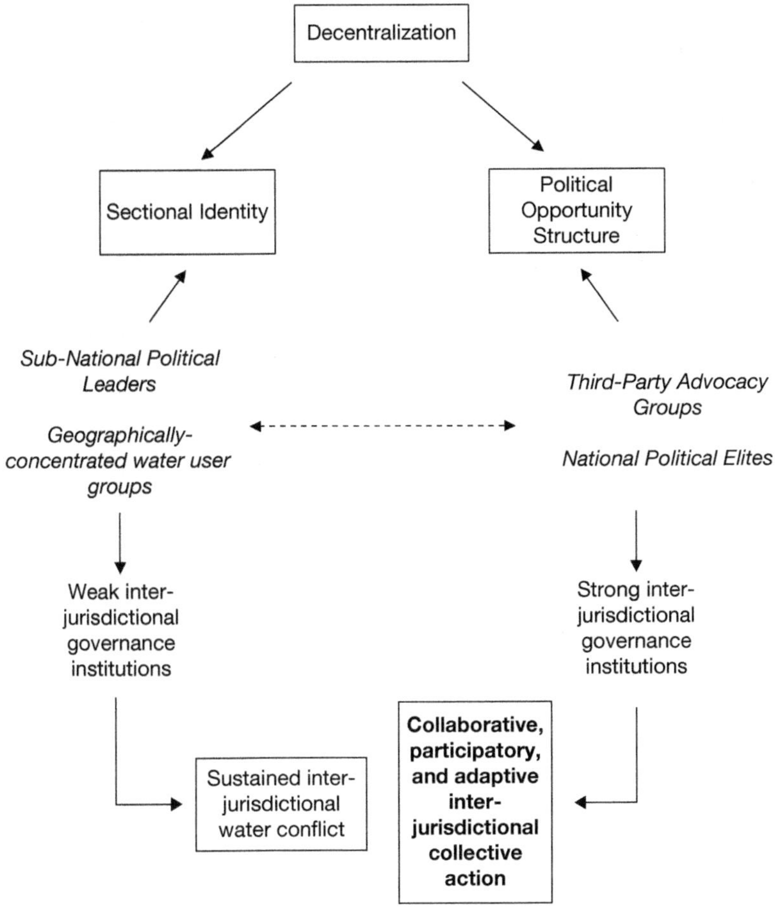

Figure 3.1 Schematic of Subnational Hydropolitics

constituents on the basis of shared sectional identity, which in turn drives them to adopt confrontational behaviors over shared water resources. In the second cluster, toward the right-hand side of the figure, third-party advocacy groups, like environmentally focused NGOs, ally with national political elites to gain support for cooperative institution-building. Throughout the figure, arrows indicate directions of influence and lobbying.

The left-hand side of Figure 3.1 shows how sustained interjurisdictional water conflict emerges. The desire of subnational political leaders to preserve their autonomy, as described in Chapter 2, tends to produce weak forms of interjurisdictional collective action that do little to restrain strategic behavior by subnational politicians. As described in the first section of this chapter, this weakness allows for interjurisdictional water conflicts to arise: in their absence, the existence of strong sectional identities creates an opportunity for subnational politicians to mobilize supporters by mapping sectional grievances onto shared water resources. Meanwhile, linking water-sharing issues to sectional identity also offers a means for geographically concentrated water-user groups, who benefit from gaining access to as much water as possible for a small number of users, to expand the sphere of conflict, potentially gaining broader support for their narrow, rent-seeking objectives.[3] This confluence of interests creates considerable pressure for subnational politicians to compete, rather than cooperate, over water. The combination of weak interjurisdictional institutions and sectionalism therefore provides both the impetus for water-user conflicts to emerge and the space for them to persist.

Where third parties fail in fostering strong institutions, three types of conflict can occur between the sectional groups that populate subnational political jurisdictions in shared river basins. The first, which I call *infrastructural,* arises from the construction of water-related infrastructure, including dams, reservoirs, and canals, and usually pits a narrow coalition of beneficiaries, such as farmers in a particular area seeking water for irrigation, against a more diffuse coalition of opponents, including residents of displaced communities, environmentalists, and economic interests, such as fishermen and those who depend on waterborne transportation, who are impacted by project construction. The second type of water-related conflict, which I call *allocative,* arises from the unequal distribution of water to different water users in a given area and often opposes agriculturalists with urbanites, environmentalists, and major industrial water users.[4] The third and final type, *qualitative,* arises when one water-user group, often an industrial sector, degrades the supply of water for other groups. Qualitative conflicts typically occur between water-user groups located upstream and downstream of one another. The common feature of all such conflicts is that they can all be readily transposed onto existing sectional tensions between groups in neighboring political jurisdictions. Shared water resources can thereby become the common currency[5] for a wide range of grievances between social groups that in combination undermine interjurisdictional collective action.

The right-hand portion of Figure 3.1 depicts how, in sharp contrast, collaborative, participatory, and adaptive interjurisdictional institutions can help prevent and resolve conflict in shared river basins. Often, these institutions are established over time, and this institution-building arises as a result of pressure from third parties. Third parties, including NGOs, are most effective when they can form alliances with political elites, particularly at the national level, who can provide resources, advice, and access to bureaucratic and legislative allies. These alliances can then, through political lobbying, side payments, provision of technical advice, and other means, induce neighboring jurisdictions to create strong, cooperative governance institutions. By providing platforms for negotiation, dialogue, and consensus-building, as well as decision-making rules that provide alternatives to conflict, such institutions promote interjurisdictional cooperation over time.

Figure 3.2 builds on this framework to explain in greater detail how these divergent processes unfold over time and how cooperation can emerge from conflict. All actors in this framework exhibit strategic behavior—that is, each acts rationally to pursue its own interests and does so on the expectation of interaction with other actors displaying similar behavior (McLean 1987). Each of these actors also operates at a different level of governance, indicated by its position relative to the x axis. The remit of water resource agencies is national, and that of subnational politicians is local and regional. For subnational politicians, their position at the meso-scale means that they must constantly balance relations with their constituents, their

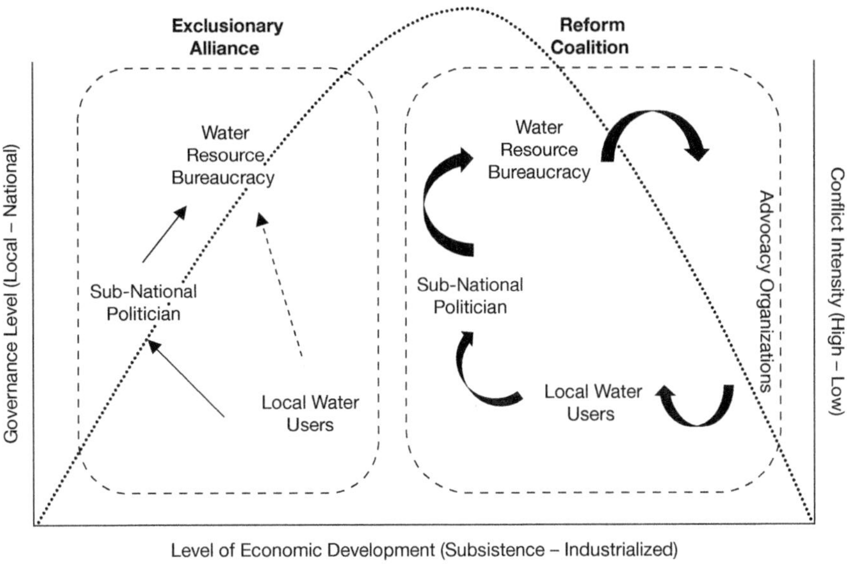

Figure 3.2 Schematic of Exclusionary and Reform Coalitions in Subnational Hydropolitics

neighbors, and the central government, creating a complex calculus in which cooperation is favored in some cases but not all. Local water users, meanwhile, are most directly tied to local and regional political leaders, particularly through elections and other forms of political representation. Because their interests with respect to water are highly dependent on local circumstances, they must find ways to broaden their support, or expand the sphere of conflict, by appealing to broader shared interests, such as sectional identity. Third-party advocacy organizations, in contrast, uniquely span different levels of government and, in their pursuit of influence, lobby politicians and officials at local, regional, national, and transnational levels (Myint 2012). As a result, third parties are uniquely capable of finding effective entry points to advocate for the development of strong institutions that restrain strategic behavior on the part of political actors at both national and subnational levels.

Relationships between these actors are animated by an underlying process of economic development, depicted along the *y* axis. As societies move from subsistence agriculture toward industrialization, three major changes take place: the number of water users multiplies, straining existing institutions for water resource allocation; pollution increases as a result of industrial waste and nutrient runoff, degrading drinking water supplies for some communities and inhibiting recreational and cultural uses of water; and finally, demand for water supply infrastructure, including dams, reservoirs, and canals, rapidly increases. The first two of these challenges creates a demand for collective action because they can be effectively addressed only through interjurisdictional and intersectoral cooperation. The third challenge, however, creates an incentive for competition. Financing for large water infrastructure projects can often only be obtained from the central government. Because such financing is always limited, the incentives to seek rents are strong, and individual regions often find themselves in competition with their neighbors. To compete more effectively, subnational politicians form political alliances with neighboring jurisdictions and with water resource agencies to gain financing for water infrastructure projects from the central treasury. Though often formed with widespread support from water users in a given basin, in their pursuit of financing these alliances subnational politicians soon exhibit classic rent-seeking behavior, seeking to maximize access to water resources while passing costs along to others, including neighboring jurisdictions and the central treasury.

I refer to these coalitions as "exclusionary alliances" because they aim essentially to maximize the transfer of as many economic rents to their constituents as possible, irrespective of need or justification to the country at large. Accordingly, these coalitions benefit from limiting water resource decision-making to the specific groups, regions, or interests that support them.[6] But because they by definition represent a narrow set of interests, they attempt to forge alliances with political leaders on the basis of some broader appeal—especially shared sectional identity, which distinguishes them from competing, neighboring water users. These conflicts are shown in the left-hand portion of Figure 3.2. Importantly, however, this model is not static; and an extension of it, depicted in the right-hand portion, shows how cooperation can eventually emerge as well. As water resource management challenges increase along with levels of economic development, reform coalitions can emerge, typically led by nongovernmental advocacy organizations. Because these advocacy organizations seek the broadest possible influence to counteract the dominance of exclusionary alliances, they engage both central and local levels of government (Mertha and Lowry 2006). Where this is the case, subnational water conflict intensity can become *U*-shaped and can decrease over time, as depicted on the *z* axis in Figure 3.2. This description of subnational water conflict therefore explains not only how and why conflict arises between the users of a shared water resource but also how cooperation can emerge in its stead.

This description of subnational conflict and cooperation over water captures the salient features of the case studies examined in this book. First, although immutable factors such as physical geography and water scarcity inform the behavior of subnational actors, institutional arrangements decisively shape their interactions. Second, these institutional arrangements interact with sectional politics and ideational differences over the value and use of shared water resources. In isolation, these underlying tensions and grievances can serve to undermine trust between water users and create mismatched incentives between central and local levels of government. However, these barriers to interjurisdictional collective action can be altered by a third factor, the political opportunity structure for third parties. Where third parties enjoy sufficient resources and access to the political process, they can be effective advocates for collaborative, participatory approaches to water resource management and can forge reform coalitions with officials at different levels to encourage interjurisdictional collective action. Put simply, third parties can help to counteract the influence of sectional identity over shared water resources, creating a constituency for durable cooperation instead of conflict. Together, these three elements provide a framework for explaining the case studies which follow. While this framework needs further testing to determine whether it applies more broadly, these cases suggest that while the conditions underlying subnational hydropolitics are common, subnational conflict over water is far from inevitable.

NOTES

1. I am grateful to Professor John Waterbury for this point.
2. I am grateful to Ken Conca of American University for this observation.
3. *Expanding the sphere of conflict* is a term employed by some scholars to describe the process by which interest groups gain support and influence in the political process. See Mertha and Lowry (2006).
4. The term *allocative conflict* reflects the Rawlsian distinction between allocative and distributive justice in that allocative conflicts stem from the outcome of water resource allocation irrespective of the distributive mechanisms employed to ration water. While water can be allocated by particular mechanisms, especially pricing, that prove contentious, allocation is usually achieved through nonmarket mechanisms.
5. I am indebted to Richard Damania, Global Lead Economist, World Bank Group Global Water Practice, for both this term and concept.
6. In their tendency to co-opt power and decision-making processes, exclusionary alliances resemble what Nathanial Matthews and Jeremy Schmidt call "closed communities," which manipulate discourses around good water governance.

4

Institutional Diversity and Interstate Hydropolitics in the United States of America

The preceding chapters have emphasized the often unappreciated extent to which subnational jurisdictions engage in behaviors that resemble those of sovereign nation-states with respect to shared water resources. The United States, the world's first modern federation, provides perhaps the clearest illustration of how institutional arrangements create the conditions for such behavior to be exercised. Even in comparison to other federal systems, the U.S. Constitution grants an unusual degree of power to state governments. This asymmetry is codified in the Constitution's Tenth Amendment, which assigns all powers not specifically granted to the federal government to the states instead. The greater power of American states, even relative to their counterparts in other federal systems, is also reflected in the fact that they maintain not only independent executive and legislative bodies but also judiciaries, a feature that has resulted in the uniquely complicated American legal system wherein different states recognize different bodies of law, especially in the case of water rights (Watts 2008). Despite this fundamental asymmetry, the power of the federal government relative to the states has grown over time, especially following the expansion of federal authority during the New Deal era (Sharansky 1970; Elazar 1984; Zimmerman 2011). The United States also lacks several of the mechanisms that ensure a greater degree of coordination and cooperation between states in other federal systems. In particular, the United States lacks the prominent intergovernmental organizations, like the Council of Australian Governments, that are a feature of many other federal systems and that help to address interjurisdictional issues like water resource management.

Hydropolitics in the United States presents a twofold puzzle. First, unlike the other countries examined in this book, the United States features a notable diversity of institutional models for governing its river basins. While many American river basins, including the Colorado, are governed either by a patchwork of institutions or by none at all, organizations like the Tennessee Valley Authority (TVA) and the Delaware River Basin Commission (DRBC) represent some of the most powerful river basin governance institutions in the world (Delli Priscoli 2007). The question of how and why these organizations arose in some basins and not in others reveals much about how interjurisdictional water conflict can be effectively prevented. Second, the performance of U.S. river basin institutions varies widely both between cases as well as across different time periods. While interjurisdictional cooperation has improved in the Colorado River basin, entities like the TVA have shed functions and powers over time. At the same time, in recent decades interstate water-sharing disputes have spread beyond the dry Southwest to historically water-rich areas like the Apalachicola–Chattahoochee–Flint basin. In this chapter, I address these puzzles by examining America's most prominent interstate water conflict, the Colorado River controversy, as a primary case and its most prominent examples of institutionalized cooperation, including the TVA and DRBC, as contrast cases. I also examine the evolution of the Colorado case over time, from a generally conflictual to a broadly cooperative dynamic.

Looking across these cases and across particular time periods within them reveals that the decentralization inherent to American federalism has decisively shaped water conflict and cooperation in the United States. In particular, the ability of each state to formulate different principles of water rights allocation creates a considerable challenge both to national policymaking and to interstate coordination. In the case of the Colorado River basin, this institutional patchwork led to the creation of alliances between geographically concentrated water-user groups, especially irrigation districts and metropolitan areas, state leaders, water resource management officials, and state congressional delegations. These coalitions then formed alliances with federal agencies to obtain federal financing for reclamation projects, which I refer to as the "reclamation nexus." The reclamation nexus is a classic illustration of the exclusionary alliances described in the Introduction, which are formed by geographically concentrated water-user groups to seek central financing for water infrastructure projects. The intensely territorial politics of reclamation corrode incentives for interjurisdictional as well as central-local cooperation. However, the influence of the reclamation nexus on U.S. water politics proved to be temporary. In recent decades, new alliances have emerged between environmentally oriented civil society organizations and the same federal bureaucracies which previously supported reclamation at all costs. These advocacy groups, joined at times by national political leaders, have succeeded in creating the constituency for collaborative and participatory institution-building, providing a powerful illustration of the importance of third parties in inducing cooperation between water users—even when they seem more disposed to conflict.

COUNTRY CASE INTRODUCTION

Water resource management lies within a gap in the division of powers between state and federal government in the United States, and a strong federal role has emerged only gradually and fitfully over time, following an expansion of powers to regulate commerce and build infrastructure of national importance. As the demands on America's rivers accelerated with the pace of economic development in the late nineteenth and early twentieth centuries, institution-building flourished, producing a number of interstate compacts and management organizations. Yet certain features of American federalism have consistently complicated these efforts. As John Thorson has written,

> The constitution . . . gives insufficient attention to the horizontal aspect of federalism—that is, the relationships among the governments of a region that share a transboundary natural resource such as water. The Constitution provides few opportunities for resolving natural resource conflicts between states or within regions, and it provides even fewer formal avenues for intergovernmental cooperation in the management of shared natural resources—particularly rivers. (1994, 4)

These distinctive features of America's federal system have created a patchwork of water resource management institutions, many of them relatively weak.

Table 4.1 depicts the major features of subnational water politics in the United States. Given the country's size and geographic diversity, driving management issues vary significantly but generally include scarcity and allocation in the west and non-point-source pollution in the eastern states. The peculiar design of American federalism creates a number of institutional collective action dilemmas, including brittle interstate water-sharing agreements and incomplete federal powers over water resource management, that integration mechanisms like interstate compacts have proven incapable of entirely resolving. In large part, this is because of the

Table 4.1 FEATURES OF WATER POLITICS IN THE UNITED STATES

Case Characteristic	**Issues**
Driving management challenges	Scarcity; allocation; non-point-source pollution; environmental protection
Institutional collective action dilemmas	Weak and contested water-sharing agreements; multiuse allocation conflicts; bureaucratic fragmentation
Integration mechanisms	Interstate compacts; political party structure
Sectional cleavages	Individualistic western political culture; historic north–south cultural divide; marked discrepancies in most states between large cities and rural areas
Political opportunity structure	Highly pluralistic; multiple points of access at federal, state, local levels

persistence of sectional cleavages both between and within states. Unlike those of some other federal systems, America's state boundaries for the most part do not reflect ethnic, religious, or linguistic divides. However, states do vary significantly in terms of political culture. Interior western states like Colorado and Wyoming have historically been highly individualistic, while wealthy coastal states feature a more communitarian political culture. In addition, in most states urban and rural areas feature marked political, economic, and social divides. This is especially true in western states like California. The United States features a vibrant civil society sector, and its federal system provides multiple points of access at all levels of government, which many nongovernmental actors skillfully exploit. Even so, formal inclusion in agency rule-making and decision-making processes is a relatively recent development, having been instituted in the case of water management primarily after the 1960s.

Ambiguous Federal Powers and Patchwork Institutions for Water Resource Management

On a strict interpretation, the Constitution's silence on most aspects of water resources places them primarily under state jurisdiction, per the Tenth Amendment. However, as the American economy and population grew into the nineteenth century, it became apparent that the Constitution's primary mechanism for addressing regional issues, the interstate compact, would need to be complemented with additional federal authority in the area of water resource management. Subsequently, the judiciary has played a critical role in establishing a basic institutional framework for federal as well as state involvement in the governance of interstate river basins. With *Gibbons v. Ogden* (1824), the Supreme Court began taking a more expansive view of federal power in the water resource management area, granting it considerable powers under the commerce clause, in particular the enumerated authority to promote and regulate navigation on internal waters. Subsequent decisions have led to a slow accretion of federal powers and authorities in water resource management under a legal doctrine known as "navigation servitude." In *United States v. Appalachian Power Co.* (1940), the court specifically authorized the federal government to exercise several new functions, including flood control and, importantly for the development of organizations like the TVA, watershed development (Taylor 1981).

The institutional framework established by the courts has greatly expanded the federal government's role in water resource management, but these authorities are exercised primarily by hydraulic engineering agencies, which historically focused primarily on building water supply and flood control infrastructure. The federal role in governing river basins is exercised primarily by two powerful bureaucracies, the U.S. Army Corps of Engineers (USACE) and the Bureau of Reclamation (BuRec), an agency of the Department of the Interior. The USACE has been responsible for aiding navigation since the mid-nineteenth century and under the 1936 Flood Control Act also became responsible for building and maintaining infrastructure on many of the country's interstate waterways (U.S. Army Corps of

Engineers 2007). The BuRec, on the other hand, was created as a result of the 1902 Newlands Reclamation Act. The act's primary objective was to promote the economic development of the western states by supplying irrigated water to arid and semiarid lands. The BuRec accordingly devoted itself primarily to the construction of large storage reservoirs to supply the necessary water. Over time, however, both the BuRec and the USACE have accumulated additional management and government responsibilities, including for ecosystem protection and environmental dispute resolution.

Despite this accumulation of functions, the role of federal bureaucracies in water resource management is constrained by the unique American institution of water rights pluralism. In the eastern states, a legal doctrine known as "riparian jurisdiction" establishes the right of landowners to use and consume water from shared rivers and streams. This doctrine proved ill-adapted to use in the arid American West, where the paucity of perennial streams led communities to adopt another principle, prior appropriation, which holds that the first user to exploit a given water resource gains a preferential right to its use. Although the institution of water rights pluralism developed organically, Congress essentially codified it by declining to exercise its power to determine how water rights would be allocated on land acquired from Mexico in 1848, which subsequently became part of newly created states in the Colorado River basin (Hundley 1975). As a result, each western state was left to adopt its own water rights doctrine, and a complex hodgepodge of doctrines coexist in different states, including riparian, prior-appropriation, and even some Spanish legal provisions (U.S. Bureau of Land Management 2001). In practice, this doctrinal multiplicity acts to greatly complicate federal involvement in multistate western water issues, such that the federal government typically has to defer to state legislation and policy regarding water use, even on federal land (Tarlock 1996; Huffman 2009).

Congress has proven very reluctant to address this asymmetry, especially in the West, where a strong individualistic ethos predominates; and in this vacuum a wide variety of institutional arrangements have been developed. The primary institution for the governance of interstate river basins is the compact, of which hundreds have been concluded since ratification of the Constitution of 1787. Compacts, a mechanism established under Article I of the Constitution, are effectively interstate treaties on matters of mutual concern and require ratification by both state legislatures as well as the U.S. Congress. Most water-related compacts simply address the division of waters among neighboring states, rather than their cooperative management. While several compacts do establish governance institutions, such as interstate commissions, most lack enforcement or dispute-resolution mechanisms (Schlager and Heikkila 2009). In two cases, those of the Delaware and Susquehanna River basins, the federal government joined the riparian states as a party to the compact. A few noncompact interstate institutions also exist, primarily for purposes of pollution control, such as the Ohio River Valley Water Sanitation Commission. From 1965 to 1980, several federal–state river basin commissions were active under the terms of the Water Resources Planning Act. Finally, a few unique river basin management organizations exist, such as the Mississippi River Commission and the TVA, which exercise at least some management functions at a basin-wide scale (Delli Priscoli 2007).

This considerable variety in institutions for river basin governance reflects the influence of a wide range of political actors, including third-party bridging organizations. The most important bridging organizations in the United States have been national political leaders, especially presidents and secretaries of the interior, and environmental and conservation organizations. While presidents have relied on informal prerogatives of their office, such as their convening power and that of the "bully pulpit" to promote regional institution-building, secretaries of the interior have acquired specific executive powers, such as the ability to allocate water deliveries to states in the Colorado River basin. At particular points in history, these powers have been effectively utilized to encourage basin-wide cooperation, but at other times political leaders have been reluctant to exercise them, usually because they have feared alienating entire states and their accompanying congressional delegations.

The role of civil society organizations, on the other hand, has grown significantly over time. The fragmentation of power between branches and levels of government offers a relatively open political opportunity structure (Meyer 1993). These favorable conditions aided the development of a strong environmental movement, including large numbers of organizations that operate at national, state, and local levels of government (McLaughlin and Khawaja 2000). The role of nongovernmental environmental organizations has moreover been codified by federal legislation such as the National Environmental Policy Act, which provides them with a voice in bureaucratic rule-making. These reforms, coupled with increasing political mobilization on the part of environmentally oriented advocacy groups, have led federal bureaucracies to become much more receptive to civil society. In combination, this influence of civil society has helped to balance the tendency of states to prefer weak forms of cooperation with the federal government as well as their neighbors. These dynamics in turn help to explain why in some cases cooperative institution-building has failed and in others it has succeeded.

Strategic Behavior and Institutional Development in American River Basins

Given the constrained role of the executive branch in water resource management and the historic deference of Congress to state prerogatives, the governance of interstate river basins in the United States has been shaped primarily by the interplay of state interests. In the early decades of American economic and political development, this relationship often took the form of a reclamation nexus in which federal agencies and state political leaders colluded to seek federal financing for water projects. This form of rent-seeking strategic behavior was dramatically described by former Secretary of the Interior Harold Ickes, who decried "this willful and expensive Corps of Army Engineers closely banded together in a self-serving clique in defiance of their superior officers and in contempt of the public welfare" (Maass 1974, xiv). The BuRec has also historically typified the rent-seeking bureaucracy. Although the Newlands Reclamation Act which created it envisioned the BuRec to be self-sustaining by means of fees paid by users of the water it provided, in fact little federal investment has been recovered. Subsequently, the BuRec has frequently been accused of funding projects which benefited small numbers of politically influential western landowners, at great expense to the treasury (Rucker and Fishback 1983). At times, competition between the USACE and the BuRec has directly hindered cooperative institution-building, particularly in basins like the Missouri, where the two agencies competed for influence (Thorson 1994). These textbook examples of bureaucratic strategic behavior have been dramatically moderated in recent decades, but they nonetheless served to reinforce the tendency of state governments to prefer weak forms of cooperation over shared water resources.

State political leaders, as described in the previous section, are generally motivated by a desire to preserve their autonomy relative to both the federal government and neighboring states. Consequently, they have traditionally resisted river basin institution-building, which they feared might diminish state powers. The result is that, although many agreements exist to apportion interstate waterways, they vary considerably in the extent of their provisions and institutions to support them are often very weak. This weakness is especially evident with respect to interstate compacts, which often lack enforcement and conflict-resolution capacities and are undermined over time by a lack of attention and resources devoted by state parties. Given these limitations, interstate compacts have in most cases failed to prevent interstate competition and disputes over shared water resources (Heikkila, Schlager, and Davis 2011; Zimmerman 2012). Instead, the interests of state-level actors are often dominated by concentrated economic interest groups like irrigation districts and water-hungry municipalities. Particularly in the American West, these interests formed alliances with state politicians and congressional representatives in order to seek federal financing for reclamation projects. Yet the very prevalence of this behavior created intense competition among the states, and in many shared river basins it has undermined cooperative institution-building.

The failure of most existing interjurisdictional agreements to prevent water conflict is most evident in the number of interstate water disputes brought before the U.S. Supreme Court. Some basins, such as the Colorado, have been subject to repeated cases of litigation, while others, such as the Yellowstone and the Apalachicola, have more recently given rise to legal disputes. Yet this growing number of interstate water law disputes has proven just as inadequate in preventing conflict as the weak compacts which gave rise to them. Although the Supreme Court has adjudicated many of these interstate water quantity disputes, it has to date resisted articulating a uniform doctrine for apportioning interstate waters, on the grounds that each case must be decided on its individual merits. As a result, the growing body of judicial decisions on interstate water disputes has failed to coalesce into a set of principles for preventing future disputes. The complaint of a leading water engineer in the 1930s regarding the Supreme Court's attempts to allocate western waters remains equally valid today: "The temporary nature of these late decisions, and their dependence on changing conditions, preclude water users of an assurance of future use" (Maddock 1932, 1663). Rather than resolving interstate disputes, the mechanism of judicial settlement frequently has sustained interstate competition over water. As one observer noted of interstate litigation in the Missouri River basin,

"What often goes unnoticed in these proceedings is the hidden cost to mutual trust and long-term relationships among the states" (Thorson 1994, 4).

Despite the prevalence of interstate disputes, the United States also features several signal examples of collaborative and participatory river basin governance institutions, and even the Colorado River basin has become the site of growing cooperation between levels of government. Perhaps the best example of such cooperative institutions, however, is the DRBC, which possesses significant planning and management authorities and represents both federal and state governments (Gentzler and Hines 2006). Each of these examples has featured an alliance of national political elites and environmental and conservation organizations, who together have played a crucial role in promoting collaborative and participatory institution-building. Historically, presidents and secretaries of the interior have favored river basin institution-building as a means to seek electoral credit for resolving long-running disputes, while environmental and conservation organizations have done so to gain greater voice in river basin governance. Yet efforts to promote cooperative, basin-wide institution-building have had to overcome strong opposition from states seeking to preserve their autonomy, as the case of the Colorado basin illustrates.

AUTONOMY, SECTIONALISM, AND INTERSTATE COMPETITION IN THE COLORADO RIVER BASIN

The Colorado River is the major waterway in the arid American Southwest, and consequently water scarcity has long been the dominant issue in its management. For the better part of a century the subnational states of the Colorado River basin have continually fought among themselves and with the federal government to reach a durable allocation of basin waters, particularly in light of emerging urban and environmental water uses. Yet despite a history of animosity, cooperation and the performance of the Colorado's patchwork management institutions have steadily improved over time (see Figure 4.1). The history of interstate collective action in the Colorado River basin spans nearly a century, which I divide into three periods for the purpose of this case study. In the first, 1918–1945, the efforts of national and state political leaders to conclude a durable compact agreement were undermined by Arizona's refusal to join in fear of surrendering water rights to California. In the second period, from 1946–1968, the role of the federal government expanded considerably, particularly in financing reclamation projects, for which the states assiduously competed. This period also witnessed important phases of institution-building, including conclusion of the Upper Basin Compact in 1948, and a landmark Supreme Court case, *Arizona v. California*, in 1963, which settled the two states' long-running dispute. In the third, most recent period, 1969–2014, the growing power of environmental movements helped to restrain both bureaucratic and subnational strategic behavior and set the stage for a period of cooperative institution-building toward the end of the twentieth century.

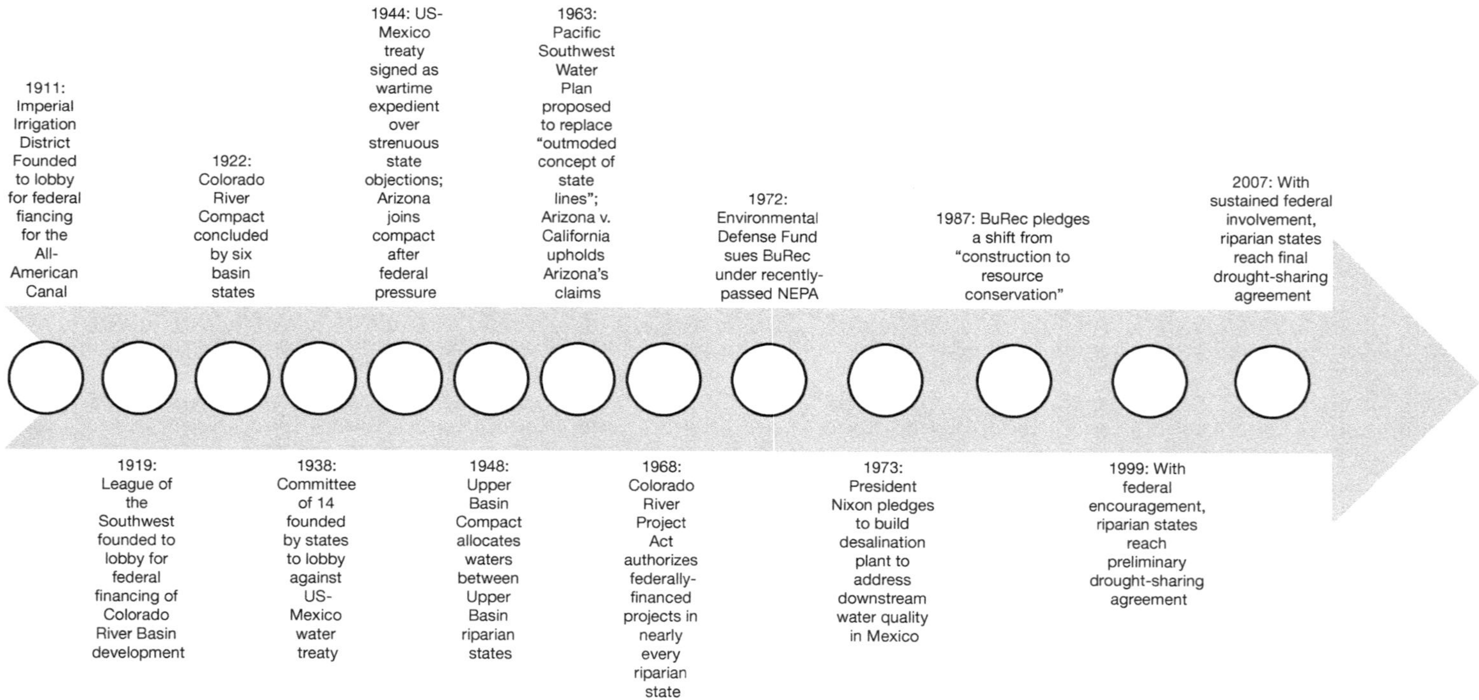

Figure 4.1 Timeline of Institution-Building in the Colorado River Basin

The Colorado River is located in the western United States and is shared between seven American states as well as between the United States and Mexico (see Figure 4.2). Large portions of Arizona, Colorado, and Utah lie within the basin, while the remaining states contribute a smaller share of the river's flow. Waters of the Colorado are apportioned by a 1922 interstate compact between an upper basin, which consists of those parts of Wyoming, Colorado, Utah, and New Mexico which lie above Lee's Ferry, Arizona, and a lower basin, constituting those parts of Arizona, Nevada, and California which lie below Lee's Ferry. The compact apportions 7.5 million acre-feet of water (MAF) annually and in perpetuity to both the upper and lower basin but does not allocate water between the individual states—an omission that has been at the heart of much subsequent litigation between the riparian states. Most of this legal conflict has centered on issues of interstate allocation, although infrastructural and qualitative issues have also come into play at different periods.

In the upper basin, a 1948 interstate compact allocates 51.75% of water to Colorado, 11.25% to New Mexico, 23% to Utah, 14% to Wyoming, and 50,000

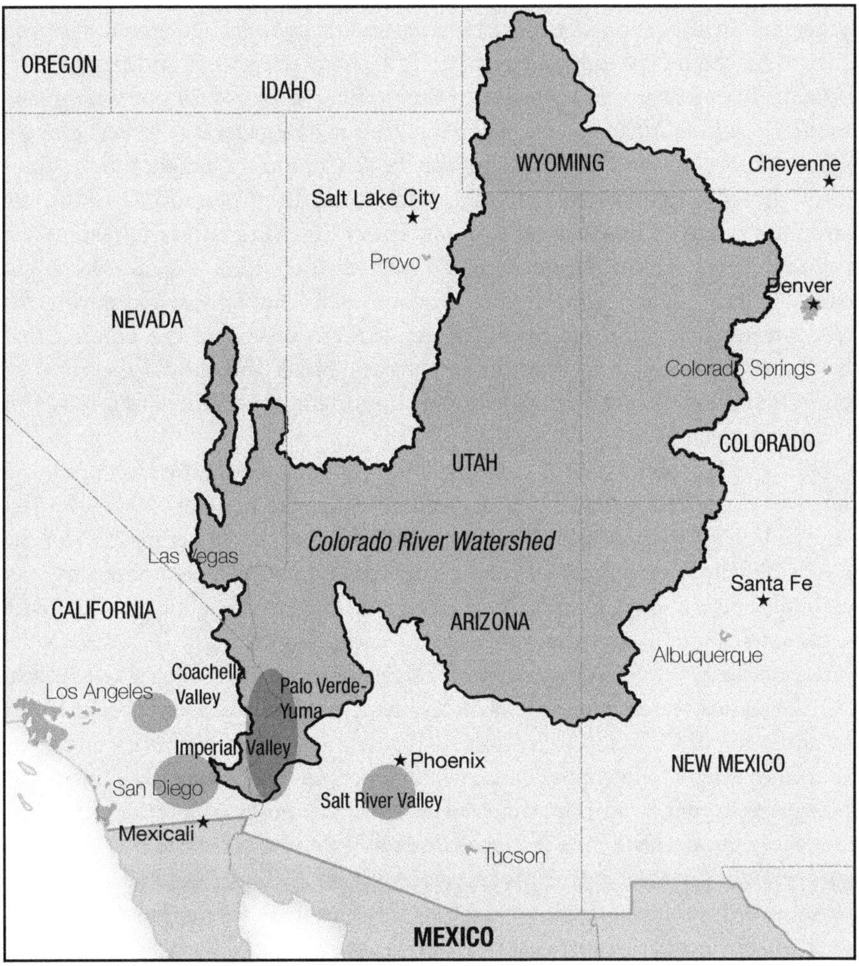

Figure 4.2 Physical and Political Features of the Colorado River Basin

acre-feet annually to Arizona. In the lower basin, a 1964 Supreme Court decree apportioned 4.4 MAF to California, 2.8 MAF to Arizona, and 300,000 acre-feet annually to Nevada. Additionally, a 1944 treaty guarantees Mexico a downstream flow of 1.5 MAF annually, subject to minor restrictions (U.S. Bureau of Reclamation 2008). Subsequent agreements under this treaty framework provide for the control of salinity in waters delivered to Mexico, as well as for the maintenance of baseline levels of environmental flow in the Colorado River delta (Bennett 2000). Collectively, these compacts and agreements are known as the "Law of the River" (LOTR). The LOTR regime has historically been hobbled by a number of structural deficiencies, particularly its failure to recognize Native American water rights and its reliance on a baseline Colorado River flow that research suggests is significantly above long-term historical averages (Price and Weatherford 1976; Tarlock 1987; Miller, Rhodes, and Macdonnell 1997). But the most significant impediment to the LOTR regime and interstate cooperation in the Colorado River basin has always come from the riparian states themselves.

For the better part of a century, elements of the federal executive branch have attempted to employ a mix of legal authority and persuasion to affect a final agreement on allocation of Colorado River waters, but final agreement has been elusive. At certain periods in the nation's history, presidents and presidential aspirants have lent the full prestige of their offices to the cause but with mixed results. President Calvin Coolidge and then commerce secretary Herbert Hoover played important roles in brokering the 1922 Colorado Compact, but subsequent efforts to employ the bully pulpit to resolve the Colorado River dispute have been trumped by electoral realities. Harry Truman's efforts to resolve the Colorado River dispute, for example, foundered when it became clear he might lose California by appearing to take Arizona's side. Similarly, when during the 2008 presidential campaign Senator John McCain expressed the opinion that the Colorado Compact favored the upper basin states and should therefore be renegotiated, reaction from the electorally important upper riparians forced him to retract his statement (Walton 2011).

Efforts by U.S. secretaries of the interior to promote cooperation have produced somewhat better, though still uneven, results. Under the Boulder Canyon Project Act, the U.S. secretary of the interior was designated as the "water master" for the lower Colorado River basin. In an unprecedented assertion of federal authority over water allocation in the Colorado basin, the act further specified that all water users in the lower basin must enter into a contract with the secretary for deliveries of water. However, this initial assertion of federal authority proved to be only partial and did not lead to sustained federal involvement in settling interstate disputes over the allocation of Colorado River waters. The secretary's legal authority under the act did not extend to the upper basin, where waters were subsequently allocated by the separate Upper Basin Compact. Moreover, for the most part interior secretaries have chosen to delegate their responsibilities as water master to the BuRec, which has operational control over dams and reservoirs on the lower Colorado, and have played an active role only in times of crisis (U.S. Bureau of Reclamation 2008). The most notable recent example occurred during the George W. Bush administration, when Secretaries Gale Norton and Dirk Kempthorne brokered a 2007 agreement between the Colorado basin states to share the burden of water delivery reductions during drought conditions.

If these attempts at persuasion have been less than entirely successful, instances where the federal government has exercised its constitutional powers, particularly in the field of foreign relations, have been more effective at achieving environmental objectives. This is best illustrated in the form of several international agreements concluded by the U.S. Department of State. The department's involvement in the basin has been limited to issues pertaining to Mexico, but because of the severity of water scarcity in the Colorado basin, its actions have had a profound effect on basin governance. Additions to the 1944 U.S.–Mexican treaty have indeed produced some of the most successful examples of cooperative Colorado River basin management, including a 1973 agreement to construct treatment plants to control salinity on downstream flows as well as a 2012 accord, known as "Minute 319," which provides for carefully calibrated flows to maintain the ecosystem of the Colorado River delta. This provision for maintaining minimum environmental flows has subsequently come to be seen as a model for promoting sustainability on international transboundary waterways.

It is especially noteworthy that while the involvement of environmental and conservationist organizations in the Colorado basin has historically been limited, their lobbying played a critical role in the environmental flow provisions of Minute 319 (U.S. Department of the Interior 2014). It is also significant that these organizations, led by the Environmental Defense Fund, chose to lobby the executive branch rather than Congress. In contrast to executive branch efforts to resolve the Colorado River dispute, legislative attention to the issue has been irredeemably warped by lobbying from individual state delegations and a consequent divide between the representatives of riparian and nonriparian states. While congressional representatives from western states have been avid supporters of reclamation projects in their respective constituencies, those from eastern and midwestern states have been largely critical of federal expenditures to benefit western states.

The interests of the riparian states themselves, especially Arizona, California, and Colorado, have been shaped by geographically concentrated water-user groups, which allied with particular branches of state governments to secure, with federal funds, the maximum quantity of Colorado River water. California exemplifies this form of behavior as two sets of subnational actors sought federal funding to bring Colorado water to arid southern California. The first, consisting of the cities of San Diego and Los Angeles, sought to supply the industrial and domestic water needs of a rapidly growing population. Los Angeles was particularly aggressive in its pursuit of water: its Department of Water and Power had, under the direction of its chief William Mulholland, constructed long-distance water-transfer projects which had been criticized as stealing other regions' water (Nadeau 1974). In addition, agricultural interests in the Coachella and Imperial Valleys sought large quantities of Colorado water for irrigation purposes. In both valleys, irrigation districts were the nexus of political and lobbying activity. The Imperial Irrigation District (IID) was particularly powerful: from the start, it was empowered to issue bonds and tax farmers and, almost from its founding in the early 1900s, employed professional lobbyists (Hundley 1975).

However, the real influence of geographically concentrated water-user groups like the IID came from the alliances they forged with members of both the California State Assembly and the U.S. Congress. These groups, led by the IID, established strong relationships with Republican Senator Hiram Johnson and Republican Congressman Philip Swing, whose constituency included the Imperial Valley. The Swing-Johnson bill, advocated by both delegates in the face of fierce opposition, eventually delivered to southern California the Boulder Dam, intended to guarantee stream flow and hydropower to the region, as well as the All-American Canal, intended to guarantee an uninterruptable supply of water to the region. In addition to these legislative alliances, southern California water users aligned themselves closely with members of the California Colorado River Commission (CRCCA), a state agency charged with pursuing the state's interest pertaining to the Colorado. Sectional tensions animated the politics of water within individual states, especially California. Not all Californians supported this water-maximizing agenda: northern California citizens, politicians, and interest groups generally opposed southern California's efforts to secure water at the expense of other states and believed its federally funded water projects to be wasteful. This tension had little effect on moderating the influence of groups like the IIDor their preference for competition over cooperation with neighboring states.

The salience of sectional tensions in shaping the politics of water in California was paralleled in Arizona. Just as California politics were marked by a sectional schism between north and south, in Arizona the opposing territorial interests of the central Salt River Valley faced off against the western Yuma area, which bordered on the Colorado River itself. The Salt River Valley, lying far from the main stem of the river, required a complex and expensive system of dams, canals, and pumps to obtain Colorado River water. This system, first termed the High Line Canal and later incorporated into the Central Arizona Project (CAP), depended on guaranteed access to large quantities of water from the main stem of the river. This geographical discrepancy led to a sharp division in the interests of water-user groups in the two regions: Yuma-area farmers strongly supported the Colorado River Compact as a means of securing adequate water resources for Arizona, while central Arizona farmers, believing that irrigation of the arid interior of the state would require far more water, held out for greater quantities of water and federal assistance. Despite opposition from Yuma's representatives in the state legislature, central Arizona water users nonetheless secured the support of successive governors and Arizona's congressional representatives, who sought federal reclamation funds with notable unity and perseverance. This support was most effectively lent by the Colorado River Commission of Arizona (CRCAZ), a state agency created in the image of the CRCCA. The CRCAZ's primary purpose was to devote bureaucratic resources to securing water rights for Arizona over those of California, and it played a crucial role in both the state's appeals to Congress as well as its litigation against California, which in 1963 paved the way for congressional authorization of CAP.

The upper basin states featured a more diffuse set of water-user groups, but I often refer to the upper basin states as a single constituency because their interests coincided for most of the Colorado River controversy. In the early phases of interstate collective action in the Colorado River basin, the upper basin states, being more rural and less heavily populated than their downstream neighbors, were united in seeking rights to Colorado River water to support future development, whereas their lower basin counterparts sought federal financing for development in the much nearer term. Much of the history of the Colorado River basin therefore revolves around sectional conflict between the upper and lower basin states. The peripheral involvement of the other riparian states like Nevada and Wyoming reflects relatively low agricultural, industrial, and urban water demands and their small contributions to total Colorado River flow. Nevada, which primarily allied with California, sought revenue from hydropower dams along its southern border, while New Mexico and Wyoming were concerned mainly with securing federal funding for irrigation projects on Colorado tributaries within their own borders. The desire to strike a balance between pursuing individual state interests and the necessity of cooperating to seek federal funding would form a tension in the basin that would take decades to resolve.

Water Rights and the Dilemma of Autonomy, 1918–1945

Interstate collective action in the Colorado River basin arose in reaction to the distinctive historical geography of the American West, which led almost immediately to sustained sectional conflict between two riparian states, California and Arizona. California was by far the most economically prosperous and politically powerful of the two. Anticipating future demand from its neighbors for the same Colorado River water on which it largely depended, California fought from the outset to secure senior water rights to in order to supply geographically concentrated water users in its arid southern regions. Inland Arizona joined the Union in 1912, sixty-five years after California, but likewise staked its future on the ability to divert Colorado River water for irrigation. Recognition of the need to prevent such competition led state leaders to attempt to create the conditions for interstate cooperation over the Colorado. In 1919, representatives from all Colorado River basin states formed the League of the Southwest, and one of the league's earliest acts was to propose an interstate compact to divide the river's waters. Arizona, though, refused to ratify the compact, on the grounds that its right to fully utilize the flow of the Gila River, an important tributary within Arizona's borders, was not fully protected. The initial period of institution-building in the Colorado River basin was thus dominated by the desire of the states to secure preferential rights to water while preserving their autonomy—a desire that left the basin's institutional architecture too weak to prevent continued conflict between water users throughout the early twentieth century.

Following the First World War, the development of irrigated agriculture in the American Southwest attracted the attention of both state and national political leaders, which in turn raised the vexing question of allocation of the Colorado River basin's scarce water resources. In 1921, the League of the Southwest proposed to allocate the Colorado's waters by means of an interstate compact (Kruckman 1921). The idea quickly gained the support of Herbert Hoover, a former hydroengineer and then the secretary of commerce, who in 1922 exclaimed that "behind all the precise and commonplace language of this compact lies the greatness and romance of the West" (*Arizona Republican* 1922c). Hoover was soon appointed as the federal government's representative to the Colorado River Compact negotiations, in which capacity he continually warned wavering states that the federal government would refuse to approve or fund any developments along the river without a compact to settle interstate "quarrels" (*Los Angeles Times* 1922, 1925). President Coolidge, meanwhile, urged "immediate completion" of the compact, lending presidential authority to the states' nascent efforts at cooperation (Armstrong 1923).

But despite the interest of national politicians in quickly concluding an interstate compact, institution-building in the Colorado River basin was from the outset complicated by a combination of physical and historical geography. The upper basin states, especially Colorado, contributed the vast majority of the river's flow. However, the greatest demand for water lay in downstream California, which was by far the most populous and economically developed part of the basin. The mountainous and relatively undeveloped upper basin states consequently sought to secure water rights on the basis of expected future, rather than current, demand. As Delph Carpenter, a Colorado water lawyer credited with being the first to propose the idea of an interstate allocative compact, noted, "The topography of the upper basin limits the extent to which each of the upper States may go in its development and its corresponding consumption of river flow" (Carpenter 1923, 6). Nonetheless, upper basin political leaders were acutely aware of the need to preserve water resources for later use. Utah Democratic governor George Dern warned that while Utah possessed "great agricultural potentialities in the Colorado River Basin," it was not yet "ready to develop them. . . . When that time comes we do not want to wake up and find that we have plenty of land but no water with which to irrigate it. We do not want our future growth stunted" (Dern 1926, 6–7). The very insistence of political leaders in each state on securing water rights for future expansion laid the seeds of sectional tension between the upper and lower basin, even as the riparian states entered compact negotiations.

Largely on the strength of Carpenter's advocacy, upper basin political leaders soon became persuaded that a compact offered the best means of securing their individual states' water rights against claims from the lower basin states, especially California (Colorado River Commission 1925). Republican governor Clarence Morley of Colorado promised his constituents the compact would provide "that protection . . . which is our just due as a State of origin of the river" (*Los Angeles Times* 1926). His successor, Democrat William Sweet, likewise promised in his inaugural address that the compact "protects our development from adverse claims on account of any great reservoir or other construction on the lower river. It removes all excuses for embargoes upon our future development and leaves us free to develop our territory in the manner and at the times our necessities may require" (Sweet 1923). Utah's governor Dern, meanwhile, perfectly captured the sectional tensions at the heart of the compact negotiations by explaining that the compact "was reluctantly accepted by the upper basin states" as a means of preventing assertion of "new prior rights against the upper states" (Arizona Colorado River Commission 1927, 11). Such claims reflected the advocacy of geographically concerned water user groups that attempted to cast their parochial interests as those of their parent states as a whole, in opposition to their water-hungry neighbors.

California was perhaps the best example. Its quest to secure in perpetuity senior rights to Colorado water was decisively shaped by powerful urban and agricultural water users in its southern portion. Beginning in the late nineteenth century, California's fast-growing and water-scarce southland region, particularly the cities of Los Angeles and San Diego and the irrigated lands of the Imperial and Coachella Valleys, turned increasingly to the Colorado for its water needs. Irrigators and city officials in these regions accordingly began to lobby for a share of Colorado River water. As a 1924 report noted, "California makes no contribution to the waters of the Colorado but . . . on account of the low-lying lands within her borders can use advantageously the greater part of the water allocated to the lower basin" (Clark 1924, 15). Irrigated agriculture quickly expanded throughout the early twentieth century, and in 1911 the IID was organized as the world's largest single agricultural unit. The IID quickly emerged as a powerful political force and soon began calling for federal assistance to build an "All-American Canal" which would divert water from the Colorado on the American side of the border, instead of relying on water from Mexico as had previously been the case (Nadeau 1974). At a 1919 conference, the IID allied with Los Angeles and Coachella farmers to seek congressional appropriations for the canal, in return for which they were guaranteed Colorado water. Less than two months later, IID lobbyists were in Washington, "handing out cantaloupes produced in the [Imperial] valley" (Hundley 1975, 41–42).

Buoyed by such tactics, geographically concentrated water-user groups led by the IID soon dominated California's water resource policy. Even as Hoover promised state legislators that signing the compact was "the first step in obtaining . . . works for California" (Palmer 1923), southern California water users, led by delegates from Los Angeles and the Imperial Valley, resolved to support the compact only on the condition that Congress finance both the All-American Canal as well as a large dam at Boulder Canyon to store water for southern California (Palmer 1925b). Northern California leaders, including Republican governor Friend Richardson, viewed such conditional ratification as a selfish attempt to block comprehensive development of the Colorado that would needlessly antagonize other states (Palmer 1926). But southern California interests, flaunting their political power, quickly mobilized to override their opposition. Los Angeles's City Council "went on record . . . against unconditional ratification in order to bolster up the efforts of [Johnson and Swing]" and dispatched lobbyists to Sacramento to persuade the state legislature to vote accordingly (*Los Angeles Times* 1926). In the end, the California Assembly overwhelmingly approved conditional ratification of the Colorado River Compact.

The promise of large new federal appropriations for construction of water projects quickly attracted the attention of federal agencies, which viewed the conclusion of the compact as a firm basis on which to seek out new clients for their expertise in building water supply infrastructure. Secretary of the Interior Hubert Work lost no time in telling Congress it should make Boulder Dam a "national undertaking" because "no agency other than the Federal Government should be instrusted with the . . . distribution of its opportunities. All uses can be coordinated and the fullest benefits realized only under centralized control" (U.S. House Committee on Irrigation and Reclamation 1928, 7). Federal agencies, while taking Work's advocacy to heart, soon realized the value of forming alliances with water-user groups within the riparian states, and these alliances rapidly coalesced into a powerful political nexus in favor of reclamation projects. This nexus found expression in a study of proposed reclamation efforts conducted by the BuRec that enthusiastically endorsed the IID's proposals. The study, which became known as the Fall-Davis report, provided a technical justification for the Swing-Johnson bill introduced by an Imperial Valley congressman and a California senator to fund construction of Boulder Dam (California Colorado River Commission 1930, 18).

The emergence of this reclamation nexus between local water groups like the IID and federal agencies provoked heated opposition from the other riparian states, which feared that the alliance might threaten their own water claims. During a conference held in 1927 to finalize the terms of the interstate compact, Governor Dern of Utah alleged that "Every State has the inherent sovereign right to control the uses of water, which is essential to its existence. . . . The theory advanced by attorneys for the Bureau of Reclamation, and embraced in the Swing and Johnson bills, that Congress has the power to allocate and apportion the waters of any western river among the States, regardless of their will, is abhorrent to our whole plan of government" (Arizona Colorado River Commission 1927, 11). Indeed, when California, prompted by the IID and Los Angeles delegates, refused to ratify the compact unconditionally, Utah briefly withdrew from it in protest (Palmer 1927). This incident caused Dern to give a speech which indicates more clearly than any other the threat that state leaders saw to their autonomy: "It is my duty as governor of my State to have vision enough to guard its future. I have no ambition to let the people of my State a generation or two hence say . . . 'back in 1925 we had a stupid governor who was asleep and allowed Utah's water to be appropriated by other States, and as a result our growth was stunted' " (Dern 1926, 17).

The delicate balance between preserving state autonomy and water rights, while also seeking federal financing, was best illustrated by the case of Arizona. Mindful of the state's acute water scarcity, geographically concentrated water-user groups lobbied fiercely for large-scale irrigation projects that would divert Colorado River water to the state's inland regions. Perhaps the most eloquent statement of Arizona's distinctive interests in the basin comes from congressional testimony in which a state official proclaimed its "wider and more general interest than that of any other state. All of Arizona lies within the drainage area of the Colorado River. Every section and locality of the state is more or less directly affected by any agreement that may be entered into with respect to the waters of the Colorado River" (U.S. House Committee on Irrigation and Reclamation 1928, 32). Obtaining access to a greater quantity of Colorado River waters thus quickly became a salient political issue in Arizona. In 1922, Republican governor Thomas Campbell staked his re-election campaign on a promise that irrigation from the Colorado would support the state's future economic development, "making profitably productive over 2,500,000 acres of now arid, desert lands" (*Arizona Republican* 1922e).

Arizona water users recognized that these ambitions could be fulfilled only with federal financing and with it some form of interstate collective action. As the *Arizona Republican* editorialized, "Arizona is so situated that it would require the expenditure of much more money to reap the full fruits of development than any other state" (*Arizona Republican* 1922d). This fact, and the desire to secure sufficient water resources for future development, compelled state leaders to initially support the idea of an interstate compact. One of Arizona's compact negotiators explained the state's participation by noting that "Unless the people of the upper states . . . can be assured of reasonable protection . . . we may expect the most violent opposition to any program [of reclamation]. . . . Four states will thus unite against us in carrying out any such program, each of whom has its delegation in Congress who may be expected to be in hearty sympathy" (*Arizona Republican* 1922b). Legislators from Mohave County, which sits astride the Colorado and therefore expected to benefit from its development, were particularly strident in demanding that Arizona sign the compact and that it "gets its full share of the cream" (St. Charles and Householder 1922).

Other Arizona water-user groups, however, feared that the compact would provide too little water to satisfy their future requirements and accordingly sought to reframe the debate over compact negotiation as one pitting Arizona's autonomy and control over its own water resources versus its riparian neighbors. A vocal proponent of such fears was George Maxwell, who helped mobilize Arizona farmers in opposition to the compact by founding and leading an organization advocating a "High-Line Canal" to bring water from the Colorado to central Arizona, in order to ensure sufficient water for the state. The compact, Maxwell charged, "was a smokescreen to cover the design of turning the waters of the Colorado over" to irrigate some 900,000 acres in California, leaving only enough water to irrigate some 282,000 acres in Arizona (*Arizona Republican* 1922a). Maxwell quickly gained allies among both Democratic and Republican state politicians concerned that Arizona faced a poor bargain in the compact negotiations, and after the 1923 legislative session began, rarely was "a day allowed to pass . . . without an added measure designed to stave off action on the Colorado River compact" (*Los Angeles Times* 1923a).

The argument that California's thirst would inevitably rob Arizona of its own claims to the Colorado gained support throughout the state, and Arizona politicians began to explicitly cast the compact negotiations in terms of Arizona's autonomy and viability as an independent entity. Arizona Democratic governor George Hunt declared that "acceptance of the Colorado River compact would be the greatest disaster Arizona could know" (*Los Angeles Times* 1923b). In the 1924 state elections, the issue of compact ratification was so salient that observers noted it "will determine the result more than the personality of the candidates, or even their politics" (*Los Angeles Times* 1924). In a reflection of these political realities, Hunt successfully ran for re-election on an anticompact platform and declared Arizona "unprepared to negotiate" to "distribute resources from the biggest asset she has" (Palmer 1925a). A particular objection, which would form the basis of Arizona's policy for another forty years, concerned the claim that accession to the compact would give California rights to the Gila River, a tributary of the Colorado arising within Arizona. A state official warned that failure to secure the Gila's waters for the state's exclusive use "means that at least a million acres of Arizona's best land . . . must forever remain a desert, instead of being transformed into thousands of farms, the products of which would add millions of wealth to the state" (Stevens 1928, 26).

Driven by this iron-clad perception of Arizona's individual interest in the Colorado River, the state's congressional representatives derided the Swing-Johnson bill as "a measure designed exclusively for California's benefit" (Arizona Colorado River Commission 1927, 6–7). Hunt meanwhile denounced California for attempting to "secure through Congress . . . those things which she could not possibly hope to secure through agreement" with the riparian states (U.S. House Committee on Irrigation and Reclamation 1928, 49). State officials also railed against California's irrigation districts, which they charged were hijacking the state's participation in Colorado River basin institution-building. "So far as California's intense interest in the Colorado River problem is concerned," an Arizona official complained, "these two districts [Imperial and Coachella], for all practical purposes constitute California" (U.S. House Committee on Irrigation and Reclamation 1928, 33). Upper basin states, too, were disturbed by the clear influence of southern California sectional interests. The breakdown of a 1927 Denver compact conference, for instance, was attributed by upper basin representatives "to the unyielding attitude of a political element from the Imperial Valley" (Jones 1927). California representatives, for their part, remained opposed to negotiating with Arizona, warning their compatriots that "It would be very unwise for California to be dependent upon water that Arizona might at any time claim and take away" (California Colorado River Commission 1930, 35–36).

In this air of sectional acrimony, interstate collective action in the Colorado appeared stalled. But tireless advocacy by Swing-Johnson bill supporters enabled it to sidestep Arizona's refusal to accede to the compact by reaching an agreement between California and the upper basin states. In 1928, Colorado Republican senator Lawrence Phipps introduced an amendment to the Swing-Johnson bill providing that only six of the seven riparian states need sign the compact for it to take effect. A New Mexico–sponsored amendment, meanwhile, helped to allay the concerns of the upper basin states by limiting California, which Carpenter continued to call Colorado's "greatest water enemy," to 4.4 MAF of Colorado River water annually (*Los Angeles Times* 1928; *Chicago Daily Tribune* 1928; U.S. House Committee on Irrigation and Reclamation 1928, 332). On the strength of these provisions, the Swing-Johnson bill passed both houses of Congress. The Swing-Johnson bill, formally titled the Boulder Canyon Project Act of 1928 (BCPA), funded construction of Boulder Dam and, in recognition of Arizona's refusal to participate in interstate collective action, provided for federal allocation of water in the lower basin by requiring that both California and Arizona contract with the secretary of the interior in order to use water stored by the dam. This unprecedented provision established, for the first time, a clear federal role in the management of Colorado River waters. By avoiding the question of Arizona's participation in basin collective action, however, the BCPA simply sowed the seeds of protracted legal conflict.

In part, the stunted state of institutional collective action in the Colorado River basin reflected the fact that despite Washington's expanded role in Colorado water allocation as a result of the BCPA, federal agencies and actors found their powers constrained by interstate conflict. National leaders, particularly successive presidents and secretaries of the interior, resolved to avoid becoming ensnared in the details of resolving a bitter controversy between electorally important states. Hoover, for instance, had strongly supported the compact, yet during the election electoral sensitivities restrained him to simply issue gentle encouragement "that a seven-state pact ultimately be made and Arizona brought into the fold" (*Los Angeles Times* 1929a). After taking office, Hoover assiduously cultivated an image of neutrality and "took the position that this matter should be left entirely to the States interested, without any interference or suggestion on the part of the White House" (*Los Angeles Times* 1929b). When the compact finally entered force in June 1929, Hoover felt compelled to support it despite Arizona's noncooperation, stating that the compact "opens the avenue for some of the settlement of other regional questions as between the states rather than the imposition of these problems on the federal government" (*Los Angeles Times* 1929c).

At the same time, California water users' attempts to assert senior water rights continued to undermine basin-wide, cooperative institution-building. In 1931, representatives from the city of Los Angeles testified before the California legislature that further attempts should be made to "dispose of a possible future adversary in Arizona" by refusing to allow it to join the compact (*Los Angeles Times* 1931). Accordingly, California pressured the other basin states to block an Interior Department proposal to deliver water to Arizona, despite the department's authority to do so under the provision of the BCPA (*Los Angeles Times* 1934). Arizona, meanwhile, filed suit in the Supreme Court, charging that Boulder Dam "attempts to deprive a sovereign State of its natural resources without its consent" (Benedict 1930). The failure of this and other lawsuits brought by Arizona against California during the 1930s provoked extreme reactions in Arizona: the Arizona Legislature resolved to condemn any state official who supported the compact, and Democratic governor B. B. Moeur deployed armed National Guardsmen to prevent construction of Parker Dam (*Washington Post* 1934). Arizona's Colorado River Commission, meanwhile, ruled out cooperation with neighboring states, declaring that "Arizona could agree with Wyoming, Colorado, Utah, and New Mexico, but the differences between Arizona and California are fundamental and cannot be adjusted by agreement" (Colorado River Commission for Arizona 1935, 33). Collective action in the basin, it was clear, remained hindered by the compact's perceived threat to state autonomy.

As in the previous decade, however, the desire to preserve state autonomy continued to be balanced by a desire to assert state interests at the federal level, necessitating limited forms of interstate cooperation. By the end of the 1930s, growing water demand in the state and the prospect of a large Mexican water claim under a proposed United States–Mexico boundary treaty persuaded Arizona's leaders that they required mainstream Colorado River water, which in turn necessitated ratification of the compact (Webster 1935; *Los Angeles Times* 1937). In this concern Arizona was joined by the upper basin states, which together sought to exploit political fragmentation among different elements of the federal bureaucracy. As it became clear that President Franklin Roosevelt and the State Department favored concluding a treaty with Mexico, Colorado basin states began to aggressively lobby agencies they felt would be sympathetic to their interests, namely the Interior Department and the BuRec (Colorado River Commission of Arizona 1940, 43–44). In 1938, all seven basin states convened a conference to petition the administration to ensure no American water would be ceded to Mexico (*Los Angeles Times* 1938). Not for the first time, the states' successful exploitation of political fragmentation undermined federal efforts to play a direct role in basin governance. As a U.S. Senate committee report put it, "From the standpoint of the United States, the existing [Boulder Canyon Project] act is extremely difficult of administration" because it entailed "the determination by the Secretary of the Interior of highly controversial issues between various sections of the Colorado River Basin" (U.S. Senate Committee on Irrigation and Reclamation 1940, 6).

The difficulty of negotiating between these divergent state interests was illustrated in the aftermath of the U.S.–Mexican treaty, which was ratified by the Senate in 1944 in an attempt to solidify Mexico's support for the United States during the Second World War, over the strident opposition of riparian states. Just as the states had long feared, the treaty granted Mexico 1.5 MAF of Colorado River water annually (Stanford 1945). Despite this loss, Arizona's support for the upper basin states' attempts to block the U.S.–Mexican treaty gained it considerable favor, and in 1943–1944 this favor was repaid when all five upper riparians defied California by voting to approve Arizona's entry into the Colorado River Compact without first compelling the state to renounce its claims to the Gila River, as they had previously agreed (*Los Angeles Times* 1943). Encouraged that these shifting state alliances might signal an opportunity to finally bring Arizona into the basin's nascent institutional framework, Interior Secretary Ickes proposed to grant Arizona a contract for delivery of 2.8 MAF annually from the Boulder Dam, on the condition that Arizona would ratify the Colorado River Compact. Arizona quickly complied, for the first time joining the interstate cooperative agreement to which the other basin states had been party for nearly two decades (*Los Angeles Times* 1944). Yet rather than presaging an era of greater basin-wide cooperation, the Arizona–upper basin alliance merely set the stage for a period of intensified competition between water users in the various states as each state sought federal funding for reclamation projects within its borders.

During the first phase of the Colorado River case, initial attempts at interstate collective action were hindered by political lobbying on the part of sectional interests in the form of geographically concentrated water-user groups in riparian states. These interests sought to maximize their allocations of water under the proposed compact and, in an effort to gain support for their positions, cast them in terms of a struggle for state autonomy and control over water against their neighbors. In response, state leaders staked their political legitimacy on attempts to secure preferential water rights. Guided by the prevalence of prior-appropriation doctrine, state leaders sought to establish claims to water as early as possible, in order to protect them from subsequent claims by other states whose water needs would only grow over time, especially in the upper basin. This competition created a zero-sum perception of Colorado River water allocation and continually weakened and undermined early attempts at institution-building in the basin. In the end, the fear by upper basin leaders that California would secure more water through an alliance with federal bureaucracies than through a compact led them to support an interstate allocation agreement. It is significant, however, that even in this climate of fear this agreement was concluded only with the strong support of third-party actors in the form of national political leaders. The desire of state officials to protect both autonomy and preferential rights to water nonetheless led them to stop well short of constructing the infrastructure to support collaborative and participatory management of Colorado River waters. Instead, all states soon appealed to the reclamation nexus to secure water for their respective constituencies, thereby sustaining competition for water in the Colorado River basin, even as the federal government continued its efforts to broker agreement among the riparian states.

The Rise of Reclamation Politics, 1946–1968

The second phase of institution-building in the Colorado River basin was distinguished by a shift in the political arena away from the basin itself toward Washington, DC, and in particular to Congress and the U.S. Supreme Court. But this shift in arena did not signal a decisively greater role for federal actors, who remained limited by both legal authority and political calculations in their efforts to mediate interstate water conflict. Instead, this new phase was characterized by a distinctive brand of reclamation politics, in which the riparian states competed to win greater congressional appropriations for water storage and delivery projects. The nexus between geographically concentrated water-user groups, congressional representatives, and the BuRec meanwhile continued to crowd out the emergence of collaboration and participatory decision-making in the Colorado River basin. The efforts of national leaders to promote comprehensive basin-wide institutions and development projects were consistently undermined by this reclamation nexus. Nonetheless, a landmark moment in the history of riparian collective action occurred in 1963, when the Supreme Court, granted Arizona's long-standing claim to an extra 2.8 MAF of Colorado River water, paving the way for completion of the Central Arizona Project (CAP) and settlement of Arizona's long dispute with California. While most players in the Colorado River drama, including the states themselves, expected this settlement to finally allow each state to build dams and canals in peace, in fact it only underscored the need for collaborative and participatory institutions in the face of growing water stress.

As the postwar era dawned, resolution of the Arizona–California conflict appeared a distant possibility. Southern California water users plotted to block the CAP, the crown jewel of Arizona's reclamation nexus, from being authorized by Congress. California's Colorado River commissioner warned state leaders that "Arizona is now aggressively promoting before Congress bills to authorize two water projects. . . . They can only be served if the Metropolitan Water District and San Diego projects in California are denied the bulk of their water" (Ainsworth 1946). In response, southern California's powerful sectional interest groups intensively lobbied the state's congressional delegation to block CAP authorization, again casting control over interstate water in terms of the state's economic future. The California State Chamber of Commerce warned the state's congressional representatives that the CAP "will stop the growth of Southern California and halt the expansion of business and industry" (*Los Angeles Times* 1947b), while groups like the Los Angeles Chamber of Commerce and the Colorado River Association relentlessly attacked the alleged cost of the CAP (*Los Angeles Times* 1948a). This mobilization proved highly effective at gaining the support of California congressmen.

By 1948, Republican representative Norris Paulson could promise his constituents that the state's entire congressional delegation, both Democratic and Republican, was "already solidly behind the Southern California stand," and with this unity California's delegation prevented a vote on CAP authorization (*Los Angeles Times* 1948b; W. Francis 1949a).

In response, Arizona's own congressional representatives forged a bipartisan alliance to build support for CAP authorization both within the state and in the upper basin. Rapid postwar economic growth combined with acute drought led Arizona leaders to view their future water supply with the same degree of concern as did California's and with a similarly uncompromising view on the prospect of sharing water with neighboring states. Arizona's senior senator, Democrat Carl Hayden, applied the sharpest insult of the times to cast California in adversarial terms, comparing Imperial Valley farmers to "the Politburo in Moscow" and pleading with Congress to "save the civilization" of central Arizona by financing the CAP (W. Francis 1949b). In pursuit of this goal, Arizona's congressional delegation presented a united front across party lines, introducing legislation intended to block California's own water projects and presenting data to congressional committees questioning their value (W. Francis 1949c). At the same time, sectional interest groups within Arizona resorted to increasingly ambitious means of portraying California as a greedy, riparian bully. An organization called the Central Arizona Project Association, representing local businesses and agricultural water users, arranged the use of Arizona Air National Guard planes to fly reporters over California, for example, providing a bird's-eye view of the neighboring state's allegedly wasteful use of Colorado River water (Johnson 1977).

In the postwar period, the U.S. federal government, meanwhile, buoyed by new economic powers gained during the New Deal and Second World War eras, attempted to advance an ambitious regional development approach to the problem of securing cooperation in the Colorado River basin. Yet the persistent influence of the reclamation nexus in basin politics greatly frustrated these efforts, as it had in the past. In 1947, a Republican-controlled Congress, citing interstate conflict, cut President Truman's proposed reclamation budget in half, prompting Interior Secretary Julius Krug to warn the Colorado riparian states that their squabbles were undermining his department's plans for comprehensive development of the basin (*Los Angeles Times* 1947a). Hopes that the Arizona–California deadlock would be broken, and reclamation funding rereleased, were again raised in 1950, when the Water Resources Policy Commission, appointed by President Truman to evaluate reclamation projects across the country, issued its report. The commission strongly favored the formulation of comprehensive development plans for each of the country's major river basins, on the principle of "One river, one plan." More pointedly, the commission also proposed canceling all Colorado River basin projects until their economic feasibility could be reviewed by an independent commission (Francis 1950). Such ambition found eager supporters in some federal agencies, particularly the BuRec, which were frustrated that the ongoing interstate dispute prevented construction of new water projects.

But despite strong federal support for a comprehensive, regional approach to basin development, each of the riparian states made it clear that it expected support for individual reclamation projects within its borders and would fight any attempts on the part of the federal government to enhance its role at the expense of state autonomy. The commission's regional plan for the Colorado basin was quickly condemned by the states as amounting to "Federal domination of the nation's water resources" (Ainsworth 1950). In California, Washington, D.C. was added to the state's growing list of perceived water enemies. Despite his partisan relationship to the Democratic administration, Representative Cecil King of Los Angeles, for instance, scored its proposals for basin-wide development by remarking how "strange that all these visionary and untried schemes originate either in Arizona, which wants California's water, or in the Department of the Interior" (L. Francis 1949a). Yet, in fact, opposition to federal involvement from both California and Arizona ensured that Washington enjoyed only limited influence in the Colorado River basin. When the Truman administration attempted to make CAP financing contingent on reaching an agreement between Arizona and California, the opposition of Arizona's congressional delegation forced it to back down after only a week (L. Francis 1949b). Instead of insisting on interstate cooperation, as it had previously pledged to do, the Truman administration bowed to state lobbying efforts by funding separate reclamation projects in both states, including a "record-breaking postwar appropriation" for Los Angeles–area water projects as well as a new Arizona aqueduct to appropriate Colorado River water (Francis 1952a, 1952b). The BuRec commissioner was soon reduced to lamenting that the 1922 compact had "served to create and to emphasize a division in the basin that does not exist physically. It also served to obscure . . . that basin-wide planning and development has demonstrated its value" (Hebert 1960).

As this episode serves to highlight, the power of the reclamation nexus between state water users and their congressional representatives ensured that institution-building in the Colorado River basin was shaped almost entirely by the states rather than the federal government. These interests continued to revolve primarily around securing financing for reclamation projects; by the late 1940s, the quest for increased reclamation financing drove Arizona into an alliance with the upper basin states against California, and in 1947 it joined an intergovernmental body called the Colorado Basin States Committee, from which California and Nevada had previously withdrawn. The five states formed a lobbying organization to oppose the proposed Glen Canyon Dam that would occupy a site within Arizona but which California supported because it would enable the storage of large quantities of Colorado River water as part of the Colorado River Storage Project (CRSP), which could then be channeled into California. The upper basin states quickly adopted Arizona's well-practiced anti-California slogans and charged that Glen Canyon was for California merely an attempt to "obtain more water from the Colorado than she is entitled to" (*Christian Science Monitor* 1947). By the end of 1947, Arizona had succeeded in dividing the upper and lower basins into distinct sectional groupings. The *Christian Science Monitor* (1947) reported that "Five states of the Colorado River basin have now perfected a unified, fighting organization to oppose the stand taken by California and Nevada in regard to dividing up the waters of the Colorado River. The new thing is not the opposition, but that at a recent meeting here the four 'upper basin' states . . . submerged all differences to team up with Arizona."

This alliance finally created the counterweight to California's political influence that Arizona leaders had long sought.

On the strength of their newfound common cause against California, Arizona and the upper basin states cemented their alliance by concluding the Upper Basin Compact, in which each state recognized one another's claims to Colorado River water in a thinly veiled rebuke to California. This novel comity led to a landmark moment for interstate collective action in the basin as Article 8 of the Upper Basin Compact created an interstate commission, with representatives from each of the four upper basin states as well as the federal government. The commission constituted the first permanent, interstate body devoted to operationalizing a water-sharing agreement. Crucially, however, the states made certain that the commission would stop well short of pursuing the sort of comprehensive regional development advocated by federal officials. Instead, its design ensured that the commission would remain a tool of state rather than federal control: it was invested only with the ability to collect data, undertake research projects, and issue reports; and the states ensured control over its functions by requiring the concurrence of four of the five members of the commission to take any action, no matter how routine (U.S. Bureau of Reclamation 1948). As in previous periods of the history of Colorado River collective action, the states' preference for autonomy produced only weak forms of interjurisdictional cooperation.

Moreover, despite the security afforded by the Upper Basin Compact, state leaders remained disturbed by growing water demand in California and they feared that without constructing water-storage projects, they would be forced to shoulder a disproportionate burden in times of drought (Fleming 1951). Because California utilized the majority of Colorado River water, the upper basin states came to view southern California water users as their primary opponent in seeking federal financing for upper basin reclamation projects. The upper basin states were accordingly determined to employ the CRSP to secure water for their constituents, even in the face of California's political power in Congress. As in the past, California's strong sectional interest groups were believed to be the largest roadblock. As a member of Colorado's Colorado River Commission told a Mountain States Water Users Conference, "potent Southern California agencies" were "again advocating reclamation policies calculated to defeat the interests of the upper-basin states" (*New York Times* 1950). These same state scrambled tocalibrate their strategies to undermine California's congressional delegation and its perceived allies in Washington. New Mexico Democratic senator Clinton Anderson duly warned Californians that if they blocked the CRSP, "no new [reclamation] projects will be approved for California as long as I'm chairman" (Fleming 1955).

This threat provoked a reconsideration of California's confrontational stance among state officials and opened a brief window for a cooperative strategy to emerge. After a Sacramento official warned that "California is about as popular in legislative circles in Washington as a good case of smallpox," California leaders sought to improve relationships with upper basin states, calling for "cohesion" at a conference of western states (Brunn 1950, 3; *Los Angeles Times* 1950, 10). But moderation in the state's position was fiercely opposed by southern California's sectional interest groups, who, as in previous phases of the Colorado River basin dispute, largely controlled California's policy. The state's geographically concentrated water users merely echoed previous calls for the state's congressmen to help "bring about the defeat of the Upper Colorado river projects" (*Los Angeles Times* 1955). The Metropolitan Water District (MWD) and CRCCA continued to portray other riparians as free-loading threats to the state's very survival, warning Congress that the CRSP amounted to "one of the greatest raids on the Federal Treasury that has ever occurred" that would force southern California to "gamble with the thirst of our citizens" (Shannon 1955b). Republican congressman Craig Hosmer of Los Angeles called the proposed projects illegal, railing that "Southern California is the most productive area of the nation and we must have the water to which we are entitled by contract, by compact and by appropriation" (Shannon 1955a). The salience of these sectional water politics was further demonstrated in the 1958 California Senate campaign, when candidates traded accusations that the other had failed to protect the state's water rights from the other Colorado River basin states, particularly those of the upper basin (Blanchard 1958). The lingering influence of California's geographically concentrated water-user groups therefore prevented it from joining the spirit of cooperation that increasingly united Arizona with the upper basin.

Even in the midst of this new spirit, however, similarly parochial interests guided Arizona's Colorado River policy to favor reclamation projects over cooperative institution-building. In reference to the CAP, Senator Hayden told the Senate that "Arizona urgently needs more water, without which she faces a slowly withering economy" (U.S. Senate Committee on Interior and Insular Affairs 1963, 3). Such claims nonetheless were supported by only some of Arizona's water users, especially irrigators in the central part of the state. The *Los Angeles Times* reprinted a letter from a legislator representing Arizona's northern Navajo County which claimed that "From the very outset the Central Arizona Project has not had, nor has it now, the support of any large segment of our people. All the fire and fury has come from a few interested landowners in a small section of the valley" (*Los Angeles Times* 1953b). Nonetheless, on the strength of these vocal few, Arizona filed suit against California in 1952, seeking adjudication of its water rights in advance of seeking Congressional approval for the CAP. As in previous cases, Arizona laid claim to 1 MAF hitherto utilized by California, but which Arizona claimed because it arose from within the state (*Los Angeles Times* 1953a). However, in its new suit Arizona modified its legal strategy, asserting that Colorado River tributaries belonged to individual states rather than to the basin as a whole as the 1922 compact had stipulated (August 2007, 80–85). This assertion of state sovereignty over the water within its borders perfectly illustrated the continuing dominance of sectional state interests and the opposition to cooperative institution-building in the basin.

Exasperated by this continuing interstate competition for water, Interior Secretary Stewart Udall, a three-term Arizona congressman and no stranger to the Colorado River controversy, attempted to finally overcome the troublesome nature of state-based territorial politics by proposing the gigantic Pacific Southwest Water Plan (PSWP). The PSWP aimed to meet the energy and water needs of the entire western United States by developing a network of large dams, canals, and transmission lines to channel water and power from the Pacific Northwest to the Southwest, thereby sidestepping the animosity between Arizona, California, and the upper basin states (Dominy 1964). This regional approach, Udall proclaimed, promised to finally transcend "the outmoded concept of state lines . . . only regional planning and action will enable us to meet the growth needs of this area" (Sherman 1963). Yet despite Udall's full-throated support for such a regional solution, he was soon forced to acknowledge the dominant influence of the very sectional boundaries he had attempted to erase. In testimony before Congress, Udall promised that "The rights of the individual states likewise must be respected and the aspirations of the states accommodated" (U.S. Senate Committee on Interior and Insular Affairs 1963, 311).

In the wake of Udall's failed assault on the autonomy of Colorado River riparian states, state political leaders and their allies in Congress quickly mobilized to thwart aspects of the PSWP which threatened state control over water. Powerful Democratic senator Henry Jackson of Washington voiced concern over proposed water transfers from the Columbia River to the more arid parts of the Southwest and demanded that the PSWP be reviewed by an independent national water commission. As the governor of Washington State warily told Congress, "we are a downstream state within a basin whose waters have consistently been looked upon and prejudged as a source of import supply" (U.S. Senate Committee on Interior and Insular Affairs 1967, 631). Meanwhile, national conservation organizations lobbied representatives of eastern states, such as Republican congressman John Saylor of Pennsylvania, to oppose the PSWP because of its threatened effects on the West's dramatic natural scenery (Fleming 1954). The combined influence of this congressional opposition outside the Colorado River basin was sufficient to dramatically reduce the scale of the PSWP, returning the congressional debate to projects favored by the individual states. The appearance of environmental advocacy groups, up to this point a marginal influence in the interjurisdictional politics of the Colorado River basin, provided a harbinger of the critical role they would later play in neutralizing the influence of the reclamation nexus and lobbying for the creation of more cooperative governance in the basin.

For the remainder of the 1960s, however, the preference of politicians to maintain their autonomy at nearly any cost, prevented cooperation and institution-building in the Colorado basin. Upper basin political leaders were just as concerned at the threat of federal control over Colorado River waters as they were of surrendering senior water rights to their downstream neighbors. Given this fear, upper basin congressional representatives proved to be some of the fiercest opponents of Secretary Udall's PSWP. State representatives continued to claim the right to secure and control a minimum supply of water for users within their respective jurisdictions. In 1963 congressional hearings, Democratic senator Clinton Anderson of New Mexico pressed Udall on planned changes to the operation of Glen Canyon Dam as a result of the PSWP, asking "I do think you recognize, don't you, Mr. Secretary, that the opening of the gates does jeopardize the bank account in the upper basin states? I would recognize we have a right to try to protect our bank account" (U.S. Senate Committee on Interior and Insular Affairs 1963, 327). Indeed, sensing a growing federal threat to their control over water resources, the upper basin states modified the role of the existing Upper Colorado Basin Commission to act as a "watchdog group . . . devoted to seeing that the upper river is developed as planned, and that its interests are protected" (Fleming 1963). The commission duly turned its attention to opposing "any part in a compulsory program of overall planning for the Colorado River basin" (*Los Angeles Times* 1965).

Even as state-based opposition in Congress doomed the regional solutions advanced by Interior Department officials, the Supreme Court began a final adjudication of the allocation of water in the lower basin between California and Arizona. California, having lost some congressional allies as a result of the upper basin's alignment with Arizona, filed its own lawsuit in an attempt to preempt Arizona's case. California's complaint, filed on behalf of the state as well as the IID, MWD, and other southern California water users, charged that "Arizona has permitted and encouraged the expansion of the irrigated area . . . in knowledge of the inadequacy of the surface and underground water supply," which impinged upon its "senior and prior" water rights as specified by the 1922 compact (Francis 1953a). Meanwhile, federal agencies sought to use California's lawsuit to clarify and expand federal powers to govern the Colorado River basin, thereby permitting planned water infrastructure projects to go ahead. When the Justice Department joined the suit in 1953, it did so on behalf of several federal agencies as well as Native American tribes (Ainsworth 1953). Its brief reflected the exasperation of federal agencies at their apparent lack of authority relative to the states: "the United States is in doubt as to its obligations and responsibilities. . . . There is a pressing need for a decree by this court declaring, determining, confirming and quieting the title of all parties" (Francis 1953b, 18). However, this attempt on the part of the federal government to assert new authorities raised still further the ire of states that feared the loss of their autonomy.

In some ways, the states' concern was amply justified by events: contained within the Justice Department's apparently innocuous filing was an audacious attempt to expand federal powers over Colorado River water. While federal authority in the basin remained largely indirect and limited by the terms of the BCPA, the invocation of Native American water rights created a novel means by which the federal government might seek to directly allocate Colorado River water between California and Arizona. The Justice Department's brief to the Supreme Court accordingly sought to secure federal authority to allocate some 1.8 MAF of Colorado River water, nearly the entirety of that claimed by Arizona. As in the case of the PSWP, however, the federal government soon backed down in the face of pressure from the states. Coordinated lobbying from eleven western states and public protests from the leaders of California, Arizona, and New Mexico forced the Justice Department to withdraw its petition (*New York Times* 1953; Francis 1953c).

In light of the effective state opposition to the attempts of federal agencies to settle the Colorado River basin dispute, the states effectively abandoned cooperation and negotiation and turned to the judicial system to determine the final partition of lower basin water. The California–Arizona lawsuit remained as contentious as in previous phases of the dispute; a pretrial conference so exasperated a court-appointed special master that he resigned, saying simply, "I give up" (*Los Angeles Times* 1956). After a twenty-six-month-long hearing period, a reappointed special master upheld Arizona's claim to an additional million acre-feet of Colorado

River water, awarding it 2.8 MAF in total and reducing California's allotment by 978,000 acre-feet annually to 4.4 MAF (*Chicago Daily Tribune* 1960). The court itself upheld this partition in its final 1963 decision in *Arizona v. California* and in the process reaffirmed the interior secretary's power to apportion water among water users in the lower basin states, particularly in times of drought (Blair 1963). Officials in Washington, expectant that the court's decision had finally solved the dispute, appeared eager to wash their hands of the matter. Secretary Udall refused to become embroiled in continued controversy surrounding California's 4.4 MAF, telling Congress, "It is our view that this is a matter between the states, and however they want to work this out is satisfactory so far as the administration is concerned" (U.S. Senate Committee on Interior and Insular Affairs 1967).

Yet as the federal government effectively withdrew from direct involvement in basin institution-building, state competition for water and reclamation funds reached new heights. Indeed, for the states, the Supreme Court's decision merely paved the way to completion for their long-delayed individual reclamation projects, completion of which was bound to create competition since in many cases they involved diverting the same flows of water. The continued influence of the reclamation nexus in this period is illustrated by the emergence in the late 1960s of a previously unimaginable alliance between California and Arizona, whereby California agreed to support CAP authorization in exchange for a guarantee of 4.4 MAF of Colorado River water annually. Northcutt Ely, a state official who had long led California's opposition to Arizona, conceded with startling frankness that "We would rather reestablish our alliance with Arizona than to get involved in open warfare with you. The cement of this alliance is the recognition of our existing investments of $600 million, just as we recognize your larger aqueduct" (U.S. Senate Committee on Interior and Insular Affairs 1967, 293). This unprecedented accord settled a nearly fifty-year dispute between California and Arizona, but the fact that it was reached only after both states' water rights claims were finally guaranteed demonstrates how completely sectional politics shaped their relationship. Only when demand from both states' powerful water users could be ensured could the two states resolve their long-standing conflict.

Indeed, the strain placed on the basin's water supply as a result of the California–Arizona accord alarmed the upper basin states, which continued to compete with the lower basin in seeking federal financing for reclamation projects. Both lower basin states asserted water claims that were in excess of the Colorado's total flow by the mid-1960s, raising fears that these claims would "bankrupt" those of the upper basin states (*Christian Science Monitor* 1965). As Wyoming's state engineer explained in 1967 congressional testimony, "As the upper basin develops, the water supply to the lower basin will progressively diminish . . . Wyoming fears that future pressures arising from the developed economies in the lower basin will be muscled with sufficient political strength to effectively inhibit the future development of Wyoming" (U.S. Senate Committee on Interior and Insular Affairs 1967, 409). Ever determined to protect CAP financing at all costs, Arizona's congressional delegation adopted a carrot-and-stick approach, on the one hand, blocking authorization of upper basin reclamation projects until CAP financing could be secured and, on the other, seeking the support of Pacific Northwest states by assuring them of Arizona's continued opposition to transfers from the Columbia River basin (Irwin 1966; *Los*

Angeles Times 1967). As before, management of the Colorado River basin was decisively shaped by interstate hydropolitics.

Arizona's appeal to the interest of the upper basin states in securing federal financing for reclamation projects within their respective jurisdictions proved to be remarkably effective in securing their cooperation. This new strategy would fund each state's proposed reclamation projects but at the cost of further straining the basin's water resources. The upper basin states' opposition to the California–Arizona agreement was quickly dropped when the Colorado River Basin Project Act was amended to include financing for projects throughout the upper as well as lower basin. Finally, and crucially, Arizona agreed to supply California with 4.4 MAF annually in perpetuity, meaning that Arizona would bear the brunt of reduced allocations in times of drought. The resulting Colorado River Project Act of 1968 (CRPA) authorized a total of $1.3 billion in project funding for CAP as well as dozens of other projects throughout the western states. The act passed with the sole but telling opposition of an eastern congressman, the conservationist Republican John Saylor of Pennsylvania, and national environmental organizations (*Los Angeles Times* 1968a). Upon signing the bill, President Johnson gushed that "good and reasonable men have put aside their differences in favor of regional progress" (*New York Times* 1968).

Nevertheless, the CRPA illustrated more than anything else the extent to which state territorial politics proved capable of undermining cooperative regional institution-building. The act's provisions concerning basin governance were notably weak, as were those of the Upper Colorado Compact Commission. Neither the act nor the compact incorporated rules or procedures concerning the allocation of basin water resources under drought or scarcity conditions, despite the fact that it was already apparent that completion of all planned reclamation projects would divert or impound virtually all water in the basin. The *Los Angeles Times* (1968b) noted that, despite CRPA, "within a decade the West must face up to the hard fact that the Colorado River without augmentation simply doesn't contain enough water for all of the [statutory] commitments." The second phase of the Colorado River case thus illustrates how, in the absence of credible third-party involvement, the dominant influence of individual state interests produced weak forms of cooperative governance that failed to provide a forum for solution of emergent challenges like water scarcity. Fortunately, this trend has been largely reversed in the third and most recent phase of the Colorado River case.

The Growth of Civil Society and Cooperative Governance, 1969–2014

The third and most recent phase of institution-building in the Colorado River basin has been marked by a profound shift toward a more cooperative set of relationships both between levels of government and among the riparian states themselves, and by extension towards more collaborative and participatory decision-making. The signal event of this most recent period was the broadening of the political opportunity structure for social movements, particularly those relating to Native

American rights and environmental protection, which weakened the reclamation nexus between territorial economic interests and federal agencies that had previously underpinned interstate competition for Colorado River water. The growth in scope and influence of national environmental organizations, combined with wider changes in American political discourse, led federal political leaders and agency officials to shift their focus to managing rather than controlling the Colorado River. These same movements also helped to reduce the influence of geographically concentrated water-user groups within states. Finally, toward the end of the twentieth century, increasing physical water scarcity provided a powerful impetus for interstate cooperation as it became clear that the Colorado River flows on which the 1922 compact had been calculated were unlikely to persist in a changing climate. During this third phase, institution-building in the Colorado River basin for the first time became defined more by cooperation than by competition. This transformation was enabled by the profound influence of national environmental organizations, whose activism led to far more constructive federal engagement in basin governance—and created unprecedented constraints on the sectionalism that had long sustained interstate competition rather than cooperation over the shared waters of the Colorado.

Environmental issues had begun to influence institution-building in the Colorado River basin in the postwar era, and concern regarding the environmental impact of reclamation projects featured prominently in congressional debate over the Colorado River Project Act in 1968. However, the potential for environmental issues to reshape riparian politics became clear only in 1969, when President Johnson used his last ninety minutes in office to place substantial areas along the Colorado River under federal protection, thereby preventing their use for ranching, mining, and other extractive activities. While conservation groups like the Sierra Club praised the move as "true conservation in the public interest," state officials were furious and protested on the familiar grounds of a threat to state autonomy. Utah's director of natural resources expressed hope that the next administration would "undo the damage Johnson has done," while Utah Republican senator Bob Bennett called the federal designation "the most blatant type of greed I can imagine" (Goodman 1969). Despite this tired objection to perceived federal overreach, the growing political salience of environmental issues induced Washington to take a more active and constructive role in governance of the Colorado basin after 1970s than it had in previous decades.

In part, the federal government's increased role in basin governance relative to the states during the 1970s was the result of its long-standing prerogatives in the field of foreign relations. Soon after taking office, President Richard Nixon made improving ties with Mexico a top priority. The Mexican government made it clear that this would require Nixon to solve the growing problem of salinity in the lower basin, which Mexico claimed made Colorado water essentially worthless by the time it entered its territory, in violation of the 1944 treaty (Semple 1970). The issue became so politically heated that a 1972 article noted that "When Mexicans discuss relations between their country and the United States, they tend to mention first the level of mineral salts in the Colorado River" (Severo 1972). Continued discussion of the salinity issue bore fruit when in 1973 President Nixon announced that the United States would commit to building what was then the world's largest desalinization facility to ensure that water flowing into Mexico remained of sufficient quality to support irrigation (Childs 1973). Water-user groups within the riparian states, however, grew concerned that this commitment would reduce water availability in the southwestern states. In response, they demonstrated their lingering political power in Congress by persuading it to designate the Department of the Interior, over which groups like the IID and the MWD held more influence, as the lead agency on the desalinization project, rather than the State Department, as the administration had proposed (McElheny 1974).

Despite this setback, the success of the salinity-reduction effort marked a new period of commitment in Washington to promote environmental protection in the Colorado River basin. This new commitment was in turn spurred and catalyzed by the rapidly growing political influence of the environmental movement. The decade of the 1970s opened with passage of several landmark pieces of environmental legislation that greatly improved the political opportunity structure for environmental advocacy organizations, most notably the National Environmental Policy Act of 1970 (NEPA) and the Endangered Species Act of 1973. Both acts obligated federal agencies like the BuRec to take environmental issues and impacts into account in the planning and construction of reclamation projects. Nonetheless, as state opposition to Johnson's conservation measures attests, the development of collaborative and participatory decision-making was far from inevitable and stemmed from activism on the part of environmental organizations, which skillfully utilized the provisions of new environmental legislation. As early as 1972, barely two years after the NEPA created the environmental impact assessment process, organizations like the Sierra Club and Friends of the Earth induced a distinct change in orientation among federal agencies like the BuRec and the USACE. Where once these agencies had been obsessed with "quick technological fixes," news reports noted that by 1972, prodded by environmental groups, they were actively considering conservation-oriented approaches (Ripley 1972a). The same year, the Environmental Defense Fund illustrated the growing power of environmental organizations by using the NEPA to sue the BuRec for proposing to sell water allocated by Congress for irrigation for use by coal companies, a notable challenge to what had previously been an unassailable bureaucratic prerogative (Kenworthy 1973).

At the same time, environmental and Native American rights movements combined to help undermine the legitimacy of reclamation projects in both Arizona and the upper basin. The issue of Native American water rights had been left unresolved in the Supreme Court's 1963 *Arizona v. California* verdict, which was silent on tribal water claims. Moreover, as part of CRPA passage in 1968, Colorado Democratic congressman Wayne Aspinall, chair of the powerful Reclamation Subcommittee, persuaded Arizona's Navajo Nation to surrender its senior water rights to Colorado River water in exchange for federal financing of infrastructure projects, thereby freeing additional water for use in planned upper basin reclamation projects (Greider 1969). Congress's general reluctance to recognize Native American water rights increasingly became a focal point for social activism. In a well-publicized incident in the early 1970s, Arizona's White Mountain Apache Tribe posted armed guards to prevent state officials from serving a court order to pump water from within the tribe's reservation in order to irrigate farmland in the Salt River Valley (Slocum 1973). At the same time, the growing influence of environmental conservation projects was illustrated by the simultaneous founding of an Arizona organization called Citizens Concerned About the Project, which quickly drew attention to the CAP's perceived environmental impact and filed suit to force the BuRec to complete an environmental impact assessment (Awkesasne Notes 1975).

Perhaps no state more clearly embodied the shift in political power from the reclamation nexus to third-party advocacy organizations than Colorado, where environmentalists joined with residents of the state's West Slope region to oppose completion of the proposed Fryingpan–Arkansas Project, strongly favored by the BuRec and Colorado congressional representatives, especially Aspinall. One of the most powerful congressmen of the postwar period, Aspinall represented what was at the time the largest single congressional district in the continental United States and one that featured little manufacturing and even less appetite for the environmentalism increasingly influential in other parts of the country. Supported by conservative constituencies, Aspinall doggedly sought federal money for reclamation and resource extraction activities, while opposing the conservation measures now called for by environmental organizations with some regularity. In many ways Aspinall embodied the old sectional tensions that had impeded cooperation between the riparian states; a 1971 *New York Times* profile quoted a congressional colleague as saying that "In many ways he is very, very provincial. It is fair to say that he has repeatedly put his district above the national interest" (Fradkin 1971). But such sectionalism had by the early 1970s ceased to be the modus operandi in the Colorado River basin. The changing balance of power between environmental and rights organizations, on the one hand, and the BuRec and state officials, on the other, was illustrated at the 1972 dedication of the Fryingpan–Arkansas Project, when Interior Secretary Rogers Morton acknowledged "diverse opinions" concerning its wisdom and Colorado Republican governor John Love declared that the project represented "the end of an era" of large water infrastructure projects (*New York Times* 1972).

The demise of the reclamation politics embodied by politicians like Aspinall was quickened by the release in December 1972 of a report by the National Water Commission that strongly criticized the federal government's approach to reclamation projects. The commission had first been appointed by President Johnson, and part of its mandate included a re-evaluation of the federal financing of western reclamation projects. Its highly anticipated report was fiercely critical of the interlocking system of state and federal agency interests, which had sustained competition in the Colorado River basin since the early twentieth century, and recommended wholesale reform to the process by which the benefits of such projects were assessed. Under the Newlands Reclamation Act, reclamation projects were supposed to be financed by user charges that would enable costs to eventually be recouped, but in practice rates were almost invariably set well below cost recovery levels. The report specifically questioned the practices of agencies like the BuRec which, thanks to their alliances with powerful congressional committee chairs, had previously been largely above public reproach. As a news report noted at the time, the report was "particularly rough on flood control, irrigation and waterway development by the

Army Corps of Engineers [and] the Interior Department's Bureau of Reclamation" (Ripley 1972b).

Despite the commission's damning report, the final collapse of the reclamation nexus had to await a noisy public debate later that decade, when the new Carter administration began to publicly denounce the old pork-barrel politics of reclamation. In 1977, President Carter, in an attempt to deal with the nation's fiscal crisis, proposed to cut federal funding for nineteen reclamation projects from a budget already approved by the Ford administration. Despite warnings from senior congressional leaders that the proposed cuts would imperil the president's legislative agenda, the administration at first pressed the attack, calling reclamation projects "blatant boondoggles" (*New York Times* 1977). The strength of the long-standing reclamation nexus provoked a furious backlash; as a news report noted at the time, "Western governors and members of Congress are lobbying furiously to restore funds for the projects already in trouble." The familiar tropes of state sectional politics were quickly deployed as Arizona Democratic governor Raul Castro warned Congress that any reduction in CAP funding would result in "catastrophe" for the state (Rawls 1977). The BuRec, too, initially attempted to maintain its long-standing alliance with state water-user groups. In 1977, BuRec officials warned Colorado farmers that "California would get the water if [they] didn't use it" and pressured them to sign a petition demanding federal financing for a new dam, which the BuRec then presented to Congress to justify its construction (Knudson 1987). In the short term, this old reclamation nexus prevailed, and the Carter administration eventually agreed to sanction most of the funded reclamation projects, including the CAP (Madden 1977).

Yet it was already clear from the statements of state political leaders that conventional reclamation had become politically untenable. Even as Colorado Democratic governor Richard Lamm fought to save three proposed dams within his state, he found it prudent to channel the message of conservationists and acknowledge that "there's a limit to the carrying capacity of the West." As a result, Lamm warned his constituents, future reclamation projects were unlikely to be approved (Lichtenstein 1977b). California Democratic governor Jerry Brown likewise sought to justify his appeals for billions of dollars of federal reclamation funding by highlighting the state's unprecedented water conservation efforts in the aftermath of a severe drought (Lichtenstein 1977a). A critical element of such shifts in position to conservation rather than continuous supply-augmentation was the diminished political power of southern California water-user groups. This shift is illustrated by a 1978 newspaper profile of San Diego's MWD, which it called "the second most powerful legislative body in the state," in part because the MWD employed full-time lobbyists in both Sacramento and Washington, DC. Nonetheless, the profile noted, the MWD had come under attack during a crippling recent drought for its water pricing policies, which substantially favored agricultural rather than urban water users (Hill 1978). Unlike in previous phases in the history of Colorado basin institution-building, geographically concentrated water-user groups could no longer count on success in advancing narrow, absolute claims to water. Partly as a result of this unprecedented criticism, sectional interest groups like the MWD began to lose their disproportionate influence in both California and congressional politics.

At the same time, provisions of the Endangered Species Act of 1973 began to fracture the previous alliance of convenience between federal agencies and state governments in support of reclamation projects. The Endangered Species Act established a new federal responsibility to protect threatened wildlife, which in turn led the federal government to delay, modify, and sometimes block construction of water infrastructure projects favored by the states. In 1980, the Interior Department blocked the construction of a dam and storage reservoir on Utah's White River, leading Republican governor Scott Matheson to protest that "we ought to have a modicum of jurisdiction on our future" (*New York Times* 1980). But instead of supporting reclamation projects favored by states, as they had hitherto done almost without exception, some elements of the federal bureaucracy began to ally themselves with environmentalists. In 1981, the Colorado River Water Conservation District, which supported additional dam construction, protested to the Interior Department that "As viewed by the federal courts and agencies, the Endangered Species Act . . . stands supreme in the land no matter what it costs the country and its people in lost opportunities" (Associated Press 1981). While this objection would at one time have reached sympathetic ears in the Department of the Interior, pressure from a national conservation organization, the National Wildlife Federation, compelled Interior Secretary James Watt to cancel a planned expansion of hydropower facilitates at Glen Canyon Dam, despite his well-known opposition to environmental pressure (Shabecoff 1981). Led by this example, federal agencies began increasingly to turn to advocacy organizations rather than state leaders for political support.

The demise of the old politics of reclamation was further evident in the struggles of Arizona political leaders to shepherd the CAP, which had been authorized in 1968 as part of the CRPA, to completion during the 1980s. The perception that the reclamation nexus pursued narrow sectional interests at the expense of the nation at large shaped debate over CAP authorization. Legislation introduced by Arizona Democratic congressman Mark Udall to settle Native American water rights, a potential obstacle to securing CAP authorization, was vetoed by President Reagan, who called it "a multimillion-dollar bailout of local public and commercial interests at the expense of federal taxpayers" (*Washington Post* 1982). Although CAP supporters eventually prevailed thanks to a 1983 Supreme Court ruling striking down Native American water claims, the project nonetheless became a focal point for political controversy at the national level (Greenhouse 1983). CAP financing remained a point of contention between Arizona and the Reagan administration, which repeatedly signaled its belief that the project amounted to a waste of taxpayer money (Schmidt 1983a). At the official dedication of the CAP in 1985, in many ways the apotheosis of reclamation in the American West, Arizona Democratic senator Dennis DeConcini felt compelled to insist that "It's not pork barrel, it's an investment in America" (Peterson 1985). But reclamation politics, as embodied by the CAP, had lost the support of too many important constituencies for such claims to be convincing. By the time the CAP was finally dedicated, the antireclamation alliance included "Midwesterners unhappy about federal dollars subsidizing Sun Belt economic competitors; environmentalists concerned about the project's threat to the rare desert bald eagle, and Arizona critics who say the Central Arizona Project is unneeded" (Boyarsky 1985a). Indeed, by the time water started flowing through the CAP, even some farmers protested that the water rates were too high and that they would prefer to continue using groundwater (Kenworthy 1994a).

These transformed politics of reclamation led to gradually a new emphasis on cooperation and conservation at the state level, in contrast to the strategy of competition and supply-augmentation previously favored by state leaders in the face of growing water scarcity. In 1983, in a notable shift from the competitive strategies favored by state leaders in the past, Arizona Democratic governor Bruce Babbitt called for creation of the Colorado River Commission, with representatives from the BuRec as well as the seven riparian states, to jointly allocate the river's flow (Schmidt 1983b). Meanwhile, in response to criticism from environmental groups, the state of Arizona enacted some of the most stringent water conservation initiatives in the world until that time, including an almost complete prohibition on groundwater pumping in some parts of the state (Stall 1985). The growing political salience of conservation was even more evident in California. As the CAP neared completion in the mid-1980s, southern California water-user groups sought to build support for transferring water from the water-rich northern portions of the state but quickly encountered bitter opposition from northern legislators in the California Assembly. In the 1985 gubernatorial campaign, in an attempt to gain the support of northern Californians, Los Angeles mayor Tom Bradley chided his own constituents for wasting water—in stark contrast to his previously unabashed support for additional Colorado River transfers (Connell 1985). Such calls to conserve water clearly illustrated the fracture of the reclamation nexus, which relied on increasing the supply of available water rather than reducing demand for it.

As more statewide political leaders began to emphasize conservation over supply-augmentation, enthusiasm for reclamation and water transfer projects began to wane even in southern California. In 1988, in the midst of acute drought, San Diego County's Water Authority adopted new conservation measures; and the city announced plans to build a large-scale wastewater reuse system (Kraul 1988; Bernstein 1989). The most significant development in California, however, was growing support for the compensated transfer of water usage allotments, known as water transfers. In the mid-1980s, the IID proposed the transfer of 100,000 acre-feet annually to the MWD in exchange for investment in water conservation projects within the IID, but the deal languished as a result of farmers who feared the permanent loss of their senior water rights (Boyarsky 1985b). In 1992, however, an important local economic interest group, the Bay Area Economic Forum, endorsed the idea of compensated water transfers; and the state for the first time established a mechanism to sell water from agricultural districts to urban areas on a large scale. This innovation was attributed by a newspaper report to a "breakdown of the historic alliance between urban and agricultural interests in pressing for more dams and reservoirs," on the one hand, and support from environmentalists and urban economic interests, on the other. Although expanded water transfer had to await further federal action, since congressional legislation specifically provided that water-user associations could veto water sales, the stage was set for a dramatic transformation from sectional politics to cooperative governance in the Colorado basin (Reinhold 1992).

A key factor in this transformation was the continued growth in the influence of environmental organizations within the executive branch, which helped to shift agency priorities in the Colorado River basin in favor of conservation and management rather than reclamation. A watershed moment occurred in 1987, when BuRec officials announced that the agency would cease initiating major construction projects and would instead change its "orientation from construction to resource management" (Coates 1987). The Environmental Protection Agency, which hitherto had played little role in western water issues, solidified this trend by blocking construction of a dam in Colorado strongly favored by state political leaders, citing its disruption to a number of threatened species (Coates 1989). Although tensions between environmental, flood control, and irrigation priorities persisted, especially in the lower basin, federal agency leaders increasingly identified with the objectives of conservation and environmental advocacy organizations and gradually incorporated them into agency regulations and processes (Lancaster 1990). In 1994, the Clinton administration announced a major change to the rules governing releases from Glen Canyon Dam to minimize variation in flows downstream, a move that the BuRec's commissioner said heralded "the difference between the old way we have treated the river and the way of the future" (Kenworthy 1994b). Interior Secretary Bruce Babbitt, meanwhile, proclaimed a "new era for ecosystems" and "a new era for dam management" (Associated Press 1996).

Even more significant than this change in policy, however, was a newfound willingness on the part of federal leaders to constructively engage in Colorado River basin governance. By the late 1990s, lower basin states faced growing demands for water as Arizona's CAP became operational. In 1997, Secretary Babbitt announced a new federal rule designed to encourage interstate water transfers as an alternative to litigation and political reallocation of increasingly scarce water. Nonetheless, the success of this new interstate water market remained, as Babbitt warned, dependent on California's ability to wean itself off the use of water which belonged to Arizona under the terms of the CRPA (Purdum 1997). Babbitt, breaking with a long tradition of federal neutrality in interstate water disputes, therefore warned southern California water users that "the time has come for me as the River Master to play a more active role" and threatened to impose limits on deliveries of Colorado River water to the state (Perry 1997). On the strength of this unprecedented personal involvement, the seven riparian states of the Colorado River basin in 1999 reached a preliminary agreement to resolve the remaining issues concerning allocation on the Colorado. Under the agreement, the states agreed to a process under which the federal government would begin reducing water deliveries, beginning with California. California, for its part, committed to an ambitious effort to line canals and construct underground storage to accelerate transfers from agricultural to urban areas (Purdum 2000).

The federal government's newfound commitment to collaboration and participatory decision-making in the Colorado River basin further intensified during the Bush administration. Beginning in the mid-1990s, multistate adaptive management working groups were formed representing federal, state, and nongovernmental stakeholders. In the early 2000s these groups scored notable successes, including building support for plans to adjust streamflow to protect endangered species (Blakeslee 2002a, 2002b). The importance of the federal government's role in inducing cooperation became even clearer when a vote by the IID board threatened to imperil the 1999 interstate agreement. In response, Interior Secretary Gale Norton did not hesitate to act on her authority to threaten reduced deliveries of Colorado River water in order to compel the IID to approve the sale of water to San Diego (Murphy 2003a). Despite the IID's protests that this unprecedented threat constituted "an example of heavy-handed and unwarranted federal interference" (Murphy 2003b), Norton quickly gained the enthusiastic support of the other six basin states, environmental groups, and California state legislators. The IID finally bowed to this combination of federal, state, and nongovernmental pressure by agreeing to the transfer, thus signaling a dramatic shift in the stance of the Colorado River basin's most important single economic actor and the erstwhile linchpin of California's reclamation nexus. In response, a Nevada water official proclaimed that "with cooperation so much better over the past 10 years, no one wants to lose the momentum. It is a mindset change that has developed among the states" (Murphy 2003c).

The shift in the IID's stance in favor of cooperation set the stage for another, even more productive, phase of institution-building in the Colorado River basin, even as physical water scarcity pressures grew acute. In 2007, with the determined personal involvement of Norton's successor, Dick Kempthorne, a set of guidelines was concluded which superseded the 1999 agreement and specified how the lower basin states would share the burden of water-use reductions during drought (Archibold 2007). As drought grew more acute in the mid-2010s, the lower basin states, along with the BuRec, agreed to further curtail water use by implementing conservation measures in order to preserve water levels in Lake Meade and to increase storage by 1.5 MAF over time, helping to buffer the effects of future shortages (Brean 2014). California's willingness to accept further water use restrictions marked a striking change from the prior history of basin competition. As one participant recalled, the state's "original attitude was 'We have the priority and those look like difficult problems—good luck.'" However, the very severity of the drought led California officials to calculate that senior water rights would not be sufficient to protect the state from rationing in the event that Lake Mead water levels dropped further. As an MWD official stated, "We absolutely agree we're better off having the lake higher. We can't only rely on our priority and say we have our priority and we're protected" (Davis 2014).

Rather counterintuitively, the threat of severe water scarcity, and the rationing of state water use, served to catalyze greater interstate cooperation. By the mid-2010s, cooperative governance of the Colorado River basin appeared poised for success, largely on the strength of robust institution-building beginning in the 1980s. Yet this recent development was precipitated by the gradual changes induced by social movements, especially the environmental movement, in the federal government's role in the Colorado River basin after the 1970s. Spurred by the more favorable political opportunity structure created by the NEPA and other environmental legislation, federal agencies broke with their traditional allies in Congress in support of ever greater numbers of reclamation projects and became increasingly receptive to third-party advocacy for institutionalized collective action that advanced environmental and conservation priorities. Simultaneously, geographically concentrated water-user groups like the IID began to lose the support of state political actors. Lacking their former allies at both the federal and state levels, these sectional interests were largely eclipsed, setting the stage for a more cooperative and less competitive set of both central–local and interstate relationships in Colorado River basin governance. Yet the success of Colorado River basin institution-building remains tentative. As a former BuRec commissioner has warned, "We need more urgency, more definitive commitments and more collaborative agreements . . . I fear the Colorado River Basin will return to its past and replicate what still goes on in many basins, which is conflict, litigation and gridlock" (Snider 2014).

FEDERAL–STATE COMPACT ORGANIZATIONS, COMPREHENSIVE DEVELOPMENT, AND COOPERATIVE MANAGEMENT IN THE EASTERN UNITED STATES

In contrast to the long and halting struggle to establish regional governance institutions in the Colorado River basin, basins in the eastern United States proved to be the site of more robust, cooperative, and effective institution-building during the first half of the twentieth century. Unlike in the west, America's eastern rivers generally face no acute problems of water scarcity, and collective action arose primarily in response to flooding and water quality issues. The individualistic political culture of the west, too, was lacking in regions like New England, where smaller states were accustomed to dealing with their neighbors on a wide range of issues (Foster 1984). But despite these differences, institution-building in America's temperate eastern river basins, most notably the establishment of the TVA and federal–state compact commissions in the Delaware and Susquehanna River basins, was shaped by a similar blend of interstate and federal–state tensions. The greater success of river basin governance institutions in the East is due to the fact that, unlike in the Colorado River basin, national political leaders and third-party environmental advocacy organizations were engaged from the outset in pressuring state leaders to participate in regional river basin institution-building.

Partly because state authority developed earlier in the East and partly because eastern rivers were more important to the economic vitality of the nation as a whole, eastern river basins attracted much earlier and more intensive attention from federal agencies and third parties. In 1879, Congress created the Mississippi River Commission (MRC) in response to arguments between civilian and military engineers over how best to control flooding and promote navigation along the Mississippi, the most economically important river in the country at the time. The MRC included several federal agency representatives as well as civilians appointed by the president, but it quickly became dominated by the USACE and was often criticized for its failure to consider outside advice. Under the Flood Control Act of 1928, passed in response to the 1927 Mississippi flood, one of the worst in American history, the MRC became responsible for conducting inspection tours and hearings with farmers and civic leaders in the riparian states in order to ascertain flood-control issues and requirements, essentially convening all relevant local leaders as advisors for the USACE's levee and dam construction projects (Camillo and Pearcy 2004). These stakeholder-consultation features set an important precedent both for the involvement of third parties in river basin management in the United States and for the use of an inclusive, participatory approach to decision-making in interstate river basins.

At around the same time, however, the nascent American environmental movement began a political campaign that would do even more to promote institution-building in shared river basins. In the 1920s, the newly founded Izaak Walton League, created by outdoor enthusiasts concerned by the deterioration of water quality in many parts of the country, mounted a public campaign to shame state governments for their failure to control pollution. These attempts met with varying levels of success, and the league, adopting a tactic that would later be used by other advocacy organizations, expanded its lobbying to Washington, pressing Congress to adopt national water quality legislation (Paavola 2006). The league soon became convinced that much of the problem lay in the absence of effective institutions to control water pollution across state lines. Although several interstate river basin management institutions already existed, including the Interstate Sanitation Commission, established as part of a tri-state compact between New York, New Jersey, and Connecticut, the league argued that these bodies were too weak to effectively control pollution (Crohurst 1936; Paavola 2006). In response, Congress created a commission charged with investigating water quality, which duly concluded that states possessed too few incentives to control pollution from powerful economic constituencies within their jurisdictions (Crohurst 1936). Federal agencies, spying an opportunity to enhance their authority, embraced the recommendation to strengthen governance at the river basin level rather than rely on state governments, driven as they were by parochial interests. A 1934 report, for example, cites "conflicts over the use of available water," "the custom of using streams for disposal of sewage and waste," and "interference by local authority with navigation" as reasons that river basins "must be developed not in accordance with purely local interests, but according to a comprehensive plan" (U.S. Public Works Administration 1934, 21).

The concept of regional governance gained increasing favor during the New Deal era, when it appeared to offer a means of sidestepping state resistance to the federal government's rapidly expanding powers. During this period, federal officials' enthusiasm for strong river basin governance institutions reached new heights. River basins, the Roosevelt administration asserted, "were more important than State lines in planning for the use and control of streams, and suggest more convenient means of cooperation between states . . . and the Federal Government" (Duffus 1936). This belief found its greatest manifestation in the creation of the TVA. This unique institution was created to achieve comprehensive economic development of the multistate Tennessee River basin, including the construction and operation of hydropower facilities to aid in regional economic development and to perform environmental management activities such as the control of soil erosion and water quality. The TVA was and remains a uniquely powerful entity. Formally established as a state-owned corporation, the TVA possesses the power to seize land and conduct law enforcement functions, even while being largely immune from congressional oversight (Chandler 1984). Its unusual degree of autonomy can be directly credited to the advocacy of President Roosevelt and crucial congressional allies like Independent senator George Norris of Nebraska, who believed that only a publicly owned and operated enterprise could be trusted provide power cheaply to the rural poor. This model of regional governance was one Roosevelt hoped to emulate in other river basins. In 1937, the president requested Congress to create seven regional river basin planning and development entities modeled on the TVA (*Christian Science Monitor* 1937a). Roosevelt explained the need for such organizations by proclaiming that "It is time to extend planning to a wider field, in this instance comprehending in one great project many States directly concerned with the basin of one of our greatest rivers" (Roosevelt 1933).

Yet despite Roosevelt's technocratic vision for a supraregional institution, the subsequent history of TVA and other efforts at river basin institution-building were shaped more by state than federal interests. Despite the inclusion of provisions in the TVA act intended to preserve most state authorities, the dilemma of autonomy loomed large in the minds of state leaders and their congressional representatives, who lobbied fiercely to block creation of TVA-like institutions in other river basins. Republican senator William Borah of Idaho even proposed a constitutional amendment banning TVA-like entities and making regional issues like river basin management the sole responsibility of state governments (*Christian Science Monitor* 1937b). Mindful of these sensitivities, TVA leaders were careful to avoid the impression of interference in state and local government affairs and increasingly focused on technical functions like flood control which did not duplicate or threaten those of the states (Lilienthal 1942; Selznick 1949; Derthick 1974). Even so, state opposition to regional institution-building persisted. In 1948, the *Chicago Daily Tribune* savaged the TVA for being "a gigantic federal enterprise" which "takes its orders from Washington. Its philosophy is federal and the states and cities it serves have no part in its operation" (*Chicago Daily Tribune* 1948). A Truman administration proposal for a Missouri Valley Authority to alleviate rivalry between federal agencies and state governments succeeded only in uniting them in opposition to this perceived assault on their authority, demonstrating once again the preference for autonomy among state leaders (Espy 1950).

Instead of a "monolithic federal entity operating in their jurisdictions," state leaders sought to pursue patchwork institution-building, replacing powerful, comprehensive basin management organizations with weaker, less formal institutions that preserved the primacy of state governments relative to the federal government (Voigt 1972, 59). "Prompted by a fear that the basin might get an all-powerful authority from Congress," representatives of nine Missouri River basin states met in 1949 to issue proposals for an "interstate agency" in place of the proposed unitary authority (Blair 1949). "Such a program," explained a Nebraska representative, "would avoid ultimate centralization of all economic power . . . in the Federal Government in perpetuity" (Blair 1952). The states were ultimately successful in forestalling the creation of powerful new basin management organizations: after prolonged lobbying, the Eisenhower administration accepted the recommendations of a report which killed the idea of replicating the TVA. Instead, it proposed a weaker interagency coordinating body which explicitly would "not . . . replace existing agencies nor would it assume full and complete authority for every phase of Federal activity in the basin" (Kennedy 1953). Following the demise of the proposed Missouri Valley Authority, river basin institution-building in the United States departed from the TVA model, although it would remain influential abroad, including in India.

Despite the demise of the TVA model thanks to state opposition, the growing influence of reclamation politics in the 1950s and 1960s, with their attendant conflict between federal agencies and the states themselves, convinced federal leaders of the need for river basin governance institutions. In 1961, President Kennedy requested Congress to pass a bill providing for the creation of a federal water resources council coupled with regional river basin commissions, each to be appointed by the president, in order to coordinate water use and reclamation projects. However, the legislation languished as a result of protests from both federal agencies and the states themselves, both of which expressed the now familiar fear of a loss of autonomy and prerogative. Only after the bill was amended to permit states to appoint their own representatives to the commissions and a specific clause was added pledging neither to expand federal authority nor diminish state autonomy was the bill passed during the Johnson administration as the Water Resources Planning Act of 1965 (WRPA). Under the act, federal funding was made available to establish river basin commissions to coordinate federal, state, and local water projects and to create a Federal Water Resources Council to review development plans in each basin (Congressional Quarterly 1965).

Following passage of the WRPA, several such river basin commissions were established, along with two very different federal–state compact organizations, the Delaware and Susquehanna River Basin Commissions (DRBC and SRBC, respectively). The DRBC and SRBC represented a new, more cooperative approach to federal–state coordination in river basin management. In contrast to the weak institutions established by most interstate compacts, the compacts which created the DRBC and SRBC included the federal government as well as the states as legal parties, making them equals in the institutions created to implement each compact. Even this limited power-sharing with the states was nonetheless fiercely resisted by the executive branch. Interior Secretary Stewart Udall warned that the DRBC would constitute a "third form of government" and that "It is our bedrock position that cession [to the DRBC] is the very essence of your proposal and that this would end up by operating river basins under sort of an Articles of Confederacy and that the constitutional system that we have simply does not permit that" (U.S. Department of the Interior 1961). When the DRBC was established over these objections, the USACE lobbied intensively to diminish the power of its sister organization, the SRBC. These efforts succeeded in stripping the SRBC's authority to review and approve USACE flood-control projects, thereby removing any threat it posed to the agency's prerogatives (Voigt 1972). Instead, federal agencies strongly favored the weaker entities authorized under the WRPA, which, as a federal report noted, "tend to be heavily oriented to federal water resource planning interests" rather than those of state or local governments (Advisory Commission on Intergovernmental Relations 1972, 206).

Yet as in the case of the Colorado, state rather than federal agency interests proved the decisive factor in shaping institution-building. Even in the humid East, state sectional interests were opposed to the idea of greater interstate cooperation, which they feared would surrender control over vital water supplies. Both the DRBC and SRBC came close to defeat as a result of opposition from geographically concentrated water-user groups like the Pennsylvania Farm Bureau and Grange. Only the steadfast support of environmentalists and sympathetic advocacy organizations like the League of Women Voters persuaded the riparian states to ratify the Delaware and Susquehanna Compacts (Voigt 1972). As in other cases of river basin institution-building, then, third parties played a crucial role in lobbying for interjurisdictional collective action. State political leaders, on the other hand, sought to create institutional arrangements which would provide them with voice in influencing federal reclamation projects but would pose no threat to their individual powers. Instead of viewing the DRBC and SRBC as platforms for expanded interstate and federal–state cooperation, riparian state governors sought to give themselves"a more definitive role in determining how, where, and to what extent federal funds would be spent in the basin," as well as an alternative to "the long, tedious, and uncertain court route to a decision in a time of interstate dispute" (Voigt 1972, 65).

Notably, while third-party advocacy was crucial in securing establishment of the DRBC, SRBC, and WRPA commissions, none of these bodies provided any direct representation for third parties. The absence of a mechanism to sustain their involvement in river basin governance over time limited the effectiveness of these bodies. Instead, despite some successes, ultimately only the DRBC exercised true basin-wide governance functions, and even it never entirely escaped the dominance of state interests. An assessment of the DRBC concluded, for instance, that

> the DRBC has been most successful in pursuing the purpose for which the states came together in the first place: settling their disputes over water allocation. . . . The shared interest of the states in avoiding resort to the Supreme Court helped them reach agreement through the medium of the commission, as it had led them to create the commission. But the states rarely launch initiatives of any kind through the commission; initiatives originate with the staff, which must cajole the states into cooperating. (Derthick 1974, 72)

State participation in the WRPA commissions, meanwhile, was limited by a desire solely to "scrutinize Federal regional water resource planning and operations," thereby maintaining state autonomy (Advisory Commission on Intergovernmental Relations 1972, 97). This continued prevalence of the state concern for autonomy led to a slow corrosion in the function of many of the commissions. By the early 1980s, the U.S. General Accounting Office issued a report concluding that because state officials "believe that river basin commissions should serve as a forum for State interests" and that these should "take precedence over those agreed on collectively in river basin commissions," they had ceased to serve a useful purpose (1981, 13). The commissions were subsequently defunded by the Reagan administration.

The contrasting cases of the TVA, DRBC, SRBC, and WRPA commissions illustrate that, as in the primary case of the Colorado, river basin institution-building has been crucially shaped, and often undermined, by the interests of subnational states, which are motivated by the dilemma of autonomy and prefer weak forms of collective action. Such positions are often advanced by narrow, but politically powerful, sectional interest groups. But just as variation in the degree of interjurisdictional cooperation in the Colorado case yields lessons for more effective institution-building, so too does variation in particular outcomes among the contrast cases. The relative success of the DRBC in containing interstate conflict over water has been credited to the relatively strong powers of the commission, indicating the importance of creating robust operational rules that cannot be weakened by the states (Delaware River Basin Commission 2007). The New England River Basin Commission, generally viewed as the most effective of the WRPA commissions, likewise highlights the importance of repeated interactions between state officials (Foster 1984). The greatest lesson, however, is that sustained, constructive engagement by third parties is necessary to foster robust regional institution-building. The sustained involvement of third parties was crucial to the initial success of the DRBC, as well as other effective inter-jurisdictional institutions, such as the Potomac River Commission (Gentzler and Hines 2006, 81). However, the federal–state compact organizations also show that successful collective action in shared river basins must be sustained by alliances between third parties, whose interests lie in creating strong, inclusive, and collaborative governance institutions, and political elites, who provide them with the access to effectively lobby for such goals. Such alliances were evident in the case of the Colorado but not with respect to the DRBC, SRBC, or WRPA, which now constitute the patchwork remains of what was once envisioned to be a comprehensive system of river basin governance institutions.

COUNTRY CASE CONCLUSION

Subnational hydropolitics in the United States illustrates the complex interaction of three primary factors, decentralization, sectional identity, and political opportunity structure, as summarized in Table 4.2. In its complexity, the U.S. case demonstrates the limits of unicausal factors like water scarcity in explaining interjurisdictional water conflict. Indeed, acute scarcity in the Colorado basin has stimulated rather than impeded interstate cooperation in recent decades. In the case of the federal–state compact organizations, meanwhile, the absence of scarcity did not lead directly to cooperation, nor did it overcome the tendency toward weak forms of collective action. In contrast, the case of the Colorado evinces the influence of sectional politics, particularly between geographically concentrated water-user groups in California and Arizona. In particular, the more rapid development, higher rates of urbanization, and consequently higher water demands of the lower riparian states sustained competition with their upper riparian neighbors for claims on the Colorado's scarce water resources. Finally, factors emphasized by the collective action literatures, like repeated interaction between state political leaders, are evident in the cases of the DRBC and SRBC. However, only the involvement

Table 4.2 Summary of Subnational Hydropolitics in the United States

Factor	Characteristics	Effect on Interjurisdictional Collective Action
Decentralization	• Ambiguity of federal powers favors states in water policy issues • Political representation favors state-based rather than partisan agendas • Water rights pluralism complicates federal action, encourages litigation	• Patchwork institution-building: mix of strong (DRBC) and weak (LOTR) interjurisdictional institutions; disputes often settled through litigation • State congressional delegations, governors allied with BuRec to form rent-seeking alliances for irrigation financing before the 1970s • National political leaders (presidents, secretaries of interior) periodically involved, but when they did engage, as in 1990s Colorado River scarcity agreements, effective in brokering consensus
Sectional identity	• Individualistic political culture encourages state-based sectional identities • Historical geography in west creates upstream–downstream divide	• State governors equated interests of geographically concentrated water user groups (IID, San Diego MWD) with states themselves, driving upstream–downstream conflict
Political opportunity structure	• Federal, state, local government offers many points of access but also makes coalition-building more difficult • Third parties favored by legislation like NEPA after the 1960s	• Groups like Isaak Walton League crucial in expanding federal role in water policy in early twentieth century • After the 1970s, legislative reforms and growing influence of environmental movement created reform coalitions between NGOs like Sierra Club and BuRec/USACE

Abbreviations: BuRec, Bureau of Reclamation; DRBC, Delaware River Basin Commission; IID, Imperial Irrigation District; LOTR, Law of the River; MWD, Metropolitan Water District; NEPA, National Environmental Policy Act; NGO, nongovernmental organization; USACE, U.S. Army Corps of Engineers.

of third-party actors explains a critical feature of subnational hydropolitics in the United States, namely variation in the degree of cooperation over time. Cooperation between subnational water users, evident in the latter part of the Colorado River case and in the initial phases of the TVA, DRBC, and SRBC, resulted from alliances formed between third parties, especially environmental advocacy groups, and national political elites, including members of Congress and agency officials. Where such alliances were formed, as in the most recent phase of the Colorado River basin case, institution-building advanced, and where they were not, as in the case of the compact and WRPA commissions, it has faltered.

As we shall see in Chapters 5–7 which follow, these dynamics mirror those of subnational hydropolitics in other countries. But at the same time, interstate collective action in shared American river basins has been crucially shaped by three particular features of the U.S. federal system and its characteristic decentralization of political and economic power. First, the lack of explicit constitutional authority in many policy areas has stoked bureaucratic strategic behavior by driving federal agencies to seek political legitimacy, either through alliances with states seeking federal financing or with bridging organizations that have accumulated political influence. Second, the structure of political representation, in which unified state congressional delegations can exert disproportionate influence on the legislative process, encourages subnational strategic behavior by providing states an arena in which to compete for federal financing and form alliances like the reclamation nexus. Third and finally, the strength and prominence of the American judicial system provide an outside option to interstate collective action, by providing a forum for states to pursue competing claims to water. The decentralization of the judicial system, in which each state maintains its own separate and mostly independent judiciary, sustains the multiple water rights doctrines that underpin many interstate water disputes in the first place.

On the other hand, institutional innovations like the NEPA have expanded the political opportunity structure for civil society, permitting bridging organizations to overcome these forms of strategic behavior and advocate, sometimes successfully, for interstate collective action. Yet where this has been the case, it has been achieved only by overcoming the pervasive influence of sectional interests and identities. As the U.S. Advisory Commission on Intergovernmental Relations noted, "most regional problems of an economic or natural character have been adapted to a mode of unifunctional administration. . . . This has presented less of a threat to the traditional levels [of government] and produced commissions that are controlled by these levels" (1972, 4). Where it has succeeded, regional institution-building has been enabled by bridging organizations, including both national political leaders and civil society organizations, especially those concerned with the environment. The critical role these actors play in restraining bureaucratic and territorial strategic behavior and promoting regional cooperation is further illustrated through contrast with the case of India, whose federal system imposes many of the same challenges to collective action in shared river basins but where bridging organizations have to date played a far less salient role in the nation's river basins.

5

Ethnolinguistic Cleavages and Interstate River Disputes in the Union of India

India is one of the world's most centralized federal systems, and its Constitution grants the federal government unusually broad powers to control the actions of state-level political leaders. At the same time, however, India's state-level politics are highly acrimonious, particularly after the emergence of state-based ethnic political parties since the 1960s (Diksit 1975). These fractious subnational politics are mirrored in numerous interstate river disputes which the center has, despite its considerable constitutional and political powers, proven unable to resolve. As John Wood remarks, "One would think that these powers would be adequate to enable the central government to play an active mediating role in an interstate river water disputes. . . . But the central government's maneuverability is often no greater than that of the states" (Wood 2007, 40). Indeed, the country's interstate disputes are so acute that the former head of India's water resource engineering agency, the Central Water Commission, warned in an opinion-editorial that "hydro-politics is threatening the very fabric of federalism" (Menon 2003). Virtually all of India's major river basins play host to long-running interstate water disputes, primarily related to water quantity allocation. Notable disputes include most of the principal peninsular rivers, including the Mhadei, the Kaveri (Cauvery), and the Krishna basins, which predate independence. Other sites of interstate conflict include the Narmada, racked by construction of a large dam; the relatively water-rich Mahanadi; and the Sutlej–Yamuna Link Canal, which supplies much of New Delhi's drinking water. Unfortunately, few of these disputes show signs of resolution in the near term. Yet as India's economy has grown and the demands on its major rivers have multiplied, these disputes increasingly constrain the development of large sections of the country.

The central puzzle of the Indian case is why interstate river disputes are both so numerous and so persistent, especially given the central government's constitutional authority to resolve them. India is distinctive among federal political systems in that its constitutional architects anticipated the problem of interstate river disputes and included specific institutional mechanisms intended to prevent them. Unlike the United States, for example, India's central government does not lack for explicit legal authorities to encourage and even compel interstate collective action in shared river basins. But New Delhi has consistently declined to exercise its powers to establish regional river basin governance institutions, and like Washington has chosen instead to remain mostly neutral in interstate river disputes. Furthermore, this curious reluctance to resolve its many interjurisdictional disputes stands in contrast with India's success in cooperatively managing some of its international transboundary rivers; the Indus Water Treaty has proven to be one of the most successful areas of bilateral cooperation between India and Pakistan. This chapter accordingly examines why interstate river disputes have arisen and persisted in India despite the existence of explicit legislative provisions designed to prevent them. It does so with reference to a primary case study, the Krishna River basin dispute in peninsular India, and a secondary case, the Damodar Valley Corporation in northeastern India, one of the country's few comprehensive river basin governance institutions.

This chapter contends that India's interstate water disputes show, perhaps more clearly than any of the other cases explored in this book, how sectional tensions sustain long-running disputes over shared water resources. Because irrigated agriculture is so critical to the economy of many Indian regions and because it enjoys commensurate salience as a political issue, shared rivers have become focal points for ethnolinguistic tensions and cleavages, which moreover have become more pronounced with the rise of state-based political parties. Interstate river disputes arise and persist in India because, against the background of sectional tensions, its subnational politicians derive electoral benefit from seeking additional water resources at the expense of their neighbors. As in other cases examined in this book, including the Colorado River in the United States, geographically concentrated water-user groups, especially irrigators, are often the first to call for obtaining greater allocations of water through judicial or political conflict with neighboring states. But over time interstate river disputes have become a convenient means for state leaders to mobilize supporters and partisans in their own jurisdictions against neighboring states, which usually and not coincidentally feature different linguistic and cultural identities. The central government, meanwhile, has little political incentive to intervene since it typically requires the support of state-based political parties to remain in power. Against this political salience and sectional resonance, the voice of bridging organizations, especially third-party nongovernmental organizations, has hitherto been largely muted. As a result, the development of collaborative and participatory governance institutions in India's river basins has been stunted. The case of India illustrates thus more clearly than any other how institutional and ideational factors interact to shape conflict, cooperation, and institution-building in shared river basins.

COUNTRY CASE INTRODUCTION

India is above all characterized by both physical and social diversity, both of which have decisively shaped its subnational hydropolitics. Even among large countries, India features an unusual range of topography and climatic regions, which results in important regional discrepancies in the availability of water relative to concentrations of population and economic activity. Most parts of India are dominated by a monsoonal climatic regime, resulting in substantial intra-annual variability in water availability. In addition, the legacy of imperial-era irrigation development has resulted in marked disparities in water demand in rural areas (Singh 1997). The significance of these discrepancies is enhanced by the fact that, among the cases examined in this book, India is the most dependent on agriculture, making water availability for irrigation a greater concern than in other countries. Furthermore, the imperial system of administration, which included jurisdictions under both direct and indirect rule, created a lingering divide between coastal and inland regions, creating pre-existing sectional tensions in many major river basins between upstream and downstream jurisdictions. These cleavages have subsequently been codified by the delineation of state boundaries along ethnolinguistic lines. Both geography and scarcity are ultimately less important causes of interstate river disputes, however, than are the institutional features of India's complex federal system. In particular, the ethnolinguistic cleavages reflected in state boundaries have created incentives for state political leaders to co-opt interstate water issues as a means of inflaming sectional tensions. As the remainder of this chapter explains, the course of India's interstate water disputes cannot be explained by scarcity alone; while it provides the spark for conflict in some basins, like the Kaveri, in others, like the Mahanadi, sectional tensions can cause conflict even where water is relatively abundant.

Table 5.1 depicts the major features of subnational water politics in India. Across the country, water scarcity as a result of climate change and economic growth is the dominant water resource management issue, along with allocation among a

Table 5.1 FEATURES OF WATER POLITICS IN INDIA

Case Characteristic	**Issues**
Driving management challenges	Scarcity, allocation, pollution, water supply and sanitation
Institutional collective action dilemmas	Weak central government institutions, weak and contested water-sharing agreements
Integration mechanisms	River boards, interstate agreements, political party structure
Sectional cleavages	State boundaries drawn around ethnolinguistic lines, state political parties largely appeal to ethnolinguistic identity
Political opportunity structure	Closed elite decision-making, state-based parties less receptive to nationwide political agendas

growing number of water users. Other issues, especially water pollution and water supply and sanitation, are also salient. The division of powers inherent to federalism creates a combination of vertical, horizontal, and functional institutional collective action dilemmas, most notably with respect to a weak central government water resource management bureaucracy and the lack of robust interjurisdictional institutions. India's constitution provides for a number of integration mechanisms to address these issues in interjurisdictional river basins, notably the creation of river boards and interjurisdictional river basin management organizations. As in other countries, political party structure offers another means of solving institutional collective action dilemmas. However, unlike other cases examined in this book, these dilemmas are exacerbated by deep sectional cleavages in the form of state boundaries drawn around ethnolinguistic lines, which have fostered the growth of state-based political parties that appeal primarily to sectional interests. Although India's democratic system has fostered a vibrant civil society, this state-based party structure, along with the historical prevalence of closed elite decision-making, has so far constrained third party–induced cooperation in India's interjurisdictional river basins.

Historical Geography and Interstate River Disputes

The legacy of the British Raj has shaped India's country's political and economic geography by creating a marked coastal–inland discrepancy in economic development. During the period of British East India Company rule, India was divided into administrative regions centered on large port cities such as Bombay, Calcutta, and Madras, which were governed directly by the British, and a large hinterland that was mostly left under native administration (Gerring et al. 2011). These cities remained economic and administrative centers throughout the imperial era even as British rule spread deep into inland areas, creating a divide between a more economically advanced and urbanized coastal zone and a poorer, more rural, and more agrarian interior (Singh et al. 2003; Basu 2009). This discrepancy, though not as large as in the case of other developing nations such as China, nonetheless is reflected today in a marked divide in water use, with urban and industrial uses predominating along the coasts and irrigated agriculture in inland regions. This coastal–inland divide has in many parts of India stoked interstate river disputes by overlaying conflict between water uses with sectional tensions between geographic regions.

At first blush, India's interstate river disputes appear to break down primarily on familiar upstream–downstream riparian lines. This phenomenon has long been noted, including in a 1972 Union government report: "Because of the fact that rivers take their birth in the highlands of an area, the natural lie of the land implies that the resource . . . is usually with the upper basin, though the inhabitants of the lower basin too are heavily dependent on it. It is this situation which frequently lies at the root of inter-State river disputes" (Ministry of Irrigation and Power [India] 1972, 340). This geographically oriented pattern of conflict replicates itself throughout India's major interstate river basins. States situated downstream of others almost inevitably develop antagonistic postures toward their upstream neighbors (Ramana 1992). As one legal scholar notes, "The harnessing of the waters of such an interstate river, always remains a difficult proposition, because every State, through which the river flows, claims to be the owner of its waters. The States which are located along such a river on its downstream, do not allow the upper States to create storages in their territories, fearing that the upstream State will not release water as and when required by the downstream State" (Garg 1999, 5).Perhaps the most dramatic illustration of the extent to which sectional interests drive interstate water disputes in India lies in a long-running dispute between the colonial-era provinces of Hyderabad and Madras over the Tungabhadra River. The dispute ended only after a modification of state boundaries placed the river entirely within the new state of Karnataka, finally removing the need for interstate cooperation (Garg 1999). But in India, boundaries between states reflect deep linguistic and cultural, as well as riparian, divides. It is the existence of these deeper sectional cleavages, rather than simple riparian geography, that has sustained India's many interstate water disputes.

Unlike in the United States, the boundaries of India's states were consciously delineated based on ethnolinguistic group lines. In most countries, subnational administrative unit boundaries reflect in part physical geography and topography, including features such as major rivers and mountain ranges. In India, on the other hand, the choice to delineate state boundaries on ethnolinguistic lines represented a deliberate attempt to break with the colonial-era pattern of administrative areas, which were seen to have "erase[ed] old frontiers and . . . creat[ed] new provinces which ignored natural affinities and common economic interests" (States Reorganization Commission [India] 1955, 1). India's constitutional architects made it clear that they did not consider historical administrative divisions, particularly those which existed under the British Raj, to be a reasonable basis for the constitution of states following India's independence. Instead, the States Reorganization Commission was charged with considering a variety of factors to determine how and where state boundaries should be drawn. Among these the commission, with considerable foresight, included "the requirements of river valley projects" and "the optimum utilization of natural resources" (States Reorganization Commission [India] 1955, 39).

In the end, however, such geographical considerations were dismissed in favor of an ethnolinguistic basis for boundary delineation. In explaining its rationale, the commission expressed its belief that India's federal system required such an arrangement: "The constituent States in a federal republic must each possess a minimum degree of homogeneity to ensure the emotional response which is necessary for the working of democratic institutions. . . . Linguistic homogeneity, therefore, provides the only rational basis for reconstructing the States, for it reflects the social and cultural pattern of living obtaining in well-defined regions of the country" (States Reorganization Commission [India] 1955, 35). Nonetheless, in making this choice, the commission also helped to codify sectional cleavages between the states. Indeed, the commission itself appeared to recognize the danger, noting that existing legislative provisions "do not appear to confer on the Central Government adequate powers in respect of projects which are of an inter-State character" (States Reorganization Commission [India] 1955, 56). This institutional lacuna rapidly proved problematic in the subsequent decades as water demands increased and the newly organized states jockeyed for central government financing for dams, canals, and other water supply infrastructure to meet them.

Indeed, India's interstate water disputes, like those of the western United States, were in their early phases linked chiefly to infrastructure development and the rapid growth in water demand for irrigation, industry, and urban use during the twentieth century. The salience of these factors was recognized by the Krishna Water Disputes Tribunal, which observed that "Until the conclusion of the Second World War, the engineering works for development of water resources were few in number, the water supply was ample in relation to the demand upon it and no use of water seriously affected other uses. There was, therefore little scope for disputes regarding the use, control and distribution of the Krishna waters. British India was subject to the unitary control of the Government of India. . . . There were minor disputes relating to the Tungabhadra waters but they were amicably settled" (Krishna Water Disputes Tribunal 1973a, 1). Subsequent disputes, including the very conflict between Krishna basin riparians that the tribunal was established to resolve, would prove far more intractable, due in large part to the prevalence of irrigation throughout most parts of India

India's agricultural sector accounts for a declining but still large proportion of the national economy, at about 17% in 2009. As in most countries, however, its impact on the country's water resourcesis disproportionate to its economic contribution. No less than 91% of total water withdrawals in India are for irrigation, making geographically concentrated groups of irrigators a potent political constituency. Historically, the government accorded high priority to expanding irrigation, especially in peninsular India; and irrigated acreage has increased by 0.6–0.7 million hectares annually since the 1950s. Conventional surface irrigation accounts for about 97% of total irrigation use, and outside the arid and semiarid areas of the northwest, irrigation allocations are demand-based, creating little incentive to improve efficiency. At the same time, water prices, set by the various state governments, remain "abysmally low" across India (United Nations Food and Agriculture Organization 2011).

The relative inefficiency of irrigated agriculture in India is at least partly reflected in its complex water management infrastructure. Governance of irrigation systems, and of agricultural water use more generally, is increasingly dominated by water-user associations, which have enjoyed official encouragement since promulgation of India's 2002 Water Policy. By 2011, India featured over 50,000 water-user associations. However, the formal role of water-user associations varies significantly and is determined by state governments, which each establish regulations determining its authorities and responsibilities (United Nations Food and Agriculture Organization 2011). The dominant feature of water governance in India, especially in its important agricultural sector, is therefore the leading role played by state governments instead of the center.

In the absence of a strong mediating role played by the center, India's states compete for water to provide to key geographically concentrated constituencies, especially irrigators. According to one of India's authorities on irrigation, "Each State government is keen to be seen as an active promoter of irrigation for the benefit of as many of its constituents as possible," and as a result, "there is naturally a desire on the part of every State to appropriate as much water as it can" (Vaidyanathan 1999).

As in the case of the Colorado River basin in the United States, the influence of these sectional interests is often magnified by state politicians so that competition for water becomes institutionalized in the form of protracted interstate water-sharing disputes. This view was echoed by Indian legal scholar Dodda Rao in the 1990s: "In view of the influence of the water resources on economic/industrial development of a state . . . [water] disputes become disputes between the Governments of the basin states" (1998, 19). Unlike in the case of the United States, however, the involvement of civil society in water resource decision-making has historically been limited relative to that of state governments, providing little alternative to interjurisdictional competition in shared river basins.

Among the democracies examined in this book, India indeed stands out for its relatively closed political opportunity structure, at least with respect to water resource management. Protest movements have played an influential role in delaying or defeating some planned large dams and water transfer projects, but only recently have they become a distinct voice in water resource management (Wood 2007). This recent emergence mirrors the gradual evolution of Indian civil society over time. The most recent phase of civil society involvement began only in the 1990s in the wake of the *panchayat* reforms, which included provisions such as reserved assembly seats for women that were designed to expand participation in local government decision-making (Dubochet 2012). Many civil society groups and organizations have yet to formulate effective strategies to utilize the multilevel political opportunity structure afforded by India's federal system (Harriss 2007). Moreover, the predominance of hierarchical and elite decision-making has historically limited actual civil society influence in government policymaking, and nongovernmental organizations have frequently been marginalized when they oppose official policy (Schiff 2014). The net effect is that the involvement of civil society has only recently and tentatively begun to challenge the primacy of governments in water resource decision-making. However, while interstate water disputes typically arise among elites and were initially confined to formal, intergovernmental mechanisms, in recent decades they have entered the popular political sphere and have become touchstones for wider sectional tensions between neighboring states.

State Primacy and Constrained Regional Institution-Building

The proliferation of interstate sectional tensions has occurred against an institutional landscape that nonetheless provides avenues for both central and local government involvement in river basin management. In contrast to those of the United States and many other federal political systems, India's constitution-level institutional arrangements explicitly allocate powers for water resource management between central, state, and local levels of government. This elaborate allocation of power is a legacy of colonial-era legislation intended to encourage each part of British India to expand the area of land under irrigation, thereby increasing crop yields and government revenue. The Northern India Canal and Drainage Act of 1873 established the state's eminent domain over water resources and provided a legal basis for British-ruled provinces to pre-empt the water rights claims of the princely states (Gulhati 1972, 31–32). This central power persisted until promulgation of the Government of India Act 1935, which shifted most responsibilities related to rivers and water resources to provincial governments (Government of India Act 1935, 304–307). Throughout this period, significant attention was paid to the prospect of interjurisdictional disputes. As early as 1942, an imperial commission laid down the principle that "a Province cannot claim to do whatever it likes with the water of a river regardless of the injury which it might inflict on other Provinces or States lower down" (Gulhati 1972, 31–32).

India's present Constitution incorporates this principle and provides the center with general power to address regional-scale governance issues, particularly those which obtain within river basins. The central government is imbued with the power both to declare rivers as "national waterways" subject to its jurisdiction as well as to "provide for the adjudication of any dispute or complaint with respect to the use, distribution, or control of the waters of, or in, any inter-State river or river valley." The center is also granted the authority, under the Constitution's Seventh Schedule, to promote "regulation and development of inter-State rivers and river valleys to the extent to which such regulation and development under the control of the Union is declared by Parliament by law to be expedient in the public interest." States, in contrast, are granted more pedestrian powers regarding "water supplies, irrigation and canals, drainage and embankments, water storage and water power." In 1992, the addition of the Eleventh Schedule to the Constitution bestowed the *panchayats* with responsibilities regarding "Minor irrigation, water management, and watershed development," as well as drinking water (Ministry of Law and Justice [India] 2011). But despite possessing these powers, the Indian federal government has proven to be curiously reluctant to act on them to resolve interstate water disputes.

This reluctance is made even more puzzling by the fact that the Union government initially displayed great enthusiasm for using these powers to kick-start the comprehensive development of India's major river basins. The best example of this enthusiasm was the River Boards Act of 1956, which empowered the center to establish special bodies known as "river boards" to coordinate development of interstate rivers. The act was closely tied to the ambitions of the Nehru government to promote regional economic development in the country's river basins. As India's Third Five-Year Plan, promulgated in the early 1960s, explained, "River basins, especially those of the larger rivers, naturally extend beyond the boundaries of individual states. For integrated and economic development of water resources, arrangements for interstate cooperation are therefore essential. The setting up of River Boards for important river basins, as envisaged in the River Boards Act, 1956, would enable a coordinated view to be taken of the needs of a river basin as a whole" (Rao 1998, 38). The center's commitment to facilitate cooperative management of river basins even extended to legislation intended to help resolve interstate disputes. A separate Inter-State River Water Disputes Act, also passed in 1956, enabled the central government to set up tribunals empowered to adjudicate interstate water resource disputes and to allocate river waters among the various states (Gulhati 1972, 41).

This considerable legislative infrastructure, though painstakingly constructed, nonetheless soon foundered against the determination of state governments to preserve their autonomy. The States Reorganization Commission, established to reconsider the colonial-era boundaries of the Indian states, noted that "The provisions of the Water Disputes Bill and the River Boards Bill also do not appear to confer on the Central Government adequate powers in respect of projects which are of an inter-State character" (States Reorganization Commission [India] 1955, 56). The Union government faced such difficulties in obtaining the cooperation of state governments that the commission considered amalgamating states expressly to reduce interstate water resource conflicts. The commission noted, for example, that in the case of two rivers in peninsular India, "if one independent political jurisdiction . . . can be eliminated, the formulation and implementation of plans in the eastern areas in these two great river basins will be greatly expedited" (States Reorganization Commission [India] 1955, 104). Despite this concern, the commission ultimately determined that linguistic homogeneity should be the primary consideration in the drawing of state boundaries (Singh 1966, 29). Consequently, interstate river disputes intensified rather than ebbed in the decades following Independence (Dwivedi 2006, 65; Singh 2008, 35).

In the wake of what amounted to the central government's abrogation of its powers over water resource management, efforts to resolve interjurisdictional water resource disputes have been approached in an ad hoc manner, with little success. Official accounts have consistently attempted to cast this failure as a rationale to further enhance the central government's constitutional authority. In the late 1980s, the Commission on Centre–State Relations lamented, for example, that "the Constitution does not itself provide for a machinery for adjudication of water disputes. It leaves it to Parliament. . . . In contrast, the Government of India Act 1935 itself provided that if a Province (or a State) felt that it was likely to be adversely affected due to distribution or control of water from any natural source, it could complain to the Governor General" (1988, 487). A report from the Central Board of Irrigation and Power likewise alleges that "Had there been some basic provision of constitutional law that would have applied directly to the contesting claims on the subject of appropriation or apportionment of the inter-State waters . . . the conflicting subnational passions that rage at times over our inter-State river waters would have been smoothened and pacified" (Valsalan 1997, 21). But these complaints, while not unfounded, obscure other, more complex features of Indian federalism that in effect constrain the federal government's role in interstate disputes.

One particularly important such constraint is the fact that in India, as in the United States, no single legal regime exists to guide allocation of water between either water users or administrative jurisdictions. However, the problem stems not so much from the legal pluralism which exists in the United States as from the lack of any dominant, well-developed water rights regime at all. Neither states nor the federal government explicitly control the country's water resources (Gulhati 1972). Because of this fundamental lack of clarity concerning water rights, "Various alternate doctrines . . . causean anarchic situation which means that in most cases there is no clarity about who can use what amount of water" (Briscoe and Malik 2006, 21). Under this anarchic situation, upstream states have no general legal obligation to take account of water use in downstream states (Gulhati 1972, 63). Upstream state governments frequently exploit this legal lacuna to refuse to recognize downstream water rights, claiming absolute "territorial sovereignty" over their water resources (Dwivedi 2006, 83).

In light of this legal uncertainty, the central government has consistently preferred to attempt to settle interjurisdictional disputes through negotiation rather than judicial proceedings. Judicial resolution, the center soon declared, "may not be sensitive to the socioeconomic situation involving the welfare and the intense reactions of vast numbers of people" and might subvert the Union government's attempts to develop major river basins (Administrative Reforms Commission 1968, 225). The problem with this approach is that it also robs the courts of the opportunity to clarify water rights entitlements between the states and the central government, further neutering the ability of the Union government to definitively resolve interstate disputes. But even when pursuing negotiated settlements, India's Union government has proven itself to be only a periodic and reluctant player in preventing or resolving interjurisdictional water disputes.

This fundamental reluctance to intervene in matters seen as primarily under state jurisdiction is reflected in India's notably weak national water resource management agencies. The center's water resource management role is meant to be exercised through a ministerial-level hydroengineering bureaucracy, formerly known as the Ministry of Irrigation and Power and since 1985 as the Ministry of Water Resources. The Ministry of Water Resources is responsible for formulating general water resource policies, while its technical arm, known first as the Central Water and Power Commission (CWPC) and more recently as the Central Water Commission, provides support and advice to state governments, primarily in the realm of irrigation and water project development. India's Planning Commission, meanwhile, is responsible for assigning budgetary resources for water project development (United Nations Food and Agriculture Organization 2011). As such, the Planning Commission has played much the same role as the Department of the Interior in the American reclamation nexus, forming a locus for state lobbying in pursuit of central government financing for irrigation projects (Singh 1997; D'Souza 2006a).

The CWPC, meanwhile, was established under the Ministry of Irrigation and Power in order to formulate water resource development plans for the major river basins and, thus, forms a rough functional equivalent to the U.S. Bureau of Reclamation and Army Corps of Engineers. Unlike the Bureau of Reclamation, however, the CWPC has proven to be an exceptionally weak bureaucratic actor and has shown little interest in engaging in interstate politics or forming coalitions with state-level politicians and interest groups. During the first few decades after its establishment, the CWPC, a later report concluded, simply "lost interest in the matter" of relations with and between state governments (Ministry of Irrigation and Power [India] 1972, 353). In part because of this puzzling disinterest, the center has never fully exercised its powers under the River Boards Act and the Inter-State River Water Disputes Act to address regional river basin issues. The 1988 Commission on Centre–State Relations concluded, for example, that although "It was expected that these Boards would help in coordinated and optimum utilization of the river waters . . . the provisions of this Act have not been put to use all these years and the Act has remained a dead letter" (1988, 488). The center's failure to exercise its considerable powers therefore represents a lingering puzzle, and one that cannot be fully accounted for simply by the division of powers over water between state and central governments..

Instead, the center's puzzling reluctance to exercise its constitutional and statutory powers to prevent and resolve interstate water disputes is better explained by the calculus of electoral coalitions under Indian federalism. An early assessment of the problem of interstate river water disputes in India concluded, for example, that the center's tendency to refer disputes to tribunals rather than become directly involved represented "the path of lowest political costs" as it could thereby "escape having to make unpopular decisions in the disputes and avoid the risk of alienating some or all of the states involved" (Gebert 1983, 33). Two more recent observers conclude that in the Indian federal system, "The wide scope for discretion, the extensive bargaining, and the multiplicity of potential vetoes work in tandem to undermine the clarity and transparency needed for speedy dispute resolution" (Rao and Singh 2005, 230). In concert, these aspects of the Indian electoral system make it politically costly for the center to intervene in interstate disputes and to reconcile their underlying differences. These costs have moreover risen in recent decades as the states have accumulated greater political power relative to the center.

Indian federalism, like its counterparts elsewhere, has been characterized by a changing balance of power between central and state governments over time. In India, however, the net result has been a more acrimonious brand of subnational politics. In the first few decades following independence, both central and state governments were dominated by the Congress Party, a fact that led to considerable centralization in policymaking and the relative minimization of central–local tensions. According to one prominent observer, the broad-tent Congress Party system "was a mechanism of integrating new elites . . . which facilitated resolution of disputes." Under this broad-tent system, state governments were considered "mere appendages of the center" (Chakrabarty 2008, 99–100). Nonetheless, even during the period of Congress Party dominance and relative centralization, the states remained important political arenas and the basis of political patronage (Rao and Singh 2005, 96). This patronage helped to contain interstate water disputes, which the government presciently feared would grow more acute following the demise of Congress hegemony. By 1968, India's Administrative Reforms Commission predicted that interstate water disputes were "likely to become more acute in the future for two reasons: 1) the great increase in irrigation involving large scale exploitation of river waters and 2) the disappearance of the phenomenon of uni-party control over all India which has so far provided an extra-constitutional means for settling disputes" (Administrative Reforms Commission 1968, 222).

In recent decades, the demise of Congress Party dominance has indeed shifted the balance of power in Indian federalism toward state governments less inclined to cooperate with their neighbors. In particular, the rise of state-based political parties drawing support from particular ethnolinguistic groups has changed the political calculus in favor of states' rights. Because these parties draw their support from ethnolinguistic groups concentrated within individual states, their platforms typically call for extensive decentralization at the state level (Bhattacharyya 2012, 27–37). Over the past thirty years, the central government has granted many of these demands. The year 1989 proved to be a watershed in Indian federalism, when the

National Front made states' rights a core element of its platform, causing the government to grant extensive decentralization in response (Saez 2002, 93; Chakrabarty 2008, 85). The power of state governments has also increased because of the economic liberalization policies adopted in India after 1989, which have reduced the central government's role in the economy, combined with an increasing deference to state political authority (Bhattacharyya 2012, 37).

In a situation familiar from the case of the Colorado River, India's states therefore play the pre-eminent role in subnational hydropolitics. Their behavior, like that of the Colorado basin states, is characterized by a preference for autonomy and a reluctance to engage in institutionalized forms of cooperation with their neighbors, instead preferring forms of patchwork institution-building. Because there are essentially no effective regional river basin governance institutions, states have every incentive to promote the interests of geographically concentrated water-user groups within their jurisdictions. As a World Bank report concludes, "The lack of clear, permanent allocations means that states often spend more time and resources over 'securing our future rights' than they do to using what is theirs" (Briscoe and Malik 2006, 22). Again as in the case of the Colorado River, this resource-maximizing strategic behavior is rooted in the states' divergent sectional interests. The Ministry of Irrigation and Power noted in the 1970s, for instance, that "Many States compete with their neighbors in the industrial, commercial and agricultural sectors. This competition often lies at the root of disputes regarding inter-State rivers" (Ministry of Irrigation and Power [India] 1972, 346).

Absent a strong central government role in shared river basins, these interests have played the primary role in shaping cooperative institution-building. State resistance played an especially important role in undermining early attempts by the central government to establish river boards in several interstate river basins in the 1950s and 1960s (Gautam 1976, 1064; Iyer 1994, 193). As the Ministry of Irrigation and Power reported, "Eventually it was decided to set up River Boards for a number of inter-State rivers. But differences had already arisen between some States over sharing river waters and several of them withheld their consent to the formation of the Boards." In several cases, state governments simply refused to respond to central government requests for information on state water use. The CWPC faced such difficulties in obtaining water-use data that it abandoned its attempts to craft comprehensive river basin development plans as early as the 1950s (Ministry of Irrigation and Power [India] 1972, 88). By the time of the Ministry of Irrigation and Power's report in the 1970s, the situation had scarcely improved: "Whatever data is collected by the States is maintained for their own purposes, and is not published by one authority, or in one place. Some data is not published at all" (Ministry of Irrigation and Power [India] 1972, 282).

As these accounts indicate, even during the period of Congress Party rule, the states effective resistance to central involvement in major river basins created an institutional vacuum that persists to the present day. In particular, the river boards which were meant to function as river basin governance institutions have proven almost entirely ineffective in addressing basin-wide issues: "River boards have been set up only for a few rivers, and even these are ad hoc bodies. They have scarcely any role or responsibility for actual planning, and the various studies prepared by them do not enter in any significant way in determining the strategy, priorities, or project selection of the concerned States" (Vaidyanathan 1999, 126). Within this vacuum, the prevalence of interstate conflict over water imposes high economic and environmental costs. According to one assessment, interstate water resource disputes "foster the inefficient use of water and . . . encourage inefficient, non-cooperative investments in dams" which, while "rational from the perspective of a state that has failed to reach an agreement with another state," also "diverts investment resources and thereby retards economic growth" (Richards and Singh 2002, 612, 618). Sectionalism and subnational politics have indeed imposed serious economic, social, and environmental costs on India's river basins, as amply illustrated by the cases of the Krishna basin dispute and the demise of the Damodar Valley Corporation.

ELITE POLITICS, POPULAR MOBILIZATION, AND FEDERALISM IN THE KRISHNA RIVER BASIN

Compared with the other cases examined in this book, the Krishna basin most closely resembles that of the Colorado in that it is the site of a long-running interstate dispute that has defied multiple attempts by the central government to resolve (see Figure 5.1). Like the case of the Colorado, the Krishna basin dispute has been characterized by both allocative and infrastructural conflict, with interstate discord focused both on water-sharing agreements and on the construction of planned dams and storage reservoirs. At first glance, the interjurisdictional politics of the Krishna basin, like those of the Colorado, also appear to be shaped largely by a combination of geography and water scarcity. The Krishna basin states reflect the same basic historical geography as their Colorado basin counterparts, in which the downstream riparian was substantially more developed than its upstream neighbors. During the course of the dispute, two new states, Karnataka and Telangana, were created; and both have adopted positions contrary to those of their former parent states, instead reflecting their new geographic positions as middle riparians (see Figure 5.2). This shifting geography, the Krishna Water Disputes Tribunal would later note, caused the dispute to become increasingly bitter over time (D'Souza 2006b). In addition, increasing water scarcity, particularly in western Maharashtra and Karnataka, coupled with Andhra Pradesh's extensive need for irrigation, sustained increasingly acute competition for Krishna basin waters (Krishna Water Disputes Tribunal 1973a).

While both historical geography and water scarcity have played a role in stoking conflict at points in the long history of conflict in the Krishna basin, its persistence can be explained only by the interaction of institutional and ideational, as well as physical, factors. The course of the Krishna basin dispute has been shaped by distinctive institutional features of the Indian federal system. The first of these is its codification of ethnolinguistic divisions in the form of state boundaries. Whereas in the United States federalism has the consequence of creating multiple water rights doctrines, in India it exacerbates sectional cleavages by drawing state boundaries along these linguistic and cultural divides. The second feature is the division of powers specified under the Constitution. Despite the Indian federal government's specific powers over river basin management, states have taken full advantage of their power over many water resource management functions to forestall federal involvement in allocating basin water resources, instead choosing to bring judicial suits against their neighbors.

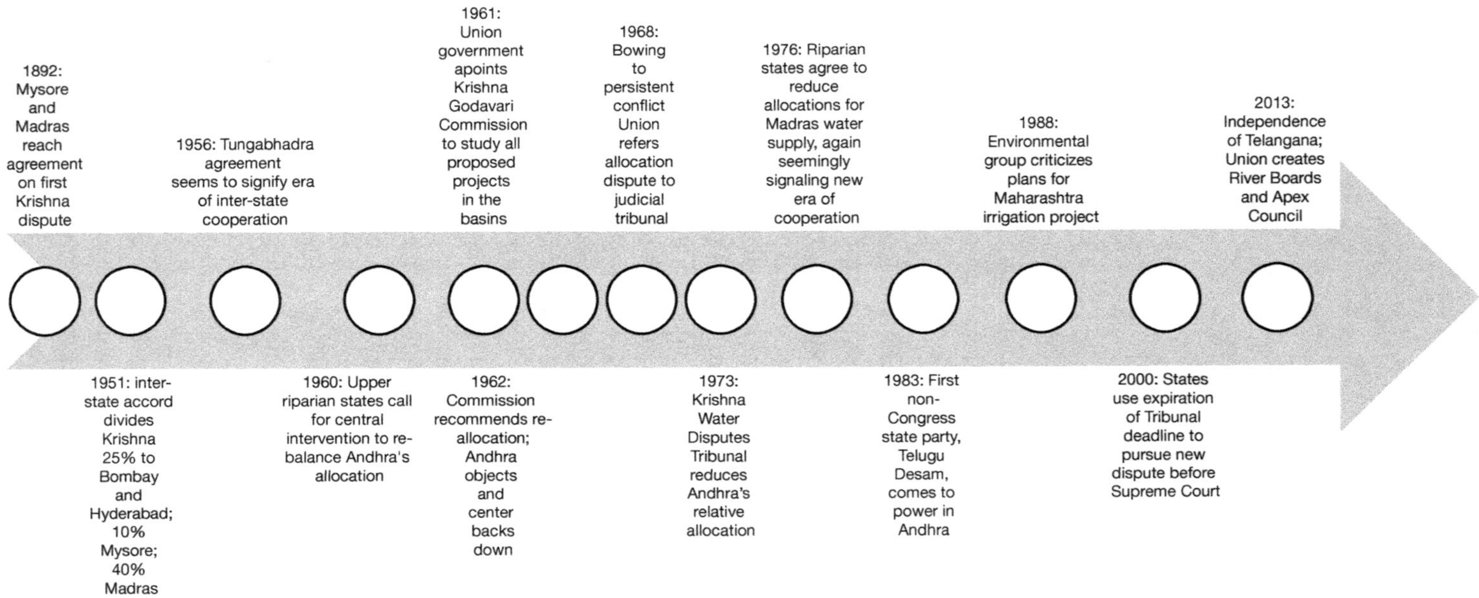

Figure 5.1 Timeline of Institution-Building in the Krishna River Basin

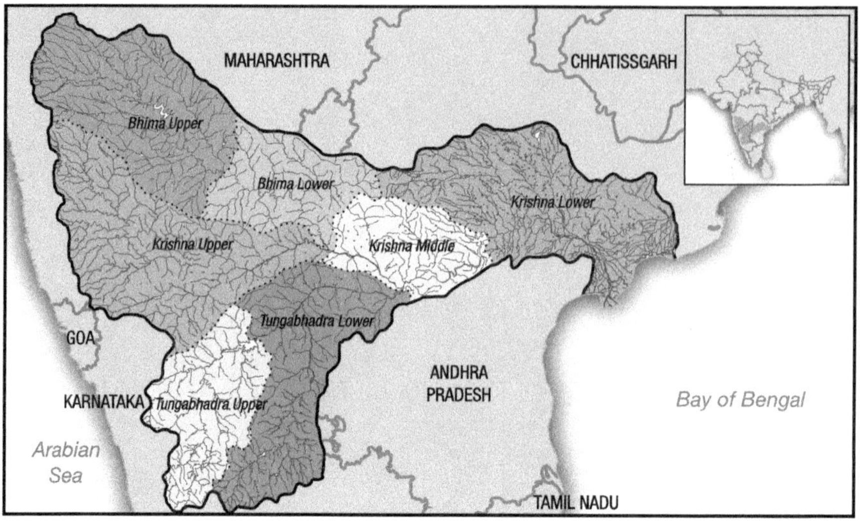

Figure 5.2 Physical and Political Features of the Krishna River Basin

As in the case of the Colorado, therefore, the Krishna basin case has been dominated by state rather than federal interests. This has been especially true since the end of Congress dominance in the 1980s and the corresponding rise of state-based politics. Finally, the increasing decentralization of power to the states after the 1970s is reflected in the growing tendency for state politicians to seek control over water resources in order to distribute them to key political constituencies.

The Krishna case can be roughly divided into three phases, each characterized by a changing balance of conflict and cooperation. The first, beginning during the British Raj and lasting roughly to 1963, was characterized by a generally cooperative relationship between the basin states and the central government and was dominated by the planning of large water infrastructure projects within the basin. However, as in the case of the Colorado, state politics and interests soon hamstrung efforts by the central government to agree on a plan for comprehensive basin development. The process of planning and funding projects supported by the individual states sowed the seeds of greater contention between them and set the stage for the second phase of conflict, lasting roughly from 1964 to 1976. During this period, the central government tried and failed to resolve the Krishna basin dispute through negotiation, finally referring the matter to a special judicial tribunal, which issued its ruling in late 1973. Both periods therefore reflect a process of patchwork institution-building, where states deliberately pursue only weaker forms of collective action to maximize access to federal financing while minimizing threats to their autonomy.

Although the tribunal's decision was expected to end the Krishna basin dispute, ethnolinguistic tensions in fact exacerbated competition between the states over Krishna basin waters, propelling the dispute into its third phase, which persists to the present day. A key feature of the dispute over time has been its translation into grassroots politics: while interstate competition for water began as a technocratic-elite concern among politicians at the state level, it has since become a rallying cry for farmers' groups, who have made the allocation of scarce water resources a political cause célèbre at the local as well as state levels. This expanded sphere of conflict has made interstate water-sharing a major political issue in most Krishna basin states and one on which it has become difficult for state politicians to compromise. Because of this political salience and sectional resonance, the institutions that proved crucial to promoting cooperative, basin-wide governance in other cases, especially national political leaders and civil society groups, have been largely shut out of the Krishna basin dispute. The long history of the dispute suggests that in the absence of the involvement of these third parties, the states of the Krishna basin are unlikely to permanently resolve their differences.

Infrastructure and Interest Formation: Origins of the Krishna Basin Dispute, 1890–1963

The origins of interjurisdictional competition for water in the Krishna basin lie in the large irrigation projects undertaken by the British Raj in the late nineteenth century. These projects were undertaken in response to both large-scale famines and the spread of British rule inland, which increased pressure to extract tax revenue from agricultural areas (Walch 1899). It quickly became clear to the rulers of both the princely states as well as the British provinces that the extent to which financing for the development of irrigated agriculture could be obtained would largely determine their jurisdictions' long-term prosperity. Because expanding irrigation required diverting water from the major peninsular rivers, geography shaped the initial formation of interests among the leaders of the various jurisdictions of the Krishna basin: the large inland princely states of Hyderabad and Mysore lay as upper or middle riparians, while British-ruled Madras assumed the status of downstream riparian. Consequently, almost from the outset the Madras government took an interest in irrigation projects in Hyderabad and Mysore that it feared might disrupt the flow of water downstream.

These dynamics are illustrated in the earliest record of an interjurisdictional dispute in the Krishna basin, which appears in the form of an 1890 letter sent by the ruler of Mysore to the British Resident in Madras in response to the "restoration and construction of irrigation works in Mysore, the right to effect which without restriction had been disputed by the Madras government." This early complaint was quickly resolved at an 1892 conference held at the behest of the imperial central government, which led the parties to come to an "amicable understanding." But much as the Colorado River dispute long revolved around an Arizona–California rivalry, this early sectional tension between Mysore and Madras, the upstream and downstream riparians, respectively, proved a harbinger of a much longer-running dispute. In a sign of future discord, the 1892 agreement specified that "in Mysore projects should be absolutely untrammeled. At the same time restrictions to be accepted by [Mysore] on certain classes of work . . . which might affect prejudicially Madras works beyond the frontier." The extent to which their relative upstream and downstream positions would underpin future differences of opinion was also suggested by Madras' insistence on a "prescriptive right" to receive unimpeded flows from Mysore (Government of Madras 1979, 169–71).

The first site of interjurisdictional dispute in the Krishna basin was the Tungabhadra, an important tributary that arises from within the boundaries of what is now Karnataka state before joining the main stem of the Krishna. The Tungabhadra was included in the 1892 Madras–Mysore agreement, with Mysore undertaking not to construct new "irrigation reservoirs" which might impede downstream flow without first securing the permission of Madras. This early agreement proved durable into the mid-twentieth century, when industrial and hydropower development plans led to new tensions between the upstream and downstream riparians. By the time of the Second World War, however, the need for a more robust institutional framework to apportion the Tungabhadra's waters had become glaringly apparent. Accordingly, a 1944 agreement between Madras and Mysore promised to "determine the apportionment between the two [governments] of the contribution to the Kistna [Krishna], giving some weightage to Hyderabad for the reason that at present such water will be used only for power by Hyderabad and for power and irrigation by Madras." The agreement served to severely restrict Mysore's right to divert water from the Tungabhadra, a provision that proved increasingly vexing to the upstream riparian (Office of the Chief Engineer 1979, 186).

Despite the seeming concurrence of Mysore and Madras with respect to these provisions, the agreement failed to specify a final apportionment of Tungabhadra waters, and as the war ended the original Mysore–Madras dispute was reignited and widened to include other states as well. By 1945, Bombay State had also become a party to the dispute, precipitating the involvement of the imperial government in New Delhi. Against this widening field of conflict, the primary riparian jurisdictions of Mysore, Hyderabad, and Madras consented to an "equitable share of waters which shall be decided by a tribunal set by the Government of India for the purpose of final appointment of the Tungabhadra water between all interested parties" (Jung 1979, 191). Nonetheless, despite continued calls for adjudication, especially from Hyderabad, no final apportionment was issued by the New Delhi government. This reluctance on the part of the central government to intervene and attempt to settle the dispute set a precedent that would often be repeated over the course of the Krishna basin dispute, even after the Raj was replaced by the independent Republic of India.

Following independence, newly installed Prime Minister Jawaharlal Nehru's enthusiasm for large river basin development projects led to a new phase of cooperation between the Krishna basin states and the center, in anticipation that construction of large new waterworks would spur regional economic development (D'Souza 2006b). During this phase, the supremacy of state-based technocratic elites, especially the chief engineers, in basin governance was matched by the rise of a new central government bureaucracy, encompassing the CWPC, the Ministry of Agriculture and Irrigation, and most of all the Planning Commission, which was charged with carrying out the national development strategy of the Nehru government. In the Krishna basin, the principal manifestation of this new emphasis on river basin development was a 1951 interstate conference concerning allocation of the Krishna and Godavari River basins, intended as a precursor to the development of large multipurpose development projects in both basins. Despite some disagreement, especially an assertion by Bombay that it had been disadvantaged by current allocations to Madras, the state parties, anxious to obtain central financing for the development of hydropower and other infrastructure, quickly reached consensus on an apportionment of Krishna basin flows. The resulting 1951 accord allotted approximately 25% of the basin's flow to both Bombay and Hyderabad, 10% to Mysore, and the remaining 40% to Madras (Planning Commission 1951). However, much like Arizona rejected the original interstate allocation specified by the Colorado River Compact, Mysore, the middle riparian, felt cheated by its lower allocation and quickly withdrew its support for the accord.

In a further echo of the Colorado case, Mysore's rejection of the 1951 agreement did not prevent the other states from eagerly anticipating what they expected to be a windfall of central government financing for reclamation projects throughout the basin. Amid this newfound comity, a joint center–state committee established to study the basin's reclamation potential recommended construction of a large storage reservoir on the main stem of the Krishna River between Madras and Hyderabad, to be constructed as a joint venture between the two states (*Times of India* 1952). This cooperative spirit was further illustrated in extensive consultations regarding construction of a canal on the Tungabhadra. At a 1956 conference of representatives from several central government ministries, including the powerful Planning Commission and state Public Works Departments, the state representatives "unanimously agreed" to the partition of canal waters at a ratio of 35:65. In an attempt to depoliticize the water-sharing issue, the state delegates also agreed that planning and design of the canal should be delegated to a Tungabhadra Board staffed with chief engineers and technical specialists. So amicable was this meeting of like-minded technocrats that, in sharp contrast to previous gatherings of political representatives, a Planning Commission delegate congratulated the state engineers on the "spirit of accommodation shown by them" (Planning Commission 1979, 197-98).

Nonetheless, this spirit proved short-lived as the looming completion of Krishna basin infrastructure projects threatened to dramatically increase demand for water for a variety of new uses, which the states feared would leave too little to satisfy their individual requirements. Faced with competing demands for irrigation and industrial uses like hydropower production, the states argued that their allotments under the 1951 agreement were insufficient. By 1960, the *Times of India* reported that "Maharashtra [formerly Bombay] and Mysore States hold that the [1951] agreement was not based on any scientific principles, and therefore, it should be revised." The report also illustrates the two states' intensifying rivalry in citing the view that the 1951 agreement gave " 'the lion's share' of the waters to Andhra" [formerly Madras] while "work[ing] to the detriment" of its own interests (Staff Reporter 1960). A 1960 editorial in the *Times of India* likewise illustrates the feeling that Maharashtra and Mysore, despite possessing most of the Krishna basin's catchment area, had sacrificed water for the sake of downstream Andhra Pradesh, which depended heavily on irrigated agriculture. The agreement, the *Times* editorialized, was "contrary to all accepted rules [in that] the States in which the river's source and upper reaches lie have been deprived of their superior riparian rights." Given this injustice, the editorial proclaimed, the only solution was involvement by the central government, including exercising its power to create an interstate river board (Observer 1960). In order to gain the center's support, Krishna basin politicians began to appeal to abstract principles of justice to bolster their respective states' claims to Krishna basin water. Maharashtra's minister for irrigation and power therefore told the State Assembly that the 1951 agreement was based on "unfairness and injustice" and blamed Andhra Pradesh's "adamant attitude of opposition" for the failure to make progress (*Times of India* 1960).

When presented with these new demands from the upstream riparian states to adjust the 1951 agreement, the central government refused to exercise its power to establish a river board, as suggested by the *Times of India*. The Ministry of Irrigation and Power tried to defuse tensions between the upper and lower riparian states by promising to explore ways to "augment water supplies" in the Krishna basin by diverting water from the neighboring Godavari River, instead of renegotiating the terms of the 1951 agreement. The ministry also attempted to delay action on the issue by establishing a study commission under noted irrigation expert N. D. Gulhati to examine water requirements in the Krishna basin states (*Times of India* 1961f). This change in tack appeared to be a recognition by the central government that differences between the states were intractable: the Union deputy minister for irrigation stated that while the central government "had tried to bring about a settlement . . . the State Governments could not agree" and that augmenting supply appeared to be the only viable option (*Times of India* 1961a). But the failure of this brief, halting attempt at developing an engineering solution to the Krishna basin dispute soon became evident as interstate sectional tensions erupted during the 1960s.

This period of intensified discord during the 1960s signaled that interstate rivers had become a focal point for political tensions. State politicians found neighboring states' refusal to abide by the 1951 agreement a convenient excuse for anemic economic growth during the early 1960s. For example, in a bridge-opening ceremony, Mysore's public works minister returned to the theme of injustice pioneered by his Maharashtra counterpart a few years earlier, blaming the 1951 agreement for the state's underdevelopment. According to the minister, the agreement's " 'unjust limitation' on the use of the Krishna and Cauvery rivers is preventing Mysore from undertaking the ambitious irrigation and power projects envisaged by it." The state's "future prosperity," the minister further warned, was "linked with the utilization of the Krishna river waters." For good measure, he went on to note that the majority of areas affected by famine in the Krishna basin lay in Mysore, vividly illustrating the extent to which it had received a raw deal (*Times of India* 1961g). Despite these appeals, the chief minister of downstream Andhra State publicly refused to renegotiate the 1951 agreement, calling the very suggestion a "conspiracy against Andhra Pradesh." Illustrating the growing political resonance of the river dispute, Andhra's chief minister in response called for "public meetings in every town and village in the State to assert the unanimous stand of the people on the 1951 agreement." In an ominous warning, the chief minister also prepared his constituents for a long battle against neighboring states, warning that giving up control of Krishna basin matters would "retard Andhra's industrial and agricultural progress" (*Times of India* 1961d).

As this warning suggested, Andhra proved to be the single greatest barrier to institution-building in the Krishna basin, much like Arizona proved to be the primary spoiler to interstate collective action in the early decades of the Colorado case. Over the course of the year 1962, the Krishna basin dispute pitted Andhra State and its desire to continue to carry out infrastructure projects without revision to the 1951 agreement against the other basin states and the central government, all of whom preferred to conclude a new agreement. The deep political salience of the issue in Andhra was highlighted by the creation of an "all-party action committee on the Krishna and Godavari waters." The committee duly presented to the Planning Commission allegations of "flagrant violations" of the 1951 agreement on the part of Maharashtra, which it claimed was illegally expanding a storage reservoir (*Times of India* 1961b). Despite these appeals, however, the central government appeared to take the side of the upstream riparians, declaring that no additional projects would be approved in the Krishna basin until a report on augmenting supply could be completed— thereby effectively ignoring Andhra's petition and supporting Maharashtra's argument that the 1951 agreement was untenable (*Times of India* 1961e). Mysore, for its part, placed little faith in the central government's plan to channel water from the Godavari, pressing instead for an "equitable distribution" of waters that would guarantee the state its "rightful share" of Krishna basin waters (*Times of India* 1961c).

The growing political stakes of interstate water allocation led state officials to redouble their efforts to lobby the central government, especially by attempting to shape the deliberations of the Gulhati Commission established to examine supply-augmentation possibilities. Madras State, for instance, "pleaded before the commission" that only the diversion of additional water at the expense of neighboring Andhra's allocation could solve a growing water shortage in Madras city (*Times of India* 1962b). Andhra Pradesh, not to be outdone, appealed to the central government's sympathy for its economic development objectives, calling its preferential allocation of Krishna waters a "life-line" for its poor and backward Telangana region, whose farmers it argued would suffer greatly if the state's water allocation were reduced (*Economic Weekly* 1962). In 1961, bowing to these lobbying efforts, the Union government appointed the Krishna Godavari Commission to study and evaluate all proposed water infrastructure projects in the Krishna basin and charged it with attempting to find a suitable compromise among the states' competing projects.

In this the commission proved to be a singular disappointment. Despite its broad brief and the hope of each state that it would support projects planned within its jurisdiction, the commission was soon hamstrung in the face of determined lobbying by each riparian state. In the end, the commission's report stated merely that insufficient water existed in the entirety of the basin to meet all state demands for water (Krishna Water Disputes Tribunal 1973a, 2). It did, however, provide some support to the lower riparian states by suggesting, against Andhra's vociferous objections, that the 1951 agreement be reconsidered. Mysore, which had been the agreement's strongest critic, attempted to capitalize on this victory by restating its appeal for a greater allocation of water to help develop the state's poorest areas. In a speech, Mysore's chief minister called for a new agreement that would not "permanently mortgage the future of the most backward areas which are to be found in our State." Andhra Pradesh, sensing a renewed threat to its water rights from its upstream neighbors, all of whom were anxious to develop water-storage and diversion projects, in turn argued that the commission's findings should be "shelved" in light of the 1962 Sino–Indian border war. In the meantime, the state argued, all previously approved water projects, many of which were within Andhra, should proceed immediately (Times of India News Service 1962).

Andhra's argument, though patently self-interested, posed a problem for Maharashtra, which as a comparatively wealthy upstream riparian had to formulate an equally compelling basis for its sweeping claims to Krishna basin waters. Rather than appeal to the center's poverty-reduction priorities, state leaders embraced a three-point argument in support of their proposed water infrastructure projects and attendant demands for Krishna basin flows. First, Maharashtra officials advanced the rather vague claim that centrally financed water infrastructure projects were needed to alleviate "scarcity conditions" in rural areas of the state. More daringly, and seemingly as a backup should its scarcity claims appear dubious, Maharashtra also argued to the Union government that allocation of Krishna River waters should be made on the basis not of equity but rather population, which would imply a 30% allocation for the state. Finally, the state sought to use its comparatively high level of economic development to its advantage, arguing that Maharashtra was in a position to put irrigated farmland to better economic use than its neighbors. In a speech in New Delhi in March 1963, a state irrigation official proclaimed that "Even from the 'benefit–cost' ratio point of view . . . Maharashtra's irrigation projects had a better record and hence its claims deserve priority." While in Andhra, the official argued, sugar cane farmers paid 10 rupees per acre in irrigation charges and in Mysore 15 rupees, in Maharashtra the comparable sum was 220 rupees. In consequence, it was argued, further developing irrigation in the upstream reaches of the Krishna basin would result in higher economic returns for the nation than in either Mysore or Andhra (*Times of India* 1963a).

Faced with these conflicting appeals for assistance from state leaders, the central government decided to postpone any action in response to the Gulhati Commission report. This refusal to intervene provoked a furious reaction in the legislative assemblies of the middle and upper riparian states, each of which had been counting on central government intervention to reverse what they viewed as the dominant influence of Andhra State. In Maharashtra, two motions were made to adjourn the state Legislative Assembly in order to recognize what the non-Congress Party opposition claimed was "the grave discontent among all sections in the State arising from the regrettable inability of the State Government to secure a favorable decision, based on sound principles and justice, from the Union Government on the dispute" (*Times of India* 1963d). In the Mysore Assembly, meanwhile, some members "urged the Government to either get a fair deal for Mysore or to resign." Such discontent spanned party lines, with a Congress member even stating that "all members of the House should be prepared to resign if justice was not done to the State" in the Krishna dispute (*Times of India* 1963c).

As such comments suggest, interstate water-sharing agreements were becoming increasingly contentious political issues and increasingly linked to deeper issues of state autonomy and prosperity. A noticeable divide began forming between politicians of the ruling Congress Party, who sought to compete for water through appeals to the central government to favor their respective states, and members of the opposition parties, who attempted to galvanize supporters by asserting each state's right to develop its own water resources. Mysore's Congress public works minister decried attempts by "certain persons who formerly held high positions to give political colour to the river waters issue by alleging that the present Ministry had failed to safeguard the interests of the State," as well as an opposition proposal that Mysore take its case to the Supreme Court rather than through lobbying the Union government (Times of India News Service 1963d). In Maharashtra, meanwhile, Congress Party legislators refused to participate in an all-party conference on allocation of Krishna River waters. The conference, they claimed, was merely an attempt at "whipping up popular feelings" (*Times of India* 1963e). In the event, the conference, convened by an economic interest group, the Deccan Agricultural Association, concluded that neither the central government nor the model of central planning could serve Maharashtra's needs. Instead, the conference proclaimed that "in a federal State, it was appropriate to leave detailed decisions relating to the use of natural resources to the States." In an echo of Mysore's opposition proposal, the conference recommended that the Maharashtra State government take its case to the Supreme Court, setting the stage for the legal battles that would define subsequent stages of the Krishna basin dispute (Times of India News Service 1963c).

The developments of 1960–1963 marked a turning point in the history of the Krishna basin dispute for several reasons. First, politicians, rather than chief engineers and other technical experts, became the primary actors, shifting the dispute from a largely technocratic into an almost entirely political sphere. Second, and in consequence, competition for Krishna basin waters became an issue both of justice and of sectional pride as state politicians came to view control over water as essential to both their states' autonomy and future prosperity. Third, and most significantly, political consensus over the best means of pursuing each state's divergent interests began to fracture. While Congress politicians in general supported continued lobbying of the central governmentin hopes of winning its support, opposition members, supported by some local constituencies, began to seek an outside option to cooperation with either the center or neighboring states and one based on seeking judicial redress for interstate water claims. This more adversarial approach relied on reasserting each state's power to control its own water resources and bringing suit against its neighbors in India's Supreme Court. These three shifts set the stage for the second major phase of the dispute, in which the central government made a concerted, but ultimately failed, attempt to end Krishna basin competition, through first negotiation and then arbitration.

Crisis and Central Response: Failed Negotiation and Litigation, 1964–1976

The persistence of the Krishna basin dispute into its second phase presents a striking contrast to the contemporaneous success of the Indo–Pakistani Indus Water Treaty to allocate the waters of the shared Indus River. This contrast bears some elaboration as an illustration of just how acrimonious India's interstate water conflicts came to be after the 1970s, and indeed proved far more contentious than its international water relations. Before independence, the division of the Indus had been the subject of at least three separate negotiations between the directly ruled provinces and princely states of British India, none of which produced a durable agreement (Salman and Uprety 2002). A tentative water-sharing agreement had been reached soon after the partition of British India into the separate nations of India and Pakistan whereby India, the upstream riparian, pledged not to disrupt flow to Pakistan. By the late 1950s, however, leaders in both countries recognized the necessity of reaching a final division of the waters so integral to their agricultural economies, and serious efforts at negotiation began. But even with a water-sharing agreement so clearly in the interest of both parties, the conclusion of the Indus Water Treaty depended almost entirely on third party–induced cooperation.

The first of these third parties was an individual, David Lilienthal, former director of the American Tennessee Valley Authority, who on a visit to India proposed that the two sides devise an agreement to divide the basin as a whole. This idea was adopted and carried through by a second, institutional third-party actor in the form of the World Bank. The bank's role in concluding the Indus Treaty was pivotal: once the two parties, at Lilienthal's urging, had agreed to divide the basin's waters between them, the bank provided technical expertise to arrive at an impartial measure of the total quantity of water available, as well as financing for infrastructure projects that enabled both countries to utilize their share of the basin's waters (Biswas 1992). This third party–induced cooperation also produced a durable basin governance institution in the form of a commission with representation from both countries. Today, the Indus commission remains one of the few bilateral fora to have survived three full-scale wars between India and Pakistan. The close involvement of impartial, third-party bridging organizations in the Indus basin suggests their importance in building institutions for interjurisdictional collective action in India, just as in the other cases examined in this book. Sadly, however, the involvement of third parties remained lacking in India's internal riparian disputes, and their absence was keenly felt as the Krishna dispute entered its second phase.

During this second phase, the increasing politicization of the Krishna water-sharing agreement, particularly in the Congress Party stronghold of Maharashtra, attracted the undivided attention of the central government. Toward the end of 1963, the center began a concerted effort to negotiate a final resolution of the Krishna basin dispute. In a speech in November 1963, Union irrigation minister K. L. Rao inaugurated this second phase of attempted Krishna basin institution-building by declaring New Delhi's determination to "bring about harmonious agreement" between the riparian states, this time by financing interbasin transfers from the Godavari to the Krishna (Times of India News Service 1963a). Yet almost immediately, this attempt at conciliation foundered on the states' refusal to accept any reduction in their claims to Krishna waters. Prime Minister Nehru acknowledged as much in a letter to Maharashtra's chief minister, in which he declared himself against the state's proposal to divert Krishna River waters westward for the purpose of hydropower generation. Water, Nehru conceded, was a "scarce commodity" that the states could not be expected to surrender without compensation. Instead, he proposed that Maharashtra consider developing thermal power sources, with the cost to be shared by other states, in exchange for guaranteeing flows to its downstream neighbors (Staff Reporter 1964).

Such proposals for compromise fell on deaf ears. Soon after Nehru's proposal was issued, the central government changed tack, and attempted to effectively bribe the Maharashtra government by proposing to sanction several of its planned irrigation projects in exchange for the state abandoning its award of approximately one-third of the Krishna's total flow. As a newspaper report noted at the time, this call for Maharashtra to abandon its legal allocation would force state officials to risk the possibility that "other claimants might get all available waters earmarked for their own projects" (Times of India News Service 1964b). Consequently, while Maharashtra's Congress legislators continued to support interstate negotiation under the center's auspices, opposition members pressed the state to pursue judicial action. One opposition politician presciently asked the state's irrigation minister whether failure to pursue arbitration would "lead to an unending process of negotiation to the detriment of the interests of Maharashtra" (*Times of India* 1964c). Certainly, the state's downstream neighbors continued to view any new project development in Maharashtra as a threat to their own water supplies: Mysore's chief minister condemned the upstream state's proposed westward diversion projects as both "unscientific and wrong" (Times of India News Service 1964c).

In the face of Maharashtra's intransigence, neither Andhra nor Mysore saw any need to compromise its own individual position, which rested on maintaining downstream flow for both new and existing irrigation projects. Negotiations convened by Union irrigation minister Rao with the two states proved inconclusive because, as Rao conceded, "Andhra feels the diversion [planned by Maharashtra] would come in the way of implementing the irrigation projects the States had already taken up," while "Mysore wants more water for irrigating some of its own backward areas." This intransigence on the part of both its downstream neighbors provoked Maharashtra's representative to threateningly note the state's "geographically advantageous position" as the upstream riparian and imply that it might unilaterally cut off the flow of water downstream (*Times of India* 1964b). Mysore ignored this threat and persisted in its previous strategy of lobbying the central government to finance water projects in the upper Krishna basin. This strategy appeared to bear some fruit in the 1966 election campaign, when the president of the ruling Congress Party toured the state with a promise to award central financing for Mysore's water projects should Congress be returned to power. Yet at the same time the visit clearly illustrated the extent to which India's politicians had come to treat Krishna water as a distributive resource. Noting Mysore's "backwardness," the Congress representative cautioned that the ruling party "alone could deliver the goods to the people" by financing expanded irrigation (*Times of India* 1966b).

Despite this promise, however, central government financing for Mysore's proposed projects was not forthcoming, and state leaders grew increasingly frustrated with the Union government's inability to modify Andhra's allocation of Krishna waters. A report from June 1966 noted that "A feeling is growing that Andhra Pradesh . . . is getting away with more than its legitimate share and this has adversely affected Mysore's interests" (Times of India News Service 1966). Shortly thereafter, the chief minister of Mysore, newly renamed Karnataka, inaugurated a new phase of the dispute by calling for mass protest to secure central government financing. The state's members of Parliament, he proclaimed, should be prepared "if necessary to observe satyagraha or go on a hunger-strike to make the Centre concede Mysore's demand for Central take-over of the upper Krishna project." Only by taking such drastic action, the chief minister claimed, could Karnataka focus

New Delhi's attention on the state's increasingly dire water situation (*Times of India* 1966a). This increasing agitation also led the Mysore government for the first time to raise the prospect of pursuing an outside option to interstate cooperation in the form of judicial action. Mysore, state leaders told their indignant constituents, might "follow Maharashtra's example by taking legal steps to ensure its rightful share of the Krishna waters" (Times of India News Service 1967d).

Mysore's shift toward a more confrontational approach to the water-sharing dispute, based on protest and legal action, signaled its transformation from a relatively narrow political question into one that pitted the state against its neighbors in almost existential terms. The extent to which shared basin waters had become a serious sectional grievance was illustrated by an opposition member's question, posed during debate in the state legislature, as to "how Mysore could expect justice from the Union Irrigation Minister, who hailed from Andhra Pradesh, and therefore could not think objectively about the Krishna dispute" (Times of India News Service 1967f). Similarly, the sense of grievance attached to the Krishna issue within Mysore was illustrated by the public works minister's insistence that the matter of water resources "concerns the whole state" and that if Andhra's claims could not be adjusted, the neighboring state's water claims would render Mysore "condemned to perpetual misery" (Times of India News Service 1967c). Indeed, such was the depth of sectional feeling on the water-sharing issue that Mysore Congress Party leaders issued unusual denunciations of their partisan counterparts in the Union government, who they alleged had demonstrated insufficient commitment to preserving the state's interests in the dispute with its neighbors. The general secretary of the state's Congress Committee, for example, accused the Union government and its ruling Congress Party of "adopting a step-motherly attitude towards Mysore" and demanded that the "Krishna issue should be taken out of the hands of the Union Irrigation Minister" (Times of India News Service 1967e).

The final collapse of the Union government's stature as a credible third party to the dispute came as a result of renewed tension over two proposed infrastructure projects, Andhra's Nagarjunasagar dam and Maharashtra's Koyna hydropower project. Despite the personal intervention of Prime Minister Indira Gandhi, who convened the chief ministers from the three basin states at a special conference, negotiations failed to produce a new agreement on the allocation of Krishna waters. Mysore and Maharashtra both objected to the Andhra dam project, which enjoyed the backing of the CWPC, because it threatened to inundate parts of both states (Times of India News Service 1967a). Andhra Pradesh, for its part, alleged that Maharashtra had diverted water for its Koyna hydropower project in violation of the 1951 agreement. The political salience of this issue within Andhra Pradesh was illustrated by the comments of one of the state's Congress Party delegates to the Raj Sabha, who complained that Andhra was "being maligned in some quarters for . . . the Nagarjunasagar dam, when Maharashtra had violated the understanding by using waters in excess of its entitlement" (Times of India News Service 1967b). Andhra's chief minister, meanwhile, signaled his disinterest in further negotiations, dismissing as "irrelevant" a Maharashtra claim that it "was in the upper reaches of the river and could do what it liked with the Krishna waters." Instead, the chief minister warned, he would "with the full backing of the State Assembly and of the people of Andhra Pradesh will fully safeguard the interest of the State" (Times of India News Service 1968b).

By the middle of 1968, the divide between the upstream states of Maharashtra and Mysore and the downstream state of Andhra Pradesh had widened to the extent that further negotiations were clearly untenable, and judicial resolution seemed to be the only solution. Both upstream states were "tired of the long-drawn-out negotiations . . . they believe that any delay only benefits Andhra Pradesh, which . . . has been using the interregnum to go ahead with new projects, pre-empting the river supplies." The Union government, in contrast, professed itself "not keen on arbitration," both because it would delay planned development projects and because "technical problems like those of river flow . . . make judicial adjudication a very unsatisfactory proposition" (Times of India News Service 1968a). Nonetheless, all alternatives to judicial arbitration were very quickly exhausted. Prime Minster Gandhi called several more meetings to attempt to resolve the states' differences, including through the Congress Party Working Committee, to which the chief ministers of all three Krishna basin states belonged. Yet neither Gandhi's personal involvement nor the intraparty forum of the working committee proved able to bridge the states' divergent interests, and even Andhra Pradesh soon professed itself in favor of arbitration. In contrast to the other states, however, Andhra viewed judicial action as a means not to modify the 1951 agreement but rather to prevent its revision in any form (Times of India News Service 1968c).

Having at last secured Andhra's consent to arbitration, the Union government in June 1968 officially referred the Krishna basin dispute to a judicial tribunal, signaling the end of its attempts to reach a negotiated solution. As all parties prepared to await a judicial resolution, persistent drought conditions throughout the basin made the partition of Krishna waters an increasingly galvanizing political issue. In Mysore, a cross-party commission lobbied a visiting Union government delegation to fund the upper Krishna irrigation project, which they stated was "the only way of ending the menace of recurring drought and famine" in the state (Times of India News Service 1970). In Maharashtra, meanwhile, a group of farmers threatened to break open dykes on the upper Krishna River in order to irrigate parched farmland. Accusing the state government of "dereliction of duty," a gathering in Sangli district gave the government an ultimatum of eight days to address the "acute scarcity conditions prevailing in the district" before initiating civil disobedience (Times of India News Service 1972). These local demonstrations further strengthened the linkage between Krishna waters and sectional tensions, driving home their salience even as state governments began to submit evidence to the newly constituted Krishna Waters Dispute Tribunal (KWDT).

The state parties' submission of evidence and testimony before the tribunal provided few surprises, but it nonetheless served to underline the riparian states' uncomproising positions with respect to the Krishna basin. Mysore demanded an "equitable distribution" of Krishna waters and requested the tribunal to block construction of Andhra's irrigation projects as well as Maharashtra's western diversion canal. Maharashtra, for its part, echoed Mysore's demand for equitable apportionment but also called for possible diversions from the Godavari River basin into the Krishna in order to avoid diminishing its own allocation of Krishna water. Andhra

Pradesh once again found itself alone in demanding that its preferential claim on Krishna waters under the 1951 agreement be upheld as "valid and binding." The state also requested the tribunal to stay construction of projects in both Mysore and Maharashtra, which it claimed would entail "utilization of more than their respective shares under the 1951 agreement." Finally, in a reflection of its insecurity as a downstream riparian, Andhra Pradesh further asked the tribunal to issue an injunction preventing its upstream neighbors from "intercepting flows to the Delta and other irrigation works of Andhra Pradesh" (Krishna Water Disputes Tribunal 1973a, 3). On the basis of these divergent positions, the three basin states submitted claims of 829 thousand million cubic feet (TMC) in the case of Maharashtra, 1,430 TMC for Mysore, and 1,888 TMC for Andhra Pradesh. The total sum of these claims, as the tribunal quickly and disapprovingly noted, proved to be substantially more water than was available in the Krishna basin as a whole (Krishna Water Disputes Tribunal 1973a, 5).

Andhra Pradesh in particular struggled mightily to have its stake to the majority of the Krishna's flow validated by the tribunal. The state presented twice as many witnesses as the other two states combined, including a preponderance of technical experts. The state did, however, appear to have the Union government implicitly on its side. The Union referred a number of additional issues to the tribunal that appeared calibrated to support Andhra's position in the dispute. Chief among these was a request that, in addition to adjudicating the substantive claims of the state parties regarding the waters of the Krishna, the tribunal grant control of water infrastructure within Mysore to an independent body, the Tungabhadra Board. The board had originally been established in 1953 to oversee the operation of waterworks on the Tungabhadra River between Andhra Pradesh and Mysore (Krishna Water Disputes Tribunal 1973a, 4). However, because its representation had historically been dominated by Andhra Pradesh, this move was widely seen as a veiled attempt on the part of Andhra and the central government to cede control of Karnataka's dams and reservoirs, whose operation might interrupt downstream flow, to Andhra Pradesh. The panel's justices, hinting at their suspicion of such a scheme, flatly denied the request and stated that there was "no ground for taking away the administration and control . . . from the Karnataka Government" (Krishna Water Disputes Tribunal 1976, 9).

This bad omen for Andhra's interests foretold the tribunal's final judgment, which roundly rejected the state's case. The tribunal found that the 1951 agreement could not be taken as valid and binding for three separate reasons: first, because the Planning Commission had no authority to conclude a binding agreement, second, because there was insufficient evidence that the state parties intended to conclude such an agreement, and finally, because Mysore's need for water had not been properly accounted for during the proceedings. The tribunal also rejected a number of subsidiary demands by Andhra Pradesh that downstream flows from its neighbors be guaranteed along with a share of hydropower generation in neighboring states. Perhaps most damagingly for Andhra, the tribunal also recognized the need for allocation of Krishna waters according to the principle of equitable apportionment, in which prior utilization enjoyed only limited recognition. On this basis, the tribunal awarded Maharashtra 565 TMC, Karnataka 695 TMC, and Andhra Pradesh

800 TMC annually. In addition, the tribunal held that each state was entitled to use a small percentage of any excess annual flow, to be calculated on the basis of prior utilization, and that the award could be reconsidered after the year 2000 (Krishna Water Disputes Tribunal 1973a, 87–88). Finally, the tribunal suggested that a Krishna Valley Authority be constituted either as a result of agreement between the states or by an act of Parliament to supervise the tribunal's allocation and, when necessary, to direct the transfer of water between states to satisfy emergency measures. This last provision signaled a setback for both Maharashtra and Mysore, which both strongly opposed such robust forms of interstate collective action as a threat to their autonomy (Krishna Water Disputes Tribunal 1973b, 183–87).

In the aftermath of the tribunal award, the Union government breathed a sigh of relief, believing that long-stalled development projects could finally get under way. Shortly after the tribunal's award, the Union Ministry of Irrigation and Power called for the Krishna basin states to submit their development plans for review so that the basin could be "harnessed to the maximum extent possible." Maharashtra officials declared themselves satisfied with the tribunal's verdict, which awarded the state substantially more than the 440 TMC provided by the 1962 Gulhati Commission (*Times of India* 1973). Both Andhra and Karnataka, however, refused to accept their respective allocations. The issue continued to be an emotive one in Karnataka, with one of its parliamentary representatives expressing outrage that the state "cannot get a fair decision" from the tribunal. Both downstream states announced they would consider appeals to the KWDT verdict, provoking a frenzied appeal from the Union government to persuade them to accept it (*Times of India* 1974). In the end, the two states' threat of judicial action proved hollow, and they instead turned their attention to completing the long list of delayed water projects. Maharashtra proved the most enthusiastic, pledging "vigorous efforts" to ensure full utilization of the state's allocation (Sabhu 1976). Yet even this apparent progress proved to constitute a truce rather than a full resolution of the Krishna basin dispute and one which would soon flare once again on a rising chorus of sectional discord.

Devolution and Discord: Ethnic Politics and Water Disputes, 1977–2014

The KWDT award coincided with the state of emergency declared by the Congress Party government of Indira Gandhi, who asserted near-dictatorial powers in the face of growing opposition to Congress rule. But despite the Gandhi government's attempts to effect an unprecedented concentration of power, beginning in the late 1970s Indian politics came to be influenced by a new generation of state-based parties that appealed to ethnic and linguistic identity, rather than to the broad Indian nationalism that had hitherto been central to the Congress Party ethos. Following the KWDT award, these new ethnolinguistic parties increasingly fashioned interstate water-sharing disputes into touchstones for sectional tensions and rivalries. These dynamics took some time to take root, however; and at first it appeared that the rise of state-based parties might instead presage a more cooperative future in the peninsular river basins. In February 1976, following the KWDT award, Prime Minister

Gandhi announced an agreement between the Krishna basin states whereby each agreed to surrender 5 TMC annually for the growing city of Madras in Tamil Nadu, which had been experiencing a drinking water crisis. Nonetheless, implementation of the agreement was derailed by a dispute between Andhra Pradesh, on one side, and Karnataka and Maharashtra, on the other, concerning the use of an open canal to convey water to Madras, which the upstream states feared would be used by Andhra to divert additional water for irrigation (Rao 1983). Conclusion of the agreement occurred only after the landmark victory of the new, sectionalist Telugu Desam Party in Andhra's 1983 state elections. As the first non-Congress party to take power in the state since independence, the Telugu Desam Party quickly agreed to support the Madras water supply canal as a "neighborly gesture" (*Times of India* 1983b). The accord between Andhra and Tamil Nadu to supply water to Madras was accordingly hailed as "a trend-setter in solving interstate river water disputes" throughout peninsular India (Rao 1983).

Such optimism, however, was soon belied by the growing linkage between interstate water-sharing agreements and deepening sectional tensions, which were eagerly seized upon by the new generation of regionalist political parties to energize their supporters. In part, this linkage reflected severe drought conditions across many of the riparian states throughout the early 1980s, which led to an increasing number of protests among small farmers angered by delays in constructing long-promised irrigation infrastructure. In Maharashtra, for example, a "mass relay hunger strike" was threatened in February 1983 by members of Shetkari Sanghatana, a populist Marathi farmers' union, if the state did not complete a long-delayed diversion from the Krishna to irrigate some 30,000 hectares in the state's Sangli district (*Times of India* 1983a). In the other basin states, similar protests from geographically concentrated water-user groups made it clear that state water claims could be a potent political cudgel, and water-sharing issues became a frequent subject of discussion in legislative sessions. In Karnataka, for instance, an opposition member of the State Assembly pressed the state's public works minister to see that enough irrigation projects be built in the state to ensure that Karnataka's allocation of Krishna water would be "fully committed for use" by the year 2000, as stipulated by the KWDT decision (Times of India News Service 1985b). Perhaps needless to say, such insistence betrayed little concern for the state's riparian neighbors or of the need to ensure downstream flows.

Perhaps nothing better illustrates the widening sphere of political conflict over interstate water-sharing than the emergence of a bitter dispute over Andhra Pradesh's proposed canal to supply Madras with drinking water, known as the Telugu Ganga project. The canal, previously hailed as a model for interstate water cooperation, had by the mid-1980s instead become embroiled in the states' broader water-sharing dispute. In August 1985, Maharashtra lodged a formal protest with the central government against Andhra's project, claiming that its true purpose was to increase Andhra's de facto share of water before diversion and storage projects in the upstream states could be completed (Times of India News Service 1985a). In response, Andhra Pradesh's parliamentary delegation aggressively lobbied the Union government to approve the project. Prime Minister Rajiv Gandhi, in an attempt to ease apprehensions among his political allies in the state, quickly announced his intension to visit Andhra. The Karnataka government, unwilling to sanction such a display of favoritism, in turn invited the prime minister to make a separate visit to "see for himself the injustice done to Karnataka" as a result of Andhra's efforts to build the Telugu Ganga project (Times of India News Service 1986). Indeed, such was the political salience of the Telugu Ganga project that in 1989 Karnataka Janata Party leader H. D. Deve Gowda felt compelled to warn both the Union Congress Party government and the Andhra Pradesh government against believing that the "people of Karnataka were weak." Karnataka, he threatened, "was in the upper reaches and the people could also take a unilateral decision" to disrupt downstream flow to Andhra Pradesh (Times of India News Service 1989a). Andhra's chief minister retaliated by asking the center to withhold clearance of any further water projects in Karnataka until it could be established that the state was withdrawing only as much water as it was permitted under the KWDT award (Times of India News Service 1989b).

The depth of interstate discord over water-sharing is illustrated by the fact that the Telugu Ganga dispute persisted even after the 1989 general elections, when all three Krishna basin states returned to Congress Party rule. This unusual alignment was expected by many to favor interstate cooperation. However, Congress' loss at the national level instead marked a retreat in the central government's involvement in the Krishna dispute, and little progress was made. While Congress members of Parliament continued to maintain that the "Centre had a duty to intervene in settling such interstate disputes," the new Union government pledged not to intervene in the dispute unless requested to do to by the riparian states (*Times of India* 1990). Such a request never arose, in large part because of an emerging alliance between both Karnataka and Maharashtra concerning the Telugu Ganga project. Under the terms of this alliance, both states proclaimed themselves "in full agreement" that Andhra Pradesh would have to supply water to Madras out of its KWDT allotment of 800 TMC, rather than receiving any additional allocation (Times of India News Service 1990a). In the absence of central involvement in resolving this new discord, each state remained focused on construction of water projects that would allow it to utilize its full KWDT allotment. A Karnataka state representative declared, for instance, that "We have to make our best efforts to complete the scheme by hook or crook. Personally, I would advocate that we should go ahead even if we have to beg, borrow or steal" (Times of India News Service 1990a). In Maharashtra, meanwhile, a report urged that despite the unresolved nature of the Krishna basin dispute, the state's "main task now [should be] to use up the entire quota of 560 TMC before the year 2000" (Date 1991).

Accordingly, for much of the 1990s the Krishna basin dispute was defined by large-scale investment in water resource management at the state level as each state party sought to develop the necessary capacity to utilize its full tribunal allocation. This period was marked by frenzied competition among the states to secure financing for water-storage infrastructure that would allow them to utilize their full allotment of water under the KWDT award ahead of the all-important year 2000 deadline. Amid this competition, the effort to fully utilize each states' tribunal allocation became cast as a matter of economic survival in which citizens of each state must unite against their neighbors. In Karnataka, a gathering of former chief ministers warned that if the state did not act to secure the necessary investment for irrigation projects within its borders, "future generations will not spare us." Instead, the former leaders proclaimed that fully utilizing Karnataka's share of Krishna waters should be a cause shared "irrespective of . . . party affiliation, case, creed and color" (Times of India News Service 1992a). The state government, acting in full agreement, sought financing for irrigation projects of its own from the Union government and from the World Bank (*Times of India* 1995; Times of India News Service 1995).

But despite state politicians' calls for unity in the face of competition with neighboring riparians, distinct geographically concentrated interest groups within each state continued to drive interjurisdictional conflict. The influence of these sectional interests on each states' approach to interjurisdictional collective action is perhaps best illustrated in Maharashtra. In a feat reminiscent of the success of southern California water users in shaping California's position in the Colorado River controversy, irrigators in western Maharashtra had by the early 1990s gained outsize influence over the state's water resource policy and lobbied incessantly for state-funded expansion of irrigation projects. Bowing to these interests, state leaders pledged to "use every drop of our allocation" by developing large-scale irrigation projects in the western parts of the state (Agashe 1992a). Yet, as in the case of many southern California irrigation projects, it was readily apparent that such investment would benefit only a small number of farmers, at considerable public expense. According to an independent assessment, for example, "the Krishna basin projects will benefit only a limited number of land-owners and sugarcane producers from western Maharashtra," while doing nothing to aid the most backward regions of a state where 80% of farmland remained unirrigated (Agashe 1992b). Beneath the veneer of state unity, western Maharashtra irrigators dominated state politics concerning the Krishna River much as their counterparts in southern California had with respect to the Colorado.

On the strength of the western water users' political advocacy, irrigation financing continued to drive Maharashtra's involvement in the Krishna basin for the remainder of the 1990s, with a position that was deeply sectional. The goal of cementing the state's control over as much water as possible, at the expense of Maharashtra's neighbors, quickly gained cross-party support. In 1996, the new Shiv Sena–Bharatiya Janata Party (Sena–BJP) government, building on previous Congress efforts, "exhibited its sense of urgency" by creating a Krishna Valley Development Corporation to raise money for some 451 planned irrigation projects throughout the state (Government of Maharasthra 1996; Agashe and Kokje 1996). In doing so, however, the Sena–BJP faced the same dilemma as had its Congress predecessors: although these projects were needed to fully utilize the state's KWDT allocation, in many ways they made little practical sense. They would only benefit one-third of the state's population and would accrue mostly to already irrigated farmland, rather than the rain-fed regions that would benefit most from expanded irrigation. The BJP's plans to spend large sums on irrigation in an already wealthy region unsurprisingly provoked "sharp reactions" from other parts of the state, but even these growing political costs did not distract the new government from the overriding importance of utilizing the state's full share of Krishna waters. So important was this objective, in fact, that the Sena–BJP government explicitly supported several projects that would almost entirely benefit Congress-dominated parts of the state while antagonizing the areas from which it drew most of its own support (Agashe 1996).

In Karnataka and Andhra, meanwhile, the politics of water-sharing were dominated by a renewed dispute over Karnataka's upper Krishna irrigation project. Andhra Pradesh, then under non-Congress rule, objected to the height of a dam planned as part of this Karnataka-sponsored project, which Andhra feared might disrupt the flow of water downstream (Jothi 1996). Financing for the project, Andhra charged, had been granted by the central government only because the Congress-led Union government's prime minister was himself from Karnataka (Times of India News Service 1996). In an attempt to maintain the center's credibility under these accusations, the Central Water Commission refused to issue final approval for the upper Krishna project, prompting appeals by the Karnataka government to the Planning Commission, which it viewed as more sympathetic to its interests (Times of India News Service 1997). As in Maharashtra, completing storage and diversion projects to fully utilize KWDT allocations remained a top priority for Karnataka politicians, who were held accountable by their own party members as well as the opposition. The Karnataka government had to fend off accusations by both opposition members as well as its own back-benchers in the State Assembly that it had dragged its feet on the upper Krishna project (Times of India News Service 1998b).

As the year 2000 deadline approached, political tensions grew, particularly within Maharashtra, which felt the most intensive pressure to develop its irrigation infrastructure. In 1998, a group of environmentalists, including the Bombay Natural History Society, issued a report criticizing one of the larger irrigation projects planned for western Maharashtra because it would fragment an important wildlife conservation area—one of the first times an environmental organization had played a notable role in the Krishna basin dispute (Times of India News Service 1998a). Unfortunately, its involvement proved too fleeting to encourage state leaders to adopt a more cooperative approach to water-sharing, as similar organizations had by this point begun to do in the case of the Colorado. Instead, state politicians continued to use water-sharing issues as a political cudgel with which to accuse opponents of insufficient fidelity to the state's interests, which they alleged to be inseparable from utilizing as much water as possible. Maharashtra's Sena–BJP government accordingly continued to rail against the "blocking" of the state's water infrastructure projects as unconscionable political interference by the Union government in the state's attempts to secure water for itself (Vaidya 1999). Despite these blatant appeals to sectional interests, as elections approached in 1999 the state's chief minister rather disingenuously attempted to discredit the Congress Party opposition by accusing it of "trying to exploit the [Krishna] issue." During its own time in power, the chief minister charged, Congress had "left 125,000 people high and dry." If, he asked, Congress had been "so concerned about the Krishna water . . . what were you doing for 40 years?" (Times of India News Service 1999). As such allegations suggest, by the late 1990s allegations that the opposing party had failed to protect state interests in interjurisdictional water disputes had become a popular, and effective, means of winning elections.

By the time of the year 2000 deadline, the development of uncompromising positions with respect to interstate water-sharing had effectively painted Krishna basin state leaders into a corner, making it essentially impossible for state leaders to compromise in order to resolve lingering disputes over either allocations or the approval of infrastructure projects. Accordingly, in January 2000, the three riparian states and the Union government took advantage of the expiration of the KWDT allocation period to reargue their claims before the Supreme Court. Karnataka restated its now familiar complaints toward Andhra Pradesh, charging that its downstream neighbor was "utilizing the total excess water available" by planning various new irrigation projects. Downstream Andhra Pradesh, meanwhile, sought an injunction against Karnataka's Almatti Dam, part of the upper Krishna project, charging that the upstream state could raise the level of the dam to effectively utilize 442 TMC of water instead of the 155 allotted to the state under the KWDT award (Times of India News Service 2000a). On the latter issue, the Supreme Court decided in Andhra's favor, directing Karnataka to limit the height of Almatti Dam to 520 meters above mean sea level. This decision quickly provoked political recrimination in Karnataka, with a senior engineer breathlessly claiming that "Andhra Pradesh has been quicker in using its share of water . . . [while] successive [Karnataka] governments, most of them Congress governments, lacked vision" (Prasad 2000).

By the year 2000, the very notion of interstate cooperation over shared water had become politically toxic, and instead state politicians sought to strengthen or expand their base of political support by adopting confrontational positions with respect to water-sharing. The extent to which intrastate politics undermined interjurisdictional collective action in the Krishna is best illustrated by Maharashtra, which had previously avoided taking sides in the dispute between its downstream neighbors over the upper Krishna project. In 2000, however, the ruling Sena–BJP government abandoned this reluctance in the face of biting criticism from the Congress opposition over delays in completing planned water-supply projects in several of the state's rural areas. Referring to the Sena–BJP's extensive investment in highway infrastructure, a Congress deputy remarked, for example, that "Drinking water should get precedence over fly-overs." Instead of responding directly to his Congress antagonists, Maharashtra's chief minister faulted the state's downstream neighbors, asking assembly members "to sink party differences and rally behind the state government to safeguard Maharashtra's interests." Amid this attempt to build intraparty unity by promoting interstate conflict, Maharashtra issued a new claim that the KWDT had underestimated the river's total flow. The excess water, the state boldly argued, "belongs to Maharashtra." Indeed, so strong was the appeal to sectionalism that the chief minister later told the press it would utilize its full share of Krishna waters "without adhering to formalities such as intimating the two neighboring states" or even seeking the Union government's permission (Times of India News Service 2000b).

As the Maharashtra example illustrates, by the beginning of the twenty-first century unilateral rather than collective action had become the modus operandi of the Krishna riparian states. The influence of this sectionalism meant that by the year

2000, "the three-decades-old dispute is almost back to square one" (Prasad 2000). In 2001, Karnataka reacted to the Supreme Court's rejection of a final attempt to prevent Andhra Pradesh from utilizing surplus water that it had not yet impounded by declaring, "This state of affairs has caused a grave apprehension in the mind of the state of Karnataka that for all times to come its right of sharing the surplus water would be gravely endangered" (Times News Network 2001). In a 2001 Lok Sabha debate, a Karnataka delegate repeated the view that very little had changed in over a century of dispute. "Madras Presidency," the delegate drily noted, "has always deprived Mysore State from optimum utilization of water resources" (Lok Sabha 2001). As this historical allusion attests, after more than a century of controversy, the Krishna basin remained defined by sectional interests and politics among the riparian states. Moreover, this dynamic has persisted until the present day as Karnataka has struggled to utilize its full allotment of Krishna basin waters, while Maharashtra and Andhra have periodically attempted to undermine its efforts. The Union government, meanwhile, has continued to make only periodic, half-hearted attempts at mediating the dispute.

The persistence of old patterns of sectional conflict is further illustrated by the recent separation of the state of Telangana from Andhra Pradesh. Telangana is an inland region and is generally more dependent on agriculture than the more urbanized coastal plains of Andhra. Telangana, unlike the rest of Andhra Pradesh, was originally part of the princely state of Hyderabad rather than the Madras presidency. This history created a long-standing cleavage with coastal Andhra, a divide reflected in the fact that water-related conflict played an important part in Telangana's agitation for statehood. In 2010, a Union government commission created to study the possibility of separating Telangana from the rest of Andhra noted that "One of the grievances of Telangana region is that, although Telangana contributes most to the catchment area of Krishna and Godavari within the state, the allocations to the region are lower and not in the same proportion." Coastal Andhra, for its part, had long expressed "apprehensions" over the diversion of approximately 20 TMC of Andhra's 181 TMC total allocation for projects in the Telangana region (Committee for Consultations on the Situation in Andhra Pradesh 2010, 205–10). As these apprehensions foretold, the Telangana example vividly illustrates how quickly the problem of sharing Krishna basin waters between the riparian states could become a focal point for sectional tensions between even a newly formed state and its neighbors.

Following passage of a parliamentary act to create the state of Telangana in 2013, the Krishna rapidly assumed its familiar role as a touchstone for sectional tension between now-upstream Telangana and downstream Andhra Pradesh. As in the case of the other Krishna basin states, differences in water use helped to make river-sharing an issue of acute political concern. As a 2013 news report noted, coastal Andhra "is better irrigated and ha[s] adopted [more] modern farm practices than Telangana" (Krishnakumar 2013). Moreover, even before Telangana formally separated from Andhra Pradesh, the historical divisions between the two regions had already been projected onto the allocation of Krishna basin waters. An editorial warned, for example, that "the relationship between the Presidency regions and the native states was not without its points of tension ... it may be much too optimistic to assume the environment of give-and-take needed for an amicable solution on water sharing" (Pani 2013). These concerns featured prominently in debate on the parliamentary bill to create Telangana State, in response to which the legislation presciently provided for creation of two river management boards for the Krishna and Cauvery, with representation from the two states and the Union government. In a unique institutional innovation, the bill also created the Apex Council consisting of the chief ministers of both Telangana and Andhra Pradesh along with the Union minister for water resources (PTI 2013). Tellingly, however, none of these institutions provided for the involvement of nongovernmental organizations; and, as of the time of writing, New Delhi has refused to actively participate in the council (Rao 2016).

Instead, the familiar sectional politics of the Krishna basin have so far prevented much in the way of cooperative institution-building. Six months after achieving separation from Andhra Pradesh, Telangana's new irrigation minister announced that the state would file suit in the Indian Supreme Court, on the grounds that "Now [the] time has come for Telangana to fight for its rights as the region has [the] largest catchment area of Krishna river." The state would pursue this strategy, the minister added, because it did not trust the Union government–constituted Krishna Water Management Board to give it a "fair allocation" (Rao 2014). As in previous phases of the Krishna basin dispute, Telangana's choice of an outside option to cooperation provided a means for the riparian states to pursue individual and divergent interests in opposition to those of their neighbors. The state's argument that the KWDT allocation should be revisited in light of its separation from Andhra Pradesh moreover created the grounds for a temporary alliance with its former parent state in opposition to the two upstream riparian states. As a late 2014 news report noted, "There is near unanimity between Telangana and Andhra Pradesh governments in opposing Maharashtra and Karnataka" (*Deccan Chronicle* 2014). Nonetheless, this Telangana–Andhra alliance failed to give rise to enduring interstate collective action. In October 2014, a row erupted between the chief ministers of the two states over the amount of water stored in dams in upstream Telangana. While Andhra's chief minister accused his Telangana counterpart of violating central government directives stipulating a minimum water height at Srisailam Dam, Telangana's chief minister charged that the Andhra Pradesh government had "taken a vow to destroy farmers in Telangana" (*Times of India* 2014).

The emergence of competition between Telangana and its former parent state of Andhra Pradesh illustrates the degree to which the allocation of Krishna River water has focused sectional tensions between the riparian states. A recent case of interstate cooperation even amid these tensions looks to be the exception that proves this unfortunate rule. In March 2016, Telangana signed a landmark memorandum of understanding with its upstream neighbor, Maharashtra, pledging to support construction of several dams and irrigation projects along the Godavari River. As many news reports noted, the agreement was one of the most significant examples of interstate water cooperation in recent Indian history (Chandrashekhar 2016). However, the optimism surrounding the agreement was quickly dampened when Maharashtra's water resources minister complained that it did nothing to address the "ongoing dispute between the two states over the sharing of water from the Krishna River" and that "our farmers suffering from a severe drought have not been able to get . . . water because of the approach Telangana government has taken" (*The Hindu* 2016b). While it is too soon to tell whether the agreement will endure, the invocation of sectional tensions suggests that durable interstate cooperation in the Krishna basin will remain elusive.

This most recent phase of the Krishna basin dispute provides a further illustration of how sectional identity and the institutions of Indian federalism conspired to create and sustain interjurisdictional competition over shared water resources. To be sure, this competition unfolded atop a historical geographic divide between coastal and inland provinces that created basic differences in culture and water use between the riparian states. But the origins of the dispute lie in the institutional arrangements of Indian federalism and its persistence in the marked influence of sectional tension. The 1956 delineation of state boundaries around ethnolinguistic lines effectively codified sectional cleavages, making it easy for state politicians to mobilize support on the basis of shared sectional identity. Water proved to be a particularly potent means of doing so—state politicians could convincingly present interstate water-sharing issues as a zero-sum contest between the citizens of one state, united by shared cultural and linguistic identity, and those of an effectively foreign neighbor. Marked divides in water use and availability between coastal and inland regions led to predictable lobbying from geographically concentrated water-user groups, like western Maharashtra irrigators, for central financing of water-storage and transfer projects. State politicians readily took up the cause. It was soon apparent, however, that it was economically and politically infeasible to finance all such state-supported projects. The central government, which for electoral reasons feared offending any one state, effectively abandoned efforts to arbitrate the various individual state interests; and no suitable interjurisdictional institutions remained to fill the void. In this vacuum, the states turned to litigation and political competition to pursue their claims to Krishna water. These elements of the Krishna basin case resemble those of the Colorado. Unlike in the Colorado case, however, the Krishna basin dispute reveals the limited political opportunity structure that has until recently obtained in most parts of India. Politics was for most of the course of the Krishna basin dispute dominated by elites, and nongovernmental involvement in the dispute has until recently been extremely limited. In the absence of third-party involvement, state interests tend to dominate water politics in India, with negative implications for cooperation, as further illustrated in the contrast case of the Damodar Valley Corporation.

DECENTRALIZATION, REGIONAL DEVELOPMENT, AND THE DAMODAR VALLEY CORPORATION

While the Krishna basin case demonstrates how, in the absence of bridging organizations, sectionalism inhibits interstate collective action, the case of the Damodar Valley Corporation (DVC) illustrates how sectional politics can undermine institution-building even after it occurs. The DVC in many ways presents an exact contrast to the case of the Krishna basin. The upstream riparian, Bihar State, was part of downstream Bengal during the British Raj, obviating the historical tensions which characterized the Mysore–Madras relationship in the Krishna basin. In addition, the principal political issues surrounding the Damodar River were not related to scarcity or irrigation, as in the case of the Krishna, but rather hydropower generation and flood control. Most importantly, the DVC represents an example of strong institutionalized cooperation in the form of a comprehensive, basin-wide governance organization created in conscious emulation of the American Tennessee Valley Authority (TVA). Indeed, the DVC expresses essentially the same ethos of basin-scale rather than jurisdictional governance as did the TVA. The DVC was created to fulfill a vision that "Our conception of the jurisdiction of existing administrative units must change, and a river should be regarded as one natural economic unit for the benefit of the whole community it can serve, in the full development of which political frontiers must not be allowed to interfere" (Franda 1968, 47). Yet despite this sweeping vision, the DVC case, like that of the TVA, illustrates that the involvement of third parties is needed to bridge the very political frontiers it was intended to transcend. In the absence of such involvement, sectional tensions fatally undermined regional institution-building in the Damodar as well as Krishna basins.

The DVC, like the American TVA, was intended to mount a coordinated and cooperative institutional response to several regional-scale river basin governance issues. For much of the colonial period, the Damodar River Valley, a subbasin of the Ganges spanning Bihar and Bengal provinces in northeastern India, had been a fashionable resort area for Calcutta's elite. However, beginning in the mid-nineteenth century the valley was devastated by a series of massive floods. Periodic flooding disasters continued to ravage the river basin for the next hundred years, until in 1943 a government commission proposed a TVA-like organization to simultaneously prevent flooding and provide a locus for regional economic development in the Damodar Valley (Damodar Valley Corporation 1956, 4; Publications Division, Ministry of Information and Broadcasting [India] 1961, 7). In seeking to explain the government's failure to prevent massive loss of life in previous flood events, the commission faulted the existing management of the river by individual states, which it claimed lacked the necessary capacity to erect flood-control structures, and called instead for a new regional organization to improve coordination between the individual states and the central government. Accordingly, the DVC was "designed to fundamentally realign the relationship between the central authority and the provinces in order to accommodate the logistics for multi-purpose river valley development" (D'Souza 2006a, 191).

As these objectives make clear, the DVC was also intended to be a prototype for cooperative center–state relations under India's newly established federal system. The central focus of this cooperative vision was comprehensive regional economic development along the lines of that pursued in the United States under the New Deal and in postwar Europe. A 1948 DVC report illustrates the scale of these ambitions: "In its industrial potentialities the Damodar Valley has the same significance for India as the Ruhr Valley for Germany and the Monongahela Valley for the USA" (Damodar Valley Corporation 1948, 6). The following year, the DVC's annual report proclaimed that the corporation's functions would extend beyond traditional water resource management tasks like irrigation, water supply, and hydropower production to include "the promotion of public health and the agricultural, industrial, economic and general well-being in the Damodar Valley and its area of operation" (Damodar Valley Corporation 1949, 1). Such wide-ranging aims reflected not only the ambition of the corporation but also the importance India's leaders attached to the development of the Damodar. Nehru himself employed no small dose of hyperbole when he remarked in a speech that "A new India is being born in the [Damodar River] valley. It is the India of our dreams" (Publications Division, Ministry of Information and Broadcasting [India] 1961, 24).

In seeking to realize this dream, India's leaders were determined that the central government exercise firm control over the DVC to prevent its mission being subsumed by interstate politics. The central government's leading role in the Damodar Valley was codified by a 1949 interministerial summit, at which it was agreed that while for minor projects "local initiative should be encouraged and provinces and states should be allowed to have a free hand in their execution," all significant decisions "should be sanctioned and controlled by the central government" (Central Waterpower, Irrigation and Navigation Commission 1949, 9). The prominence of the Union government's role was quite explicitly intended to prevent conflict and disputes between the states which might undermine the regional development objectives of the DVC. As a 1961 report explained, "It had often happened in the past that when a river valley project affecting two or more States was taken up, [a] clash of opinions and interests hampered smooth execution of work and resulted in wasteful delay. This sort of thing could happen because there was no adjudication machinery for settlement of disputes relating to the use, distribution and control of the waters of the rivers for which projects were planned" (Publications Division, Ministry of Information and Broadcasting [India] 1961, 11). The DVC, the report went on to note, was intended to avoid this situation by effectively subsuming state-level politics within a single, regional-scale organization.

The DVC accordingly took the form of a unique institution with substantial powers of its own but also one that would operate under firm central control. Like the TVA act on which it was modeled, the DVC act provides for the corporation's fiscal autonomy, grants it authority to take out and make loans and to acquire property, and bestows upon it the right to regulate most water uses in the Damodar River basin. Moreover, the act empowered the DVC to impose punishments and sanctions for individuals in breach of its regulations (Damodar Valley Corporation 2005, part V, sections 48–49, 60). The act also included a sweeping clause granting the DVC "power to do anything which is necessary or found expedient for the purpose of carrying out" its enumerated responsibilities (Damodar Valley Corporation 1956, 8). Yet in its constitution the DVC also featured several mechanisms intended to ensure it would serve national interests. As a 1956 DVC report put it, "Although the Corporation is an autonomous body and enjoys a high degree of independence in framing its own policy, plans, and programme, it is subject to the control of, and direction by, the Central Government" (Damodar Valley Corporation 1956, 10). Moreover, although funding was to come from both central and state governments, the center was to account for the largest share. Accordingly, the DVC act specified that the corporation follow policy guidance from New Delhi and that in the event of any dispute between the corporation and the central government, the center's directives "shall be final" (Damodar Valley Corporation 2005, part V, section 48).

At first, the DVC appeared to have indeed united the Damodar River under a single authority. After concluding the agreement between the central government, West Bengal, and Bihar that formally established the DVC, a member of the Union Parliament, C. H. Bhabha, congratulated state representatives for engaging "in the closest possible co-operation with one another and with no thought of scoring provincial points against the larger interests of valley development." Because of such cooperation, Bhabha went on to say, the members of the DVC would possess "no divided loyalties" between the central and state governments (*Times of India* 1947). Nonetheless, within just a few years, the Union government began struggling both to exercise control over the DVC and to achieve its objectives to develop the Damodar Valley. Despite the scale of ambitions, the DVC got off to a slow start, and progress toward building planned flood control and hydropower infrastructure soon fell below expectations.

By 1952, many Union officials had developed widespread dissatisfaction with the DVC's progress, and increasingly suspected that the DVC had become hampered by the very state politics it was established to transcend. The government accordingly appointed a special panel known as the Rao Committee that was tasked with issuing recommendations to reform the DVC, including making changes to the composition of the corporation's board (*Times of India* 1954). But even with the benefit of these recommendations, the Union government continued to struggle to assert control over the corporation. A decade later, a 1963 *Times of India* editorial could still fault the "partisan interests of the State Governments" for the DVC's subversion of central government priorities in developing hydropower generation. Instead, the editorial endorsed the growing view among central government officials that in order to compel it to serve national interests the corporation must be made "free from the State financing system" (Kamalesh 1963).

In fact, by the 1960s state interests had indeed come to dominate the operation of the DVC and conspired to undermine any ambitions of more extensive cooperation in the basin. Beginning in the 1950s, fearing that the DVC's broad powers represented a threat to their autonomy, state officials made a concerted effort to enhance their influence in the corporation. In West Bengal, state politicians grew increasingly dissatisfied at what they viewed as the corporation's failure to address basin management issues of special concern to the state, including flood control and the provision of adequate hydropower production to support West Bengal's growing industrial base. Consequently, throughout the late 1950s the West Bengal government and its representatives in the Lok Sabha attempted to enhance the state's leverage over the DVC and its decision-making processes. During 1957 debates in the West Bengal Assembly, the chief minister complained that the DVC act was "iniquitous so far as this State was concerned" and demanded amendments to the act "giving the State Government a measure of control over the Corporation's affairs" (*Times of India* 1957). Significantly, this view was held across party lines among West Bengal's parliamentary representatives, with both Congress and Communist Party members expressing "grave disquiet" at the central government's supposedly dominant role in the corporation (*Times of India* 1958; Times of India News Service 1959).

A similar process played out in the neighboring riparian state of Bihar. In 1953, the Bihar Legislative Assembly issued a demand that the DVC move its headquarters from Calcutta in West Bengal to a site within Bihar, on the grounds that Biharis were insufficiently represented among the corporation's staff (Times of India News Service 1953). This particular grievance, namely that Bihar's interests were not adequately reflected in the corporation's activities, continued to resonate for much of the next decade. Opposition members in the State Assembly declared in 1959 that the state "had derived little benefit" from the DVC, and even a Congress Party member called upon the government to "make the shifting of the headquarters of the DVC from Calcutta to Maithon a prestige issue" (*Times of India* 1959). Such discontent grew severe enough that by 1962 the central government appointed a former state chief justice to serve as arbitrator between the DVC, West Bengal, and Bihar (*Times of India* 1962a). Despite the involvement of this eminent neutral party, however, the headquarters issue remained so emotive that no agreement was reached (*Times of India* 1963b).

During the 1960s, this discontent within the states of West Bengal and Bihar led the states to strip the DVC of its multipurpose developmental functions. In a process virtually identical to that experienced by the TVA, state government pressure denuded the corporation of virtually all its statutory functions, and it was left to focus almost exclusively on hydropower generation. This effort to cut the DVC down to size was led by the West Bengal government, which was continually buffeted by charges in the State Assembly that the corporation failed to serve the needs of poor farmers. In response, the state successfully lobbied the central government to transfer the corporation's irrigation and canal system back to state control (Times of India News Service 1963b). In 1963, for example, a Congress member of Parliament from West Bengal declared that "hopes over the DVC have turned illusory and it should now be scrapped. The power and irrigation schemes should be handed over to the different states" (Franda 1968, 104). The strength of these objections eventually persuaded the central government to divest the DVC of its regional planning and governance functions and instead to reform the DVC in a way that more closely served the individual interests of the riparian states.

By 1964, pressure from the state governments further convinced the center to dissolve the DVC as a tripartite partnership between the central government, West Bengal, and Bihar. This division of authority, a central official stated, had "resulted in a plurality of authority, conflicting interests and claims" which "adversely affected the functions of the Corporation" (Times of India News Service 1964a). The center's plan for reform nonetheless clearly bore the imprint of state interests rather than its own. As a 1964 *Times of India* article records, the Union government's original proposal was to transfer the DVC's flood-control and irrigation functions to the West Bengal government and afforestation and soil conservation to Bihar (*Times of India* 1964a). Yet even this modest proposal was derailed by sectional tensions between the two states. West Bengal continued to object to the DVC's monopoly over hydropower generation and continued control of major dams along the Damodar River (Times of India News Service 1965). Consequently, the DVC was forced to surrender control of its dams, as well as its soil conservation and afforestation responsibilities. By the mid-1960s, therefore, interstate rivalry had succeeded in transforming the character of the DVC as an organization for regional, basin-scale governance into a glorified utility focused almost exclusively on hydropower generation.

Bereft of such a regional governance organization, management of the Damodar River basin devolved into a set of sectional rivalries between the states of West Bengal and Bihar. The collapse of the DVC as a regional planning and coordination mechanism created a gap in the governance of the Damodar River basin that was filled by increasingly acrimonious disputes between the two states. Within this vacuum, the sectional tensions that had begun to characterize interstate collective action in the Krishna basin following the rise of state-based ethnolinguistic parties also began to appear in the Damodar. In 1973, an "apparently innocuous question" regarding an agreement with Bihar for the supply of water from a DVC reservoir created a storm in the West Bengal Assembly, with a Communist Party member complaining that "thousands of acres in my district are going dry" for want of water used by Bihar (Times of India News Service 1973). In Bihar, meanwhile, a 1992 government commission concluded that the DVC "had not brought any benefits to the state" and that the state's cooperation with West Bengal had failed to secure Bihar's interests. Despite 82% of the Damodar River basin lying within the state of Bihar, the commission observed, the state was permitted to use only 11.9 lakh-acre-feet, against 32.1 utilized by West Bengal (Times of India News Service 1992b). By the end of the twentieth century, despite the existence of a powerful basin-wide governance organization in the form of the DVC, interstate water-sharing disputes had come to characterize the Damodar as much as the Krishna.

The DVC represents one of India's few examples of meaningful institutional collective action in interstate river basins, and close examination of its history offers two lessons for the study of conflict and cooperation over interjurisdictional water resources more generally. First, the central government played a critical role in inducing cooperation between the states and securing their support to establish the DVC as a regional-scale institution. The center could play this role, however, only because it enjoyed credibility among the critical actors, the state governments of West Bengal and Bihar. This observation leads to the second lesson, namely that once the central government became less involved in bargaining and arbitrating the interests of the riparian states, the DVC was gradually subsumed by sectional tensions between the two states. As one Indian observer concluded, "The experiment of the Damodar Valley Corporation was . . . a failure because of the strained relationship between the states and the Corporation" (Ramana 1992, 68). The result of this sectionalism was that the DVC was stripped of its management and governance functions, bereft of any capacity to perform its original mission to bridge the interests of the Union, West Bengal, and Bihar governments or the political actors they represent. In 2011, a bill was passed in the Indian Parliament to correct these deficiencies by adding three "independent experts" to the DVC's board to provide impartial input free from competing state interests (PRS Legislative Research 2011). However, at the time of writing, the impact of this reform remains unclear. In the meantime, the case of the DVC serves to illustrate that in the absence of credible third-party involvement, subnational states typically fail to create collaborative, participatory institutions to manage shared water resources.

COUNTRY CASE CONCLUSION

The puzzle of the Indian case lies in the existence of so many subnational conflicts over water and the central government's persistent failure to establish river basin governance institutions despite its explicit power to do so. Subnational politics are especially fractious in India, and as the DVC case illustrates, they have consistently frustrated efforts to manage India's river basins at the regional scale. With respect to the salience of these subnational politics, the Indian case bears a few important similarities to the case of the United States. As in the American case, geographically concentrated water–user groups, particularly farmers' associations in areas dominated by irrigated agriculture, pressed state leaders to form a reclamation nexus to seek central government financing for water infrastructure projects. These politicians, especially chief ministers, state legislators, and state representatives in the Union Parliament, eagerly took up this cause. Over time, these leaders sought to build political support by alleging that neighboring states were claiming more than their fair allotment of shared water resources. The scope of conflict over interstate water-sharing gradually widened from a narrow issue of concern primarily to elites and certain water-user groups to one that focused sectional tensions between ethnolinguistic groups in neighboring states. Under such conditions, interstate cooperation became impossible. Finally, as in the case of the United States, in the face of such emotive politics the involvement of the central government in interjurisdictional disputes proved to be episodic rather than continual, and federal apex courts were left to adjudicate them. Moreover, other institutions which may have been expected to help facilitate interjurisdictional cooperation, especially large, broad-tent political parties like the Congress, were by electoral logic rendered unwilling and finally unable to prevent or resolve interjurisdictional water resource disputes.

Despite these similarities, the case of India also demonstrates that interjurisdictional hydropolitics are shaped by a complex and country-specific blend of factors, as summarized in Table 5.2. The Indian case highlights the fact that while immutable factors like riparian geography and physical water scarcity play a role in subnational hydropolitics, institutional and ideational factors explain their dynamics over time. Riparian geography did shape relationships between the Krishna basin states to the extent that it produced lasting regional differences in culture and water use. But so long as interstate water-sharing issues remained a primarily technocratic and elite concern, riparian relations remained relatively amicable. When, following the 1956 delineation of state boundaries around ethnolinguistic cleavages and the rise of state-based regional parties, politicians cast water-sharing as a contest between groups united by different sectional identities, interstate conflict grew in both intensity and duration. Physical water scarcity, meanwhile, led geographically concentrated water-user groups to pressure state politicians to increase water supplies. But state politicians effectively co-opted this pressure, transforming it into a more expansive case of "us-versus-them" sectionalism. Once interstate water-sharing became a popular sectional, rather than elite riparian, issue, upstream–downstream divides became far less important than those between Kannada- and Telugu-speaking water users and created an effectively intractable dispute.

Finally, the Indian case vividly illustrates by dint of their absence the importance of reform coalitions that link third-party nongovernmental organizations and national political elites. In India, the influence of civil society organizations in water governance is of recent advent, in contrast to the long-standing influence of such groups in the United States and France. In the absence of such involvement, the Union government's periodic attempts to resolve interstate water-sharing disputes have consistently been undermined by the divergent interests of subnational states. The Indian case thus offers several lessons for understanding subnational hydropolitics more generally. First, when central governments act as credible brokers for divergent subnational political interests, they create a constituency for cooperative institution-building, as illustrated by the initial phases of the DVC

Table 5.2 Summary of Subnational Hydropolitics in India

Factor	**Characteristics**	**Effect on Interjurisdictional Collective Action**
Decentralization	• States have primary responsibility for most water management issues, though the Union has considerable powers also • State boundaries encourage ethno-linguistic parties • Historically weak central water management bureaucracies	• Mostly weak interjurisdictional institutions, disputes often settled through litigation • State parliamentary delegations, chief ministers allied with irrigator associations to form rent-seeking alliances for irrigation financing • National political leaders (prime ministers, water ministers) periodically involved, usually ineffective at brokering consensus
Sectional identity	• State boundaries drawn along ethnolinguistic lines • Historical geography creates upstream–downstream divide between coastal and inland areas	• Upstream–downstream dynamics usually involve sharing water with different ethnic, linguistic groups • Farmers' groups pressure state politicians to stop neighbors "stealing" water
Political opportunity structure	• Historically few formal or informal mechanisms to link grassroots advocates and political elites • Areas like water policy historically dominated by elites, little civil society participation	• Involvement of civil society third parties historically very limited in shared river basins

case. The World Bank played a similar role, against considerable odds, in the formulation of India's agreement with Pakistan concerning the Indus. Second, although historical geography and differences in regional water availability play a role in catalyzing subnational conflict over water, it is the conflation of water-sharing issues with notions of sectional identity that causes interjurisdictional disputes to persist over time. Third, and most importantly, the Indian example illustrates that institutional arrangements that structure political opportunity are more important than those that assign responsibility for water resource management. Despite the Union government's unusually expansive powers over water resource management, the dominance of state sectional interests has so far undermined cooperative institution-building in India's river basins. The following chapters, which describe the contrasting cases of China and France, provide further evidence that the inclusion of third-party bridging organizations is critical to creating the constituency for cooperative regional institution-building and avoiding the sectional conflict that characterizes the case of India.

6

Central Control, Localized Interests, and Interprovincial Conflict in the People's Republic of China

From a comparative perspective, the People's Republic of China represents perhaps the world's most distinctive combination of political centralization and fiscal and administrative decentralization. The basic unity of the state, referred to as *Dayitong* (大一统), has long been seen as the organizing principle of governance in mainland China and underpins the modern system whereby decision-making is tightly concentrated at the central government level (Wang 2009). At the same time, however, the practical challenges of governing a large and diverse territory have historically led Chinese officials to delegate substantial administrative powers to subnational levels of government. Moreover, in economic terms China is one of the most decentralized countries in the world, with revenue and expenditure powers largely in the hands of local officials (Dziobek, Mangas, and Kufa 2011). Chinese officials are thus caught in an institutional matrix known as *tiao-kuai* (条-块), in which they are responsible both to line control by functional bureaucracies, such as the various central government ministries, as well as to territorial government leaders, including mayors and provincial governors, and to equivalent officials within the parallel Chinese Communist Party (CCP) structure. The CCP effectively controls all important political appointments, creating a potent mechanism to ensure the coherence of central and local policy objectives (Mertha 2008). This matrix is intended to ensure that subnational officials pursue priorities set by the central government but also to provide them with the flexibility to implement these policies according to individual local circumstances. In practical terms, this flexibility also translates into autonomy in a wide range of policy areas, including water resource management. Like their counterparts in more politically decentralized countries, China's subnational officials therefore also confront the dilemma of autonomy, and they sometimes attempt to resolve it through conflict with neighboring jurisdictions (Moore 2014a).

The puzzle of the Chinese case is that, like France, it features a relatively high level of institutional collective action in interjurisdictional river basins, but water management outcomes remain poor, and despite its political centralization, it exhibits widespread water-related conflict between subnational jurisdictions. As in the case of France, interjurisdictional conflict in China is the product of rapid economic growth, which has created both quantitative pressures in the form of growing demand for water as well as qualitative ones in the form of severe water pollution, which has become a leading cause of tension between upstream and downstream riparian jurisdictions. Like their counterparts in France, China's major river basins are managed by regional-scale institutions known as "water conservancy commissions" (WCCs; *shuiliweiyuanhui*, 水利委员会), which possess wide-ranging powers concerning infrastructure development, water use and allocation, and water flow regulation. Unlike their French counterparts, however, China's regional river basin governance institutions are made up entirely of central government agency officials, and they do not represent subnational or nongovernmental constituencies. As a result, China's river basin commissions possess nothing like the political legitimacy or institutional success of the French water agencies and have instead produced a marked dichotomy in the success of regional institution-building. The Chinese state's technobureaucratic approach to water resource management, manifested in the narrow composition of the WCCs, has successfully addressed issues like flood control where the incentives for interjurisdictional cooperation are high. But issues requiring collaborative and participatory decision-making, such as revising interjurisdictional water quantity allocations or resolving interjurisdictional pollution disputes, remain highly contentious. Instead of cooperating to address such issues, local officials often simply shift the cost of compliance to neighboring jurisdictions, either by exceeding water abstraction quotas set by the center or by refusing to enforce water quality legislation.

This dichotomy illustrates, perhaps more clearly than in any other case, the importance of political opportunity structure and of third-party involvement in determining the success or failure of regional institution-building. Despite the existence of a strong centralized state and robust organizational machinery in the form of the WCCs, the central government and CCP have devoted scarce attention to collaborative and participatory institution-building. In the absence of such involvement, Beijing's policy of promoting decentralized economic growth has fostered the emergence of divergent sectional interests as local officials seek to maximize economic growth in their respective jurisdictions. Like their counterparts in other countries, subnational leaders in China have responded to this institutional vacuum by pursuing individual rather than collective interests. But unlike in the other country cases examined in this book, nongovernmental third parties have so far been unable to promote collaborative, participatory collective action to fill this gap. Civil society constituencies are not represented in regional or other formal governance institutions by virtue of China's authoritarian political system. As challenges like water scarcity and pollution began to strain China's water resource policy apparatus after the 1970s, the lack of participation in China's river basin governance institutions increasingly constrained its adaptive capacity. In contrast to the cases of the United States and France, where the gradual expansion of civil society involvement helped to build robust regional river basin governance institutions, in China these institutions remain stunted for lack of legitimacy and representation; and the country's rivers have suffered as a result. In this chapter, I illustrate this process with respect to the Yellow River basin and its governing institution, the Yellow River Conservancy Commission, China's oldest and arguably most important WCC.[1]

COUNTRY CASE INTRODUCTION

In contrast to more decentralized political systems, the Chinese central government retains firm and full legal and functional control over water resource management issues. Elements of water resource policy, most notably flood control, have been a core central government function in China for millennia (Chi 1963). Historical circumstances, including a disastrous 1938 flood which killed nearly a million people, moreover ensured that China's early Communist Party leaders took a keen interest in water resource issues; and this strong central influence has persisted to the present day (Gustafsson 1984). Nonetheless, following the launch of political and economic reform in 1978, fiscal and administrative decentralization has complicated the interjurisdictional management of China's major river basins. The Chinese system provides exceedingly few formal mechanisms for interjurisdictional cooperation or interest-bargaining between and among subnational officials. Instead, and in contrast to the American tendency to channel regional issues through state governments, in China such challenges are accommodated within a vertical, central–local hierarchy dominated by bureaucratic actors. As a result, while China's hierarchical system helps to ensure that policymaking is tightly controlled by the central government, it does little to encourage cooperation between subnational officials over shared water resources. In particular, central–local tensions exacerbate water scarcity issues, while interjurisdictional conflicts over water pollution are increasingly common. As water uses multiply and issues like environmental protection and recreation become more politically salient, the technocratic and bureaucratic character of river basin governance in China increasingly prevents effective adaptation to emerging water resource management issues.

Table 6.1 highlights the major features of water politics in China. As a rapidly developing economy, China faces the full range of water management issues, including water scarcity and allocation, pollution, flood control, and infrastructure development, which has displaced large numbers of people. Despite its political centralization, China also faces both vertical and horizontal collective action dilemmas. Local officials frequently warp or ignore central policy directives, while neighboring subnational jurisdictions engage in competition for water and disputes over the construction of water infrastructure. China does possess a range of integration mechanisms, including regional river basin commissions and a tightly integrated political party and bureaucratic structure. When the central government has

Table 6.1 FEATURES OF WATER POLITICS IN CHINA

Case Characteristic	Issues
Driving management challenges	Water scarcity, allocation, water pollution, flood control, infrastructure development
Institutional collective action dilemmas	Central–local conflict, interjurisdictional conflict
Integration mechanisms	River basin commissions, vertically integrated political party and bureaucratic structure
Sectional cleavages	Marked divide between coastal and inland provinces, substantial urban–rural divide within provinces
Political opportunity structure	Extremely limited nongovernmental participation

chosen to deliberately exercise these mechanisms, they have proven largely effective in preventing and resolving interjurisdictional water conflict. For the most part, however, subnational leaders have been left to pursue independent interests with respect to shared water resources, which, as in other countries, have been shaped by sectional divides, especially differences in economic structure and water use between coastal and inland jurisdictions.

However, the Chinese case is clearly distinguished from others by its authoritarian system, in which civil society and nongovernmental participation in the policy process are highly constrained within a structure of hierarchical, centralized control. This structure has so far prevented the reform coalitions that have promoted cooperative institution-building in other countries from emerging in the People's Republic. The central lesson of the Chinese case is therefore that the mere existence of interjurisdictional institutions is not enough to prevent interjurisdictional conflict. To do so, institutions must be both collaborative and participatory, a feature that reform coalitions are critical to instilling.

Fragmented Authoritarianism in China's Water Governance

To the extent that scholars have attempted to explain China's internal hydropolitics, factional-bureaucratic explanations have historically been favored, which revolve around conflict between various elements of the central government bureaucracy (Lieberthal 1992). But as this section details, a second institutional factor is even more important in animating subnational hydropolitics in China, namely the weakness of formal mechanisms for interjurisdictional cooperation, communication, and bargaining within the Chinese political system (Moore 2014a). In general, this system is structured to bind local officials to central decision-makers at the expense of encouraging horizontal cooperation (Lieberthal and Oksenberg 1988). The consensus view among both Chinese and foreign observers is that the legal foundations for interprovincial coordination are weak and insufficient to meet China's current water resource quality and quantity challenges (Xu and Guo 2007; Liu and An 2007; Zhao 2007). China's principal water resource legislation articulates only general principles for addressing intergovernmental disputes, stating only that they should be resolved through "negotiation and consultation" (RMIT International 2003, 51–52). The National People's Congress, meanwhile, reinforced this informal, ad hoc approach by developing instructions on dealing with transjurisdictional water pollution disputes that explicitly direct local officials to handle disputes themselves before seeking the involvement of higher levels of government (Guojia Renda 2002). Despite its centralization, water resource management in China, like in most countries, lies within an institutional gap that creates the space for subnational conflict to arise.

The Chinese policymaking process is premised on the notion of local subordination to national policy objectives and accordingly is underpinned by a number of central control and enforcement mechanisms that are designed to limit any divergence in the objectives of central and local officials. First, the central party leadership controls the appointment and advancement of leaders immediately below them, who in turn control personnel management for successive subsidiary levels (Burns 1994; Brodsgarrd 2002; Landry 2008, 50–57). Second, higher-level cadres are periodically rotated to different jurisdictions in an attempt to discourage them from building independent local power bases (O'Brien and Li 1991, 173). Third, an elaborate system for evaluating cadres on the basis of centrally formulated performance targets, known as *kaohe* (考核), provides a means of judging progress on achieving central priorities even in the absence of direct control (Tsui and Wang 2004). These central control mechanisms act to prevent subnational officials from openly defying policies set by the central government but do little to address the increasingly divergent regional interests which have emerged in recent decades. Indeed, the existence of widespread water-related conflict in China indicates more clearly than perhaps any other issue the extent to which horizontal, interjurisdictional political conflict occurs despite a structure of vertical central–local control.

In contrast to those of the U.S. and Indian federal governments, Beijing's powers in the field of water resource management are both broad and unambiguous: Article IX of the 1982 Constitution stipulates that water is, with few exceptions, the property of the state, and all uses of water are subject to state regulation. In order to exercise these powers, the People's Republic has moreover developed a constellation of capable bureaucratic entities. The most important of these is a powerful hydroengineering entity in the form of the Ministry of Water Resources (MWR; *Shuilibu,* 水利部), formerly known as the Ministry of Water and Power. At various times in its history the ministry has also been responsible for energy and electricity production, but its present responsibilities include the planning, finance, and construction of major water infrastructure projects, including dams, reservoirs, and irrigation systems; the formulation of plans and regulations concerning water use and allocation; providing input on water quality issues; overseeing the implementation of water pricing; collecting and maintaining hydrological data; promoting soil conservation; and flood control (Ministry of Water Resources 2012). The MWR is represented in major river basins by river basin commissions, which are classified as "dispatched organs" (*paichujigou,* 派出机构) under its direct control. The Ministry of Environmental Protection (MEP; *Huanjing baohu bu*, 环境保护部), a historically weaker bureaucratic organization, has general responsibility for pollution control, while various other ministries, including the Ministries of Agriculture and of Housing and Urban–Rural Development, play more minor roles in particular river basin management issues. Higher-level decision-making bodies, including the State Council, are involved in setting policy objectives, approving larger water infrastructure projects, and arbitrating disputes between ministries and jurisdictions (Moore 2017).

Each central ministry is represented at both the provincial and local levels, where they coincide with separate territorial administrations at provincial, prefectural/county, and municipal levels. In China's administrative hierarchy, province-level territorial governments possess the same status as central ministries like the MWR, creating considerable confusion regarding authority. Every province-level government has an entity, usually called a "water affairs bureau," whose functions correspond to those of the central MWR; and most county-level governments also possess a branch whose responsibilities typically encompass water resource conservation, protection, flood control, irrigation, and water supply (Liu and Speed 2010). In addition to these general-purpose bureaucracies, various special-purpose entities exist, including dike maintenance committees, irrigation and reservoir district committees, and water-user associations (Hubei Sheng Huanggang Diqu Shuili Dianli Ju 1977).

Yet despite these sprawling powers and the dense institutional matrix that surrounds them, Beijing's authority over water resource management is far from monolithic. The complex matrix of authority at the subnational level creates considerable tension between central objectives and localized interests, despite China's centralized political system. Asked whether he supported the central government's planned economic strategy, for instance, one Ningxia delegate to the National People's Congress declared himself to be "duty-bound to help advance Ningxia's socioeconomic development" rather than Beijing's policy agenda (*Ningxia Ribao* 2010). These localized interests have the effect of raising the cost to the central government for imposing its preferences on those of local officials, at least to the extent that they conflict (Moore 2014a). Perhaps surprisingly, the Chinese central government, despite its considerable powers, typically resolves this dilemma by deferring to subnational interests in the case of shared water resources. As David Lampton writes, "Leaders of superordinate territorial units are reluctant to impose severe costs on subordinate levels. Consequently, in water policy, where the effect of any decision . . . inevitably affects a multitude of territorial units, one finds policy and projects slowed down by the need to build a consensus among the affected units" (1987, 170). The costs of doing so are so high that, in some cases, the central government simply abandons attempts to secure consensus, delaying projects indefinitely or effectively leaving provincial governments to their own devices (Moore 2014a). Rather surprisingly, these constrains mean that Beijing can be held hostage to subnational hydropolitics, just as can Washington and New Delhi.

Unlike those of its counterparts in the United States and India, however, Beijing's difficulties persist in spite of its sustained attempts at promoting interjurisdictional collective action in shared river basins. In 1950, the State Council created China's first modern river basin management organization, the Yellow River Conservancy Commission (YRCC; *Huanghe shuili weiyuanhui,* 黄河水利委员会), and placed it directly under the central government out of a desire that it "be free of entanglements and squabbles between provincial governments" (Greer 1979, 41). Similar WCCs were subsequently established in China's other major river basins as well as in the Lake Tai basin. Originally, the primary function of these commissions was to formulate comprehensive basin-wide economic development plans. In 1954, for instance, the YRCC supervised formulation of a plan which proposed construction of nearly a dozen large dams for hydropower generation, flood control, and irrigation along the upper and middle reaches of the Yellow River (Xinjiang Weiwuer Zizhiqu Dangwei Xuanquanbu Zhuban 2009). The 1997 Flood-Control Law and 2002 Water Law assigned responsibility to the WCCs for developing basin-wide flood management and development plans, considerably expanding their remit (Zhonghua Renmin Gongheguo Zhongyang Renmin Zhengfu 2006).

However, these institutional reforms have been outstripped by the emergence of divergent interests among the subnational actors they are supposed to help manage. The central government's reluctance to delegate decision-making power means that although the WCCs are intended to coordinate regional-scale policymaking, they do not include subnational officials or representatives, and they act more as instruments of control than as instruments of cooperative governance (Shen 2009). In place of formal institutions for interjurisdictional coordination and dispute resolution, conflicts in river basin governance have traditionally been handled informally between elites (Gustafsson 1984, 153). This ad hoc, bureaucratic approach to decision-making has never been entirely satisfactory, but it has proven increasingly inadequate in the face of dramatic growth in the number of water-user groups, as well as the growing role of nonstate interests. During the past decade, water pollution and dam construction have emerged as significant social issues, and media and civil society groups have begun to influence policy decisions regarding these two particular issues (Mertha and Lowry 2006; Mertha 2008). In large urban areas, meanwhile, the private sector constitutes an important source of investment in the water sector; and in some cases, it has become directly responsible for the operations of water supply and treatment systems (Qi 2011). Amid this growing pluralization of water users and uses, China's largely centralized, technobureaucratic approach to river basin management has proven increasingly incapable of preventing and resolving interjurisdictional conflict over shared water resources.

To be sure, the central government has made several attempts to strengthen regional governance institutions in order to address interjurisdictional water resource management issues, especially those related to pollution. Joint MWR–MEP water quality bureaus were created within the WCCs in order to more effectively supervise provincial-level authorities within river basins, though many assessments indicate that they were largely ineffectual (Shen 2009). In addition, the 2008 Water Pollution Law mandated water resource planning on a regional rather than a jurisdiction-by-jurisdiction basis, in an effort to promote compliance (Wouters et al. 2004; Shen 2009). Recent attempts to develop interprovincial pollution compensation mechanisms reinforce the central government's attempts to improve interjurisdictional collective action in shared river basins (Xinhua 2012), and a system of "river chiefs," due to be established by 2018, promises to improve interjurisdictional coordination in shared river basins by making local political leaders explicitly responsible for such matters (Hou 2017). Nevertheless, environmental outcomes remain extremely poor; less than half of river water can be treated to the point where it is safe for drinking, and a quarter of surface waters are so polluted that they are unfit even for industrial use (Stanway 2012). Such conditions form the basis of disputes between subnational jurisdictions which, despite bearing responsibility for promoting economic growth, protecting the environment, and ensuring water supplies, have few institutional mechanisms to communicate, cooperate, or negotiate with their counterparts in neighboring jurisdictions. Within this institutional vacuum, conflict persists between levels of government, individual jurisdictions, and water-user groups.

Central Control, Local Discretion, and Competition for Water

China's economic reform strategy since 1978 has relied heavily on economic decentralization, and revenue and expenditure powers have largely been transferred to subnational levels of government, referred to as "cooking on separate stoves" (*fenzao chifan*, 分灶吃饭) (Li, Kenan, and Long 2015). Local governments have been granted a wide berth with respect to enterprises in their own jurisdictions, creating new revenue streams and fostering horizontal relationships outside the vertical state hierarchy (Oksenberg and Tong 1991; Oi 1995; Yang 2004). This decentralization process has created a deep symbiosis between local government officials and the enterprises within their jurisdictions. Subnational officials therefore tend to be deferential to the interests of major economic constituencies within their respective jurisdictions, often to the detriment of resource management and environmental protection objectives. This is especially true of the many municipally-owned enterprises that directly contribut to local government revenue. A news report on pollution in a large town in Anhui Province, for example, described a situation in which regulation is lax because "the government and the town's largest employer are all but indistinguishable" (Stanway 2012). Given these conflicting interests, local governments, which bear primary responsibility for environmental protection, have historically been highly reluctant to strictly enforce pollution regulations (Jahiel 1997). At the same time, local officials are often fiercely protectionist when it comes to securing natural resources for their respective jurisdictions (Wei 2000). Thus, despite the existence of potent control mechanisms binding subnational officials to central policy objectives, subnational officials frequently attempt to "cheat" national directives in order to promote localized economic interests (Moore 2017).

Such behaviors have led directly to a surprising number of interjurisdictional water disputes (*kua xingzheng quyu de shuishi jiufen*, 跨行政区域的水事纠纷) concerning both water quality and quantity issues. Perhaps even more surprisingly, given the political centralization of the Chinese state, these disputes are characterized by "a close relationship to local economies and local-benefit thinking on the part of some local governments" (*bendi liyi de kaolu*, 本地利

益的考虑) (Zhang 2009). The sectional divides familiar from the cases of the United States and India, in other words, are evident in China despite its centralization. Although there are no reliable data on the number of interjurisdictional water disputes in China, their number has by most accounts been continually increasing since 1978 (Li, Liu, and Huang 2010). Such disputes fall into several categories, including the maintenance of flood-control infrastructure, hydropower utilization, water withdrawals, and effluent flows (Zhang et al. 2010). Even the massive Three Gorges Dam was nearly derailed by a combination of factional-bureaucratic infighting and opposition from upstream provinces such as Sichuan (Lieberthal 1993). Among these sources of conflict, however, industrial point-source pollution is by far the most common cause of interjurisdictional water disputes in China, given upstream officials' "tendency towards transferring pollution costs to [the] lower reaches" (Wang 2007, 402). The growing number of such disputes stems both from both the incentive structure for subnational officials, which favors prioritizing economic growth in individual jurisdictions over cooperation with neighboring units, as well as from the lack of robust institutional mechanisms to facilitate such cooperation. The weakness of these provisions means that local authorities rarely engage in interjurisdictional cooperation, instead adopting a practice known as "shoveling only their own snow" (*zisaomenqianxue*, 自扫门前雪) (Peng 2009, 37).

The practice of promoting local economic interests at the expense of pollution control has led to a string of dramatic interjurisdictional water quality incidents. These include a 2001 Jiangsu–Zhejiang incident in which residents of a city in Zhejiang Province, tired of receiving heavily polluted water from an upstream neighbor in Jiangsu, sank old boats in the two cities' shared river, forming a dam that backed up the polluted water into Jiangsu Province; the 2005 Songhua River spill, in which 100 tons of highly toxic petrochemicals were released; a 2007 incident when pollution from Shandong Province disrupted water supplies to a city of 200,000 in downstream Jiangsu Province for some forty hours; a 2013 incident between Shanxi and Hebei regarding the release of the chemical amine; and incidents in both 2011 and 2014 between Guangdong and Guangxi (Green 2009; Li, Kenan, and Long 2015). The extent to which interjurisdictional pollution affects water quality in downstream jurisdictions is illustrated by a government investigation into pollution on the Zhangweixin River, which flows though Henan, Shanxi, and Hebei before emptying into the Bohai Gulf in Shandong Province. No less than 82% of the pollutants found in the Zhangweixin in Shandong were found to originate from outside the province (Li, Kenan, and Long 2015). As such figures suggest, China's hierarchical control mechanisms are no guarantee of a unified approach to basin governance or of subnational officials' willingness to subordinate their jurisdictions' individual interests to those of the basin as a whole.

Indeed, the sheer number of Beijing's attempts to prevent interjurisdictional disputes gives some indication of their persistence despite the center's best efforts. In 1990, the Comprehensive Utilization Plan of the Yangtze River Basin established a water management entity for the provinces of the Yangtze, though this mechanism was primarily for the purposes of building the Three Gorges Dam. In 2011, a similar mechanism was established in the Yellow River basin which brought together the Departments of Environmental Protection and Water Resources from the nine riparian provinces under the auspices of the YRCC. The mechanism, referred to as the "Yellow River Basin Emergency Pollution Incident Communication, Coordination, and Cooperation Mechanism," includes provisions related to pollution monitoring (Zhongguo minzhu tongmeng Guangdongsheng weiyuanhui 2016). In the Huai River basin, likewise, a Water Resources Protection Leading Group has been established to investigate, monitor, and propose solutions to pending pollution disputes (Li, Kenan, and Long 2015). Some similar reform attempts have also been made at subprovincial levels. In 2006, Guangdong Province issued regulations which require local governments to more carefully monitor pollution sources near municipal boundaries, establish administrative procedures for handling disputes, and require local officials to take responsibility for resolving disputes (Li, Kenan, and Long 2015).

A more systematic attempt to address interjurisdictional water conflict occurred in 2008, when the MEP issued instructions intended to prevent interprovincial water pollution disputes. The instructions stipulated that provinces strictly control point-source emissions from enterprises located near provincial boundaries and incorporate transprovincial impacts into environmental impact assessments, which would then be approved by the MEP. The instructions also stressed the importance of information-sharing, joint pollution monitoring by different provincial environment departments, "strengthening communication and coordination," and provided that if provinces could "genuinely not coordinate," the MEP could establish its own coordination mechanisms. In addition, the instructions stipulated that in the event of a dispute, the provinces involved must attempt to resolve it through mutual consultation. The central government was only to become involved upon application from the two disputing parties. If, with MEP assistance, the two parties still could not reach agreement, the MEP was directed to submit its own opinion on the dispute to the State Council for decision and approval. These instructions moreover clearly emphasize the link between pollution disputes and social stability, stating that "environmental protection departments at all levels should attach great importance to environmental problems in trans-provincial basins . . . and safeguard environmental security and social stability" (Ministry of Environmental Protection 2008).

At first, it seemed as though this new guidance from the central government would inspire a new era of interprovincial cooperation over shared waterways. In 2008, Yunnan, Guizhou, Sichuan, and Chongqing signed a "Southwest Regional Water Pollution Dispute" memorandum of understanding to establish a regional coordination mechanism. The same year, the environmental protection bureaus of several eastern cities, namely Hangzhou, Huzhou, Jiaxing, and Shaoxing, established a joint Office of Enforcement to conduct pollution monitoring and enforcement across municipal boundaries. By 2012, the joint effort had reportedly reviewed some three hundred enterprises and responded to ten interjurisdictional pollution incidents (Li, Kenan, and Long 2015). In 2009, Guizhou's Zunyi City established a similar interprovincial cooperation mechanism with Gulin County in neighboring Sichuan Province to handle disputes. At the signing ceremony, the city's party secretary pledged to "Strengthen inter-provincial cooperation, enhance friendship between the people of the two provinces, and promote exchanges" (Zhao and Wang 2009). But like their counterparts in the United States and India, Chinese officials quickly discovered that without persistent central government involvement, centrally-formulated attempts to encourage interjurisdictional coordination can quickly become warped by subnational politics.

Despite the 2008 reforms, interprovincial pollution disputes remained a significant political issue during the 2013 sessions of the National People's Congress and the China People's Political Consultative Conference (CPPCC). As a news report of the two recorded, "topics related to water pollution sparked heated debate in the two sessions," and a delegation from Anhui called for the urgent establishment of an interjurisdictional water pollution prevention and compensation mechanism that would be led jointly by the MEP and the MWR (Jiang 2013). Matters have scarcely improved since then; a 2015 news article reported that subprovincial interjurisdictional pollution disputes are "spreading like wildfire" (*cengchu buqiong*, 层出不穷) and that the issue was continuing to cause concern among members of the National People's Congress and the CPPCC (Li, Kenan, and Long 2015). A 2016 article posted by the Guangdong China Democratic League of the CPPCC likewise warned that "pollution at provincial boundaries is increasingly severe, and transboundary water pollution incidents have occurred." In order to meet the demands of future economic development in the region, the article argued, there remains an urgent need to establish an interprovincial coordinating body chaired by provincial governors (Zhongguo minzhu tongmeng Guangdongsheng weiyuanhui 2016).

Unlike in other countries, China's political opportunity structure does not permit reform coalitions to support such institutional collective action in shared river basins. In contrast to the other country cases examined in this book, the Chinese system places tight restrictions on civil society and nongovernmental participation. These restrictions are not total; forms of citizen activism have long been tolerated at local levels, especially in rural areas, for example, and the central government has encouraged non-governmental organization involvement in certain areas, such as monitoring compliance with pollution regulations (Li and O'Brien 1996). However, the Chinese Communist Party has always been careful to prevent such activity from crossing the line between independent voice and political or social collective action (Zhou 1993). Instead, the party has consistently maintained the principle that it is the sole legitimate political or social movement. The Chinese government accordingly employs a range of control mechanisms, including censorship, arrest, and detention, to neutralize nascent mobilization potential on the part of any non-Party organization. This limited political opportunity structure makes institution-building in China technocratic rather than collaborative and authoritarian rather than participatory. In the absence of collaborative and participatory institutions to support collective action, interjurisdictional conflicts characterize China's subnational hydropolitics in much the same way as they do those of the United States and India. Instead of preventing interjurisdictional conflict, China's hybrid of political centralization and fiscal decentralization simply alters how it unfolds, as the case of the Yellow River basin illustrates.

TECHNOCRACY AND AUTHORITARIAN INSTITUTION-BUILDING IN THE YELLOW RIVER BASIN

The Yellow River basin has always been central to China's economy and its politics, but since the foundation of the People's Republic in 1949 it has assumed even greater importance in the minds of China's leaders. CCP leaders like Mao Zedong and Zhou Enlai had fought many of their military campaigns in and along the Yellow and were intimately familiar with its recurring cycles of flooding and drought. Mao, Zhou, and their contemporaries accordingly developed a keen interest in hydraulic engineering and made water resource management a priority upon taking power. Almost immediately after founding the People's Republic in 1950, party leaders convened a conference on the theme of "turning the Yellow from a river of disaster into one of benefit" (*bian zaihe dao lihe,* 变灾河到利河) (Shuilibu Huanghe Shuili Weiyuanhui 2006, 82). This early investment of interest and attention from senior leaders ensured that the governance and management of the Yellow River basin would be a fundamentally technocratic elite project financed and administered principally by the central state. Yet despite making it clear that water resource management was a top priority of a centralized and authoritarian government, Beijing nonetheless acknowledged from the outset the critical importance of interjurisdictional cooperation by creating a powerful river basin management organization, the YRCC (*Huanghe shuili weiyuanhui,* 黄河水利委员会). The creation of the YRCC, administratively part of the MWR, reflected the interjurisdictional nature of most of the Yellow River basin's driving management issues. The Yellow River basin spans seven provinces and two autonomous regions and presently supports about 10% of China's population, most of its wheat crop, and approximately half of its coal production (Giordano et al. 2004, 5–6; Cheng 2008, 189). By convention, the riparian provinces are divided into regions referred to as the "lower," "middle," and "upper" reaches (*xiayou,* 下游; *zhongyou,* 中游; and *shangyou,* 上游, respectively) (see Figure 6.1).

Despite the existence of sectional interests in the basin, the political and economic importance of the Yellow River has from the outset molded a technobureaucratic approach to its management and governance. The CCP leadership quickly formulated two guiding policy objectives, first demonstrating the prowess of the new People's Republic by finally controlling flooding in the Yellow and second making the basin a showcase of socialist agricultural and industrial development policy, which they believed required centralized bureaucratic direction. An early official publication reflects the belief that a "thorough and comprehensive solution" to the problem of flooding and siltation would require "large state investments, the support of the masses, and all-round cooperation between government and people" (Teng 1955, 24–28). Accordingly, in 1954 the State Planning Commission commissioned a report on the comprehensive development of the Yellow River, which proposed the construction of forty-six dams along the whole

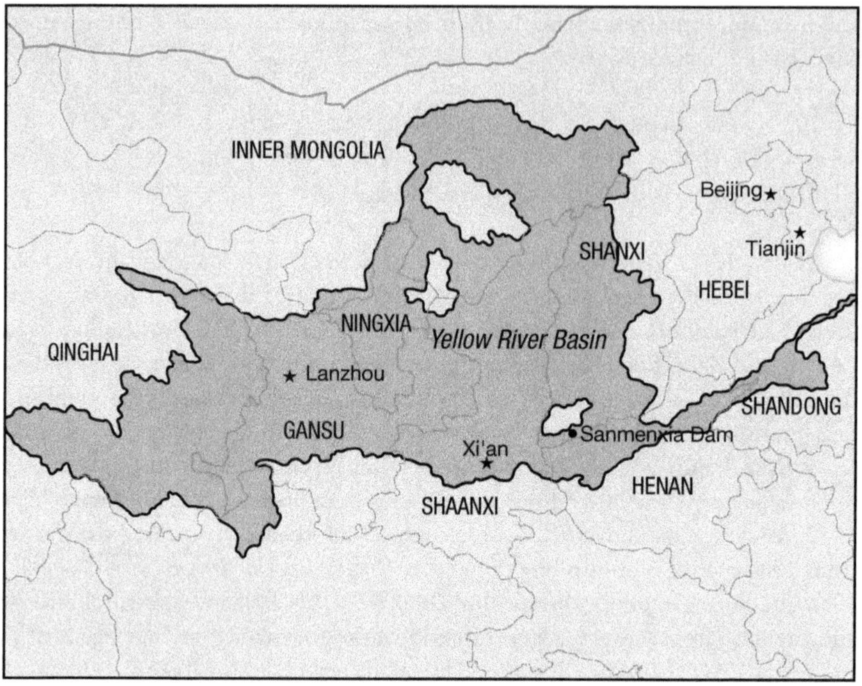

Figure 6.1 Physical and Political Features of the Yellow River Basin

length of the river for flood and erosion control, irrigation, and hydropower development. A YRCC history leaves little doubt as to the scale of these ambitions and the priority which the central government attached to them: "In short, we wanted to conquer the Yellow, and transform the Yellow Basin's natural condition . . . in order to satisfy with Yellow River resources the demands of socialist construction and the coming Communist transformation of the national economy" (Huanghe Shuili Weiyuanhui 1991, 121).

The importance China's leaders attached to the comprehensive development of the Yellow River simultaneously persuaded them to rely on a technocratic river basin management organization, much as similar ambitions had led to creation of the American Tennessee Valley Authority, the Indian Damodar Valley Corporation, and the French basin agencies. In 1950, the Ministry of Water and Power issued regulations reconstituting a hydropower development planning organization created under the Guomindang government, henceforth to be known as the YRCC. From the very beginning, this new organization, like the Tennessee Valley Authority and basin agencies, acknowledged the existence of subnational politics in the basin but sought to subsume them within a single decision-making structure. As its founding directive made clear, the newly reconstituted commission would act as a "basin-scale management institution" (*liuyuxing jigou*, 流域形机构), with the governments of the riparian provinces acting subject to its direction (Shuilibu Huanghe Shuili Weiyuanhui 2006, 83). The initial organization of the YRCC in the year 1950 reveals much about the manner in which relationships with the provinces were intended to be managed (see Figure 6.2). Particularly evident is the importance given to administrative

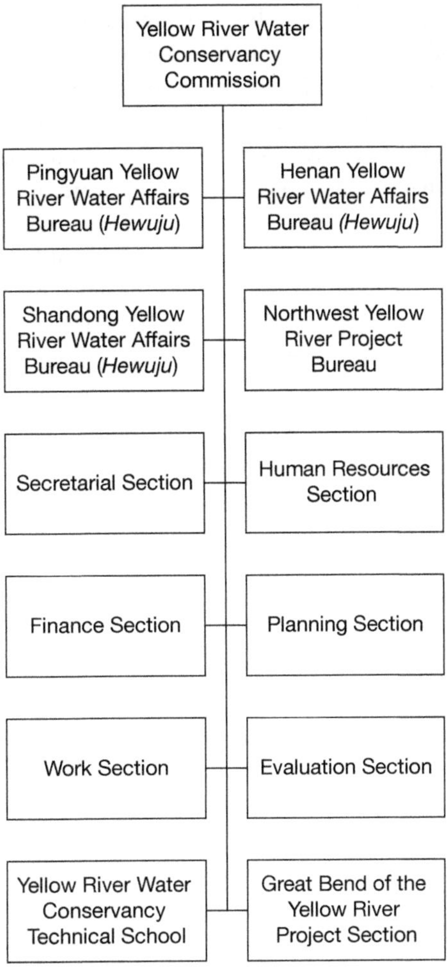

Figure 6.2 YRCC Organization Chart, 1950 (Adapted from Huanghe Shuili Weiyuanhui 1986, 390)

jurisdictions subject to flooding, with both Shandong and Henan Provinces being granted their own dedicated bureaus. Project management bureaus were created for planned major developments in the middle and upper reaches, along with a range of administrative offices and sections. Further worthy of note is the creation of a technical school, which would play a significant role in future debates over site selection and construction of dams in the middle and upper reaches. Thus constituted, the YRCC in 1950 represented a genuine attempt at basin-scale governance but also a fundamentally technocratic organization that gave little priority to addressing the interests of different constituencies within its area of responsibility.

During the YRCC's early years, these technocratic ambitions nonetheless required extensive cooperation with local authorities, especially the province-level officials to whom policy implementation was often delegated. In order to promote the central government's flood-control policies, flood-control sections (*zhihe banshichu,* 治河办事处) were established under province-level river affairs bureaus (*hewuju*, 河务局) in both Shandong and Henan (Huanghe Shuili Weiyuanhui 1995, 386–99). Such central–local cooperation was crucial to tackling the centuries-old scourge of flooding in the basin. In 1951, the *People's Daily* carried a report lauding the close cooperation between the YRCC, the CCP, provincial government officials, and Yellow River Affairs Bureaus throughout the lower reaches, which together marshaled some 200,000 peasants to aid in the critical, annual exercise of repairing and maintaining dikes along the river (Xinhua 1951). By 1955, every riparian county and township in Shandong had established a Dike Protection Committee (*hudi weiyuanhui*, 护堤委员会) chaired by the second-highest-ranking official in the county, thereby illustrating their importance (Huanghe Shuili Weiyuanhui 1995, 346). Despite the extent of this delegation to local authorities, the center also took pains to develop institutional control over local authorities, particularly regarding politically sensitive tasks such as flood control. The State Council maintained a special staff office to oversee the YRCC and established a separate National Flood-Control Headquarters with branches in each province which reported directly to the central level (Greer 1979; Shuilibu Huanghe Shuili Weiyuanhui 2006).

This emphasis on centralized bureaucratic management, albeit through a regional-scale governance institution, was strengthened by a 1952 State Council decision calling for "basin-wide planning" (*liuyu guihua*, 流域规划) (Shuilibu Huanghe Shuili Weiyuanhui 2006, 83–87). The central government also created a tiered system of responsibility for river basin development, stipulating that basins larger than 10,000 square kilometers, as well as important tributaries, were to be developed directly by the center, while provinces were to be responsible for developing smaller basins. In a further testament to the emphasis placed on basin-scale institution-building, in the Yellow and Yangtze River basins functional agencies like the Ministry of Agriculture were ordered to cede operational authority to the YRCC and its counterpart, the Yangtze River Planning Office (U.S. Joint Publications Research Service 1965). These institutions thus represented a triumph of the comprehensive, regional management model of river basin governance.

Yet even under this centralized structure, central–local conflict frustrated central policy, as illustrated during construction of the Sanmenxia Dam, the most significant storage reservoir on the Yellow in the early years of the People's Republic. The dam, begun in the late 1950s at a site in Henan Province, was intended to simultaneously generate hydropower and control siltation in the lower reaches, and it enjoyed Mao's personal support and patronage. Hydraulic engineers assigned to the project cautioned that achieving these multiple objectives simultaneously would require "all players working together" (quanju yipanqi, 全局一盘棋) in order to control soil erosion, which threatened to clog the dam's hydroelectric generators (Huanghe Sanmenxia Shuili Shuniu Zhibian 1993, 79). However, while construction of the dam remained a closely held central government priority, the critical task of preventing soil erosion was delegated to provincial authorities, who fell considerably short of the center's expectations. A 1957 *People's Daily* article admonished the provinces of the upper and middle reaches for their "extremely sluggish progress" on soil conservation work (Zhou 1957). A similar charge was repeated in 1959, when another *People's Daily* editorial singled out the middle reach provinces, urging each to improve its efforts by formulating individual strategies and convening special leading groups (*lingdao jigou*, 领导机构) to improve policy coordination at the provincial level (*Renmin Ribao* 1959).

Mutual frustration between elements of the central bureaucracy and between central and local levels of government flared further in the mid-1960s, when the failure of provincial erosion-control efforts caused sedimentation rates at the newly built Sanmenxia to rise far more quickly than had been predicted. Within four years of commencing operation, sediment had filled half the reservoir, forcing the hydropower plant to be taken out of operation (Wang 2002). In 1964, Zhou and the State Council convened a special meeting on the Sanmenxia issue that laid bare a fierce debate between MWR engineers, who favored additional soil erosion-control efforts, and Maoist acolytes, who supported the construction of additional sluice gates in order to alleviate pressure on the dam by allowing more sediment to pass through (*Renmin Ribao* 1974; Huanghe Shuili Weiyuanhui 1991, 161–62). Though seemingly technical, this debate also carried sectional implications: larger sluice gates, while favored by officials in the upstream provinces, would increase the threat of flooding downstream, provoking the vociferous opposition of downstream provincial officials. Zhou himself complained that this "local self-interest" prevented agreement on resolving the Sanmenxia Dam issue (Huanghe Shuili Weiyuanhui Huanghezhi Zongbianjishi 1986, 388; Shuixinxi Wang 2011). In the face of such opposition, Zhou was forced to compromise, establishing a "four-province-conference principle" (*sisheng huiyi yuanze*, 四省会议原则) under which releases from Sanmenxia would be determined on the basis of consultation between Shanxi, Shaanxi, Henan, and Shandong Provinces (Zhang 2008, 269).

These sectional politics, evident even in the midst of the Cultural Revolution, also frustrated the central government's development objectives in the upper reaches of the Yellow River. In the early 1950s, Qinghai Province prepared a set of "recommendations on the initial plans for the upper reaches of the Yellow River basin" calling for no fewer than six sets of terraced dams, most of which were later incorporated into the 1954 Yellow River comprehensive development plan (U.S. Joint Publications Research Service 1965). One of these, a proposed dam at a site called Daliushu near the border of Gansu Province and the Ningxia Hui Autonomous Region, became the subject of a long-running interjurisdictional dispute. Despite the 1954 plan's inclusion of a dam within Gansu, in 1958 the Ningxia Communist Party issued the "Opinion on the Development of the First Stage of Heishanxia and Daliushu," calling instead for a single high dam at a site within Ningxia instead so as to irrigate more farmland within its borders (Huanghe Shuili Weiyuanhui 1991, 268–69). The Gansu provincial government in turn objected to this proposal, insisting that the original dam was needed in order to complement several planned irrigation projects in the province (Gansu Shui Han Zai Hai Bianweihui 1996). Despite a central government directive to the two provinces to resolve their differences and produce a joint recommendation, the issue developed into a major interprovincial dispute during the 1960s as both parties sought to lobby elements of the central government bureaucracy they viewed as favorable to their respective positions (Huanghe Shuili Weiyuanhui 1991).

In 1966, the MWR's Northwest Research Institute, which was in charge of hydropower development and engineering in the upper reaches, took up Gansu's cause, issuing a report which claimed that the Ningxia Dam might induce earthquakes. A subsequent 1972 report repeated this conclusion, finding that while a dam constructed at Daliushu would have a greater water storage capacity, the cost and project-construction risk would be higher, again favoring construction of dams within Gansu. In a 1974 meeting convened by the MWR, however, Ningxia's representatives fought back, declaring that "Ningxia comrades demand consideration" of water projects within their jurisdiction. The following year, Ningxia's delegate to a meeting of upper reach provinces gave an impassioned speech in which he urged for approval of the Daliushu project on the grounds that increasing irrigation and food production was more important than hydropower. A few years later, in 1978, the Ningxia government pressed the attack by sending a letter to the MWR and the vice-premier arguing that a dam at Daliushu would "Utilize Yellow River water resources to transform the northwestern drought belt, and be the best means to develop agriculture and forestry." Sensing the political sensitivities involved, the vice-premier opted not to directly favor one side over the other and instead simply ordered the MWR and the upper reach provinces to "earnestly study" and resolve the dispute (Huanghe Shuili Weiyuanhui 1991, 270–71). Rather than enter into the complex sectional tensions involved, Beijing, like Washington and New Delhi in very different contexts, chose to stay above the fray.

As in cases of interjurisdictional conflict in the other cases examined in this book, the reluctance of national political leaders to arbitrate interjurisdictional disputes simply created a vacuum in which subnational leaders competed by lobbying different elements of the central bureaucracy. In 1981, the MWR's Northwest Research Institute issued a report in response to this renewed lobbying, restating the danger of earthquakes at Daliushu, thereby implicitly strengthening Gansu's opposition to the project. Ningxia reacted in turn with a letter to the State Council alleging that building a dam at Gansu's favored site would endanger agriculture throughout the upper reaches. "If you build a reservoir at Xiaoguanyin," the Ningxia government warned, "the principal thing it would do is to resolve the power production problem, but you can say it does nothing to alleviate the drought-affected situation of Ningxia, Inner Mongolia, northern Shaanxi, and the surrounding counties of Gansu" (Huanghe Shuili Weiyuanhui 1991, 271). Ningxia also secured the support of neighboring Inner Mongolia, whose government in 1981 sent a message to the State Council stating that "We agree with the opinion of the Ningxia government" and that "after completion of the Longyangxia project [in Qinghai], we should turn our attention to beginning the Daliushu project." This alliance between Inner Mongolia and Ningxia was sufficiently strong to persuade Gansu to modify its own position, and the provincial government soon notified the State Council that it supported a two-stage project involving construction of dams at both Xiaoguanyin and Daliushu (Huanghe Shuili Weiyuanhui 1991, 272).

With the upper reach provinces seemingly and finally in agreement, the MWR convened a work conference in 1981 to resume planning on development of the upper reaches. Once again, however, the meeting split into sectional factions, with one favoring the development of hydropower at a site farther upriver and the other favoring water storage and irrigation, with a reservoir to be built at Daliushu. The meeting failed to reach consensus, and the MWR, reluctant to resolve the dispute, once again quietly shelved the project (Huanghe Shuili Weiyuanhui 1991). Even so, Gansu and Ningxia kept up their respective lobbying campaigns. Shortly after the 1981 conference, Ningxia officials commissioned several studies questioning Gansu's two-stage dam proposal. One resulting report, prepared by the well-respected Chinese Academy of Sciences, concluded that while the two-stage proposal put forward by Gansu had some merit, on a strict cost–benefit basis the Daliushu dam represented a better investment for the central government (Huanghe Shuili Weiyuanhui 1991). The Ningxia government rather unsurprisingly registered its whole-hearted agreement with the report's conclusions, and its position was further strengthened by several additional reports which discounted the risk of earthquakes at Daliushu (Huanghe Shuili Weiyuanhui 1991). But even these additional studies failed to persuade the MWR's Northwest Institute, which instead of concurring with these findings announced in 1984 that it endorsed Gansu's two-stage proposal (Huanghe Shuili Weiyuanhui 1991). By the late 1980s, therefore, both Gansu and Ningxia could point to authoritative scientific and technical analyses to bolster their respective positions in the Daliushu dam dispute. This raised the political costs to the MWR of choosing one side over the other, which in turn ensured that the dispute continued into the 1990s. Once again, subnational politics had thwarted even Beijing's determined agenda in the Yellow River basin.

As the interjurisdictional dispute over Daliushu illustrates, after 1978 the regional governance framework established by the YRCC faced a number of strains. As the number of water users proliferated along with economic growth, new conflicts emerged among them that the sclerotic YRCC proved incapable of resolving. One of these concerned the growing challenge of water pollution. Pollution in the lower reaches was sufficiently acute that by 1978 Henan Province had established a dedicated water quality monitoring center, while intensive exploitation of fragile soils increased the problem of sedimentation throughout the middle reaches (Huanghe Shuili Weiyuanhui Huanghezhi Zongbianjishi 1986; Gao 1981). Yet even as these new challenges emerged, the YRCC remained a closed, centralized bureaucracy that had become even more technocratic over time. This evolution is illustrated by the YRCC's organization as of the year 1984 (see Figure 6.3). The creation of dam and reservoir management bureaus reflects the commission's expansion into the business of directly operating important waterworks along the Yellow, but its operation of a hospital, and even a middle school, suggests its drift into a sprawling, complex bureaucracy ill-suited to the challenges of adaptive and cooperative management.

As an organization, the YRCC was also prone to conflict with other bureaucratic agencies. In 1979, a foreign expert commented that "Administrative problems arose because the YRCC superseded provincial governments, but it [also] came into conflict with other agencies of the central government.... Its various project

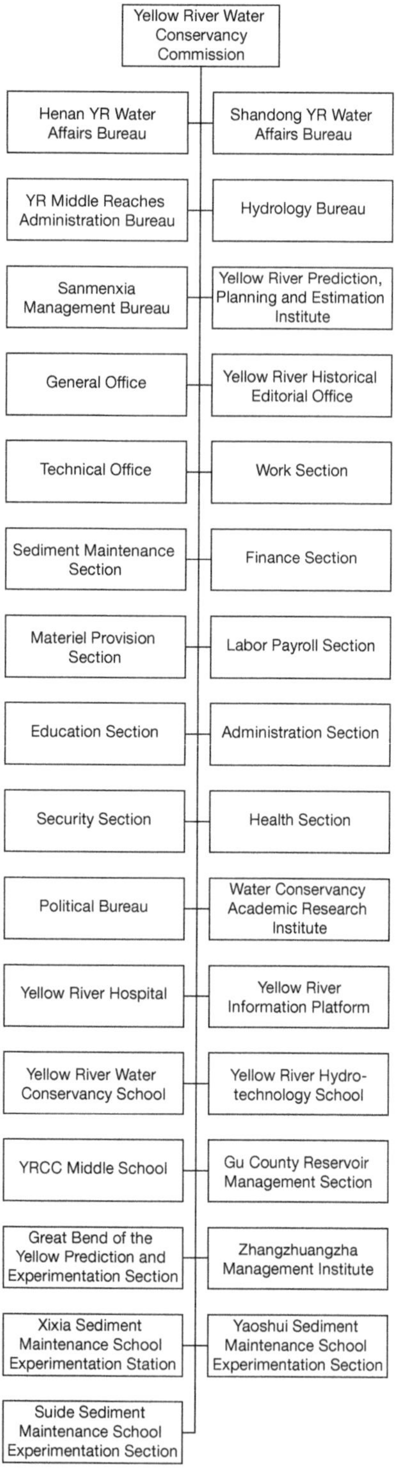

Figure 6.3 YRCC Organization Chart 1984 (Adapted from Huanghe Shuili Weiyuanhui 1986, 390)

bureaus were little more than advisory adjuncts to project bureaus of the Ministry of Water Conservancy or the State Council" (Greer 1979, 75). The deficiencies of the centralized bureaucratic model of river basin governance are even more powerfully illustrated by a brief period in which the YRCC temporarily adopted a more cooperative and inclusive form of decision-making. In the years following the inauguration of the reform and opening policy in 1978, the old central planning apparatus was dismantled, and almost immediately considerable powers began to flow to subnational and nongovernmental actors. This process had begun to exert an effect on water resource management in the Yellow River basin by 1981, when the YRCC granted the Shandong and Henan Water Affairs Bureaus full operational autonomy. Subsequently, both entities began to rapidly accumulate capacity and responsibility: the Shandong Bureau alone soon employed some ten thousand people (Huanghe Shuili Weiyuanhui 1995; Li 1999). A few years later, the YRCC announced that it would break entirely with Maoist-era practice by replacing its long-standing direct administrative control of local dike protection and flood-control committees with a long-term contract model for operation and maintenance. Instead of directly providing infrastructure engineering services, the YRCC would simply become a "supervisory" (*guanli*, 管理) entity (Li 1987).

Amid this new, more collaborative mode of management, the YRCC proclaimed that "all parts of the river are intimately related" (*jinmi guanxi*, 紧密关系). Subsequently, it focused more intently on formulating a regional solution to another growing challenge for the basin, namely allocation among the riparian provinces, many of which were experiencing substantially increased demand for water in a period of rapid economic growth (Shuilibu Huanghe Shuili Weiyuanhui 2006, 300). The commission's new spirit of cooperation was illustrated at a 1984 conference attended by all riparian provinces to agree on a new interprovincial water quantity allocation formula. Brokered by the YRCC, the provinces soon reached consensus on an apportionment of the Yellow's annual flow, based on their projected future water requirements (Li 2010). The resulting Yellow River Water Quantity Allocation Plan apportioned the waters of the Yellow roughly in proportion to the population distributions of the upper, middle, and lower reaches and received State Council approval in 1987 (see Table 6.2) (Guowuyuan 1987).

This exercise in cooperative management initially proved successful in preventing interjurisdictional conflict despite rapidly increasing water demand. On the basis of what all parties viewed as a strong consensus on the allocation of Yellow River flows, a number of new development plans were formulated, including the "Revised Yellow River Administration and Development Plan," which was adopted at a 1988 conference. In a reflection of the YRCC's new emphasis on collaboration and consensus, only after receiving the concurrence of provincial officials that the plan was "basically correct" in addressing their interests were implementing instructions for the plan issued (Shuilibu Huanghe Shuili Weiyuanhui 2006, 298–99).

However, by the mid-1990s the Yellow River Quantity Allocation Plan had come under increasing pressure as a result of skyrocketing water demand, thereby creating yet another challenge for interjurisdictional collective action within the basin. Some indication of the scale of this challenge is given by the fact that in the middle reaches, for example, industrial water use grew from 0.38 to 1.6 billion m^3

Table 6.2 Yellow River Available Water Quantity Allocation Plan (Hundred Million Cubic Meters)

Area	**Annual water quantity**
Qinghai	14.1
Sichuan	0.4
Gansu	30.4
Ningxia	40.0
Inner Mongolia	58.6
Shaanxi	38.0
Shanxi	43.1
Henan	55.4
Shandong	70.0
Hebei and Tianjin	20.0
Total	370.0

Adapted from Guowuyuan (1987).

from 1988 through 2004, while domestic consumption increased from 0.28 to 1.37 billion m^3 over the same period (Yang and Jia 2008). In addition, much of the increased water demand, especially for agriculture, was met by increasing groundwater abstraction, which had not been accounted for in the plan but which diminished the flow in the main stem of the Yellow River. As a result of these pressures, a phenomenon known as "desiccation of the Yellow" (*Huanghe duanliu*, 黄河断流) became an acute problem. From 1995 to 1998, the Yellow failed to reach the sea for about 120 days each year and did not even reach Shandong Province, devastating riverine ecosystems in the lower reaches (Giordano et al. 2004). Yet instead of repeating the successful model of central–local cooperation which had led to the 1987 agreement, the Chinese central government returned to the more familiar model of centralized bureaucratic control.

The center's primary response to the desiccation problem was the "Yellow River Framework Plan for Administration and Development," issued in 1997, which strengthened the principle of centralized bureaucratic control over water resource allocation in the basin. In particular, the new plan stressed "overall control, active administration, and strict supervision" (E 1999). On the strength of these provisions, the central government aggressively expanded its direct operational control over water resource allocation within the Yellow River basin throughout the late 1990s. In 1998, the MWR, acting under orders from the State Council, issued new regulations in response to the desiccation problem which gave the YRCC authority to assign provinces monthly water withdrawal quotas, which were then supposed to be enforced through strict reporting requirements (Guojia Fazhan Jihua Weiyuanhui/Shuilibu 1998). The system further created a strict, hierarchical control process whereby annual water usage requests had to be submitted first by provincial governments to the YRCC and then from the commission to the central MWR in Beijing for final approval (Li 1999; Shen 2012). Administrative water allocation orders issued by the MWR would then be transmitted to the state-owned dams and reservoirs, which control releases of water from the main stem of the Yellow River (Yang and Jia 2008). By adopting this system, the YRCC and its superiors in the MWR signaled a return to the technobureaucratic approach to basin governance that prevailed before the 1980s.

This renewed form of centralized bureaucratic management has had a predictable effect, successfully halting the desiccation of the Yellow River, its primary and proximate objective, but failing to address more diffuse governance challenges that require greater interjurisdictional cooperation. Chief among these challenges is that of water quality, which has deteriorated sharply as a result of rapidly growing industry and urbanization. In the Lanzhou stretch of the river alone, some 500 million tons of pollutants were released annually by the early 2000s (Gansu Dizhi Kuangchan Bao 2004). Given the scale of this pollution, only coordinated implementation of pollution control measures could address the issue. A 2000 Shaanxi Provincial Government report, for example, noted the need for "close cooperation" between administrative jurisdictions in order to protect Yellow River ecosystems (Shaanxi Zhengfu 2000, 16). Yet the YRCC structure has instead relied on increasing control over subnational authorities, rather than promoting interjurisdictional collaboration and participation. In 2006, the YRCC announced a 400 million RMB "digital Yellow River" project (*Shuzi huanghe*, 数字黄河) that relies on a network of sensors to enable the YRCC to continuously monitor water quality and water use within local jurisdictions, allowing effective tracking of local water conditions in real time (Xinhua 2006). Instead of collaboration, the YRCC turned to a form of computerized coercion to ensure compliance with central policy.

The underlying flaw in this technobureaucratic approach has been illustrated by the fact that, instead of consenting to this exercise of central control, China's subnational leaders have preferred to flaunt it. Subnational authorities have responded to this hydraulic panopticon by simply refusing to obey rules they felt were harmful to their jurisdictions. In reference to the 1998 regulations, a recent YRCC report admitted that "In implementing the Yellow River water quantity regulations, there exist some localities which do not put into practice the water quantity allocation and dispatch plan, and exceed the allocation limits in using water," while others use "administrative measures to interfere with normal management" (Huangshuihui 2011; Henansheng Huanghe Hewuju 2011). These cases of noncompliance serve to highlight the fact that by not incorporating subnational political interests in decision-making, China's water resource management institutions have frequently alienated the local officials whose cooperation is essential to address increasingly interjurisdictional collective action problems.

As a result, interjurisdictional conflict has continued to undermine Beijing's regional development objectives within the Yellow River basin. In the early 2000s, the long-standing Gansu–Ningxia dispute re-emerged, fueled by Beijing's intention to formulate a new regional development strategy for northwestern China. Both regions again attempted to lobby elements of the central government to finance projects in their respective jurisdictions. In 2001, as part of a renewed push to secure its inclusion in the new development strategy, Gansu contracted the China Hydropower Consulting Group to conduct a study in support of its proposed two-stage dam. Gansu provincial leaders appealed to Beijing's obsession with maintaining social stability by charging that Ningxia's proposed dam at Daliushu would cause "grave social problems" by inundating Gansu's most productive agricultural land, thereby stoking peasant unrest (Gansusheng Heishanxia shuidian kaifa gongcheng lingdao xiaozu bangongshi 2006).

Ningxia, for its part, began organizing tours for visiting State Council delegations to impoverished villages, which it claimed would benefit from its proposed high dam. The Ningxia government also created a website festooned with warnings from supposed experts that Gansu's favored two-stage dam would represent a "long-term error in decision-making" (Ningxia Xinwen Wang 2008). Although the State Council has subsequently called for "accelerating the timeframe for construction" of projects on the upper Yellow River, the Gansu–Ningxia dispute has so far prevented realization of its regional development plan for the Yellow River basin (Zhongguo Jiancai Wang 2011). The Gansu–Ningxia conflict thus continues to illustrate that even in a highly centralized, authoritarian political system, subnational politics can still thwart the objectives of the central government.

For over fifty years, the control, development, and management of the Yellow River basin have been important undertakings for the People's Republic. As a result of these efforts, a centralized structure and technobureaucratic approach has been developed centered on the YRCC, which directly controls water resource allocation and most water-related planning activities. This institutional approach has achieved substantial progress in addressing some river basin governance challenges, especially siltation and flooding. Some measurements suggest that siltation has been reduced by at least 300 million tons annually, for example, while the threat of flooding along the Yellow, once one of China's greatest natural hazards, has been largely eliminated thanks to the extensive construction and maintenance of flood-control works (Giordano et al. 2004). In addition, close YRCC supervision of basin water use has halted the desiccation of the Yellow's lower reaches. The problem of water scarcity has been further alleviated through the gigantic South–North Water Transfer Project, which has eased water availability concerns in the middle and lower reaches (Moore 2014b). Nonetheless, in recent years the failures of the technobureaucratic model have become increasingly apparent in the face of more diffuse basin cooperation problems.

These failures are especially evident with respect to water quality and environmental protection. Although the YRCC's administrative measures have halted desiccation, for example, they have failed to provide sufficient environmental water flows or to markedly improve water quality (Yang and Jia 2008). Even more importantly, the YRCC has failed to accumulate the political legitimacy necessary to prevent conflict between basin water users. As an International Water Management Institute report concludes, "From its founding to the present, the YRCC has not had power to act as a true basin authority . . . national, provincial and local agencies have been able to place varying claims and priorities on the river's resources and development" (Giordano et al. 2004, 35). Consequently, attempts to address basin-scale water quantity and quality management challenges remain plagued by both central–local and interjurisdictional conflicts. It is telling that the YRCC has been most successful when it has included subnational actors in decision-making and has fared worst when it has excluded subnational political leaders from decision-making and taken a dictatorial rather than collaborative approach to management of the Yellow River basin's increasingly strained water resources.

COUNTRY CASE CONCLUSION

China is virtually alone among the world's large and populous nations in not employing some form of political decentralization and instead relying on a hierarchical system in which political actors at subnational levels are tightly bound to the central government. But at the same time, China is one of the world's most economically decentralized countries, and subnational officials have considerable autonomy in deciding how centrally determined policies should be implemented. These distinctive institutional features of the Chinese political system have shaped its subnational hydropolitics in several ways, as summarized in Table 6.3. First, the hierarchical and authoritarian structure of governance in China has confined bargaining between political interests to informal negotiation between elites rather than contests in the formal political arena. In comparison with the other countries examined in this book, interjurisdictional conflicts and disputes in China are therefore pursued in a more informal manner. Instead of the open antagonism and litigation which marked the Colorado case, for example, interjurisdictional conflict in the Yellow River basin has been marked by the subtle manipulation of central government development policies and lobbying of central bureaucracies. Second, although China has established regional river basin governance institutions, it has not structured them to represent or accommodate subnational political interests; and consequently, they function more as instruments of central control rather than cooperative management. Because these institutions bear little legitimacy in the eyes of subnational officials, they are frequently ignored, and local governments simply refuse to adhere to water quantity allocations or pollution control regulations. Third and finally, the diversification in water uses as a result of rapid economic growth has placed considerable strains on China's hierarchical and technocratic approach to basin governance. Although the capacity of the central government to monitor and enforce local compliance is increasing, the costs of ensuring compliance are rising in tandem, undermining their potential for continued effectiveness. These features of the Chinese case suggest that while some

Table 6.3 Summary of Subnational Hydropolitics in China

Factor	Characteristics	Effect on Interjurisdictional Collective Action
Decentralization	• Distinctive combination of political centralization and fiscal-economic decentralization • Powerful water management bureaucracies like YRCC, but hierarchical structure means often in conflict with one another	• Individual jurisdictions have strong incentive to pass costs of water resource challenges (pollution) on to neighbors • Different bureaucracies like MWR and NDRC ally with provinces to contest power and resources between themselves
Sectional identity	• While sectional identity is limited, economic decentralization links economic self-interest of political leaders to jurisdictions	• Subnational leaders have few incentives to cooperate with their neighbors in the absence of central pressure or regulation
Political opportunity structure	• Very little formal third-party participation in political system, though opportunities for individual and some organizational input in environmental issues is increasing	• Third parties have had effectively no role in shaping interjurisdictional collective action

MWR, Ministry of Water Resources; NDRC, National Development and Reform Commission; YRCC, Yellow River Conservancy Commission.

degree of decentralization is necessary for subnational hydropolitics to occur, it need not be of the political kind. The fact that in China interjurisdictional water conflict occurs under fiscal and administrative, rather than political, decentralization merely alters the ways in which it is pursued.

At the same time, the case of China shows that subnational hydropolitics can hinder centrally formulated policy priorities even in a politically centralized system. Despite Beijing's tight control over decision-making, subnational officials enjoy considerable practical autonomy as a result of extensive economic, fiscal, and administrative decentralization. Within this more limited sphere of autonomy, China's subnational leaders pursue individual interests that often harm neighboring jurisdictions. Economic decentralization has created considerable competition between provinces, especially those in coastal versus inland areas. Geographically concentrated water users in less developed inland areas often have higher water demands and pollution discharges, which subnational officials are often content to let flow unimpeded downstream. These divergent interests relate to the distinctive economic development strategies adopted by different jurisdictions, but they are not explicitly linked to sectional identity, which the Chinese system explicitly suppresses. Interjurisdictional water conflict in China is therefore neither as profound nor as persistent as that in the United States or India. While Beijing can be just as frustrated by subnational hydropolitics as Washington or New Delhi, it has proven more able to overcome them simply by deploying its greater coercive powers. Beijing's central control and integration mechanisms, not least the CCP's control over subnational political appointments, assure that, when it chooses to do so, the center can enforce its priorities upon the local level of government. Yet the ability to compel collective action has not translated into an ability to tackle complex, adaptive management challenges like water quality and allocation. Instead, the case of China highlights the limitations of a technobureaucratic approach to water resource management and the necessity of collaborative and participatory management in response to changing circumstances.

These limitations also reinforce the key role of political opportunity structure, which in China has prevented the emergence of reform coalitions to advocate a more collaborative and participatory approach to managing interjurisdictional river basins. China's approach to institution-building has relied on a centralized bureaucratic model that, while regional in scale, does not include or accommodate the interests of water users at the subnational level. As in other cases explored in this book, this failure to bridge national and subnational levels of government has created substantial central–local and interjurisdictional conflict between riparian water users. However, the Chinese case is distinctive in featuring very few points of access for civil society organizations, while the bridging role of national political leaders is constrained by the authoritarian nature of the Chinese political system. Even as the central government has attempted to increase its leverage over subnational officials, conflict between water users has undermined its economic development and environmental protection objectives. Indeed, the case of China illustrates more clearly than any other that preventing subnational conflict over water relies not on the mere existence of a regional governance institution but rather on the degree to which it promotes collaborative and participatory decision-making. This point is further illustrated by the case of France, where a reform coalition played a crucial role in the development of collaborative and participatory river basin governance institutions that have prevented interjurisdictional conflict while successfully adapting to emerging management challenges.

NOTE

1. This chapter is derived in part from articles published in *The China Quarterly*, September 2014 (Moore, S., "Hydropolitics in China: The Pursuit of Localized Preferences in a Centralized System"), Copyright 2014 Cambridge University Press, reprinted by permission (reproduction by the author of a single journal article), and *The International Journal of Water Resources Development* 2017 (Moore, S., "China's Domestic Hydropolitics: An Assessment and Implications for International Transboundary Dynamics"), available online: http://www.tandfonline.com/doi/abs/10.1080/07900627.2017.1313157.

7

Elites, Civil Society, and Inclusive Institution-Building in the Republic of France

The Republic of France is in many ways the archetype of the centralized, unitary state, and its political institutions contrast sharply with those of the federations of India and the United States. Following the Revolution of 1789, the new republic undertook a series of political reforms intended to strip power from the landed nobility and vest it instead with a new set of egalitarian institutions, the basis of which was both centralization and uniformity. The revolutionaries believed that "justice requires the republic to be one and indivisible" (Berger 1974, 8). Inherent to this new model was a concentration of political authority, as well as political, legislative, and judicial powers, in the hands of the central government. In contrast to more decentralized and federal political systems, the French system is intended to tightly bind officials at both central and local levels and to minimize conflicts between them. Consequently, a defining feature of French political institutions is the relative cohesion of elite decision-making. According to one prominent observer, France "provides the prime example of a highly coherent administration, whereas the United States and Switzerland constitute the typical cases of lack of such coherence" (Kriesi 1995, 171). However, during the past thirty years even the French state has become more decentralized, and powers and responsibilities for some policy areas, including water resource management, have been devolved to regional governments. In comparative perspective, the outstanding feature of the French political system is in fact the presence of strong regional governance organizations, including several organized around river basin boundaries, that are among the world's most successful interjurisdictional management institutions.

France's system of river basin governance organizations, called "water agencies" (*agences de l'eau*), is by many accounts the most collaborative and participatory in the world. A global survey of river basin governance institutions concludes, for example, that the French system "is remarkable for its longevity, in how it tries to formalize representation . . . and perhaps most important, how it has attempted local and decentralized water management within the centralist state tradition in France" (Delli Priscoli 2007, 17). France's water agencies have served as the model for a core provision of the European Union's Water Framework Directive, which requires all member states to establish governance institutions around the topographical boundaries of major river basins, as opposed to national political jurisdictions (Bouleau et al. 2009). The present network of water agencies evolved from a set of previous institutions called "basin agencies" (*agences des bassins*), which were first established in the 1960s. The water agencies have since played a central role in dramatically improving water quality, incorporating relevant nongovernmental stakeholders in water resource management and policymaking, and largely preventing conflict among water users. The present system of six regional water agencies covers the entirety of metropolitan France as well as several French overseas territories. The puzzle of the French case is therefore why and how the development of collaborative and participatory decision-making at the basin scale proved so successful and so comprehensive in France compared to the other countries examined in this book. This chapter examines how the water agency system arose and how it has evolved over time.

In many ways, the evolution of the French water agencies presents a distinct contrast with the development of the river basin commissions established in the United States under the Water Resources Planning Act. Like the river basin commissions, the water agencies were established in the mid-1960s amid postwar interest in regional economic planning. Both the river basin commissions and water agencies were intended to embody a new spatial scale in governance, namely that of the region rather than the municipality or province, in the belief that water resource development required a holistic approach that would be best served by creating a new institution rather than relying on implementation by existing municipal and local jurisdictions. Like their American counterparts, the water agencies also arose from a period of sustained lobbying by third-party bridging organizations, especially national political leaders and environmentally oriented organizations. Just as in the case of the United States, French civil society groups, by advancing an environmentally oriented political agenda, helped to upend long-standing bureaucratic interests opposed to collaborative and participatory water resource management. Unlike in the American case, however, in France national rather than subnational leaders proved the dominant force in shaping institution-building. Indeed, in the French case the creation of regional governance institutions was initially viewed by national leaders as a means of thwarting the growing power of local officials. The outstanding feature of the French water agencies is that they both survived and thrived by obtaining widespread political legitimacy—a feat they accomplished by becoming genuinely participatory and simultaneously accommodating the interests of national, subnational, and civil society constituencies. The centralized French system made it easier for third parties and political elites to create a reform coalition that successfully overcame opposition from subnational interest groups.

COUNTRY CASE INTRODUCTION

By the middle of the twentieth century, water pollution had become a major political issue in France. Much like the Izaak Walton League's survey of water pollution in America had drawn attention to inadequate local pollution legislation a few decades earlier, the political debate over water quality in France soon centered on the failure of local officials to stringently implement water quality regulations. Embarrassed by what was portrayed by the national media as a loss of control over local authorities, the French government modified a system of regional planning adopted after the Second World War to create a new set of water resource management organizations based on river basin boundaries, the basin agencies (see Figure 7.1). The 1964 Water Law created one such agency for every major river basin in metropolitan France, and their powers and responsibilities have been progressively expanded through subsequent legislation. Though initially opposed by local governments as a threat to their customary authority in the field of water management, the basin agencies were eventually implemented nationwide and have become powerful, multipurpose institutions in their own right—in stark contrast to the experience of the United States, where the development of river basin management organizations has

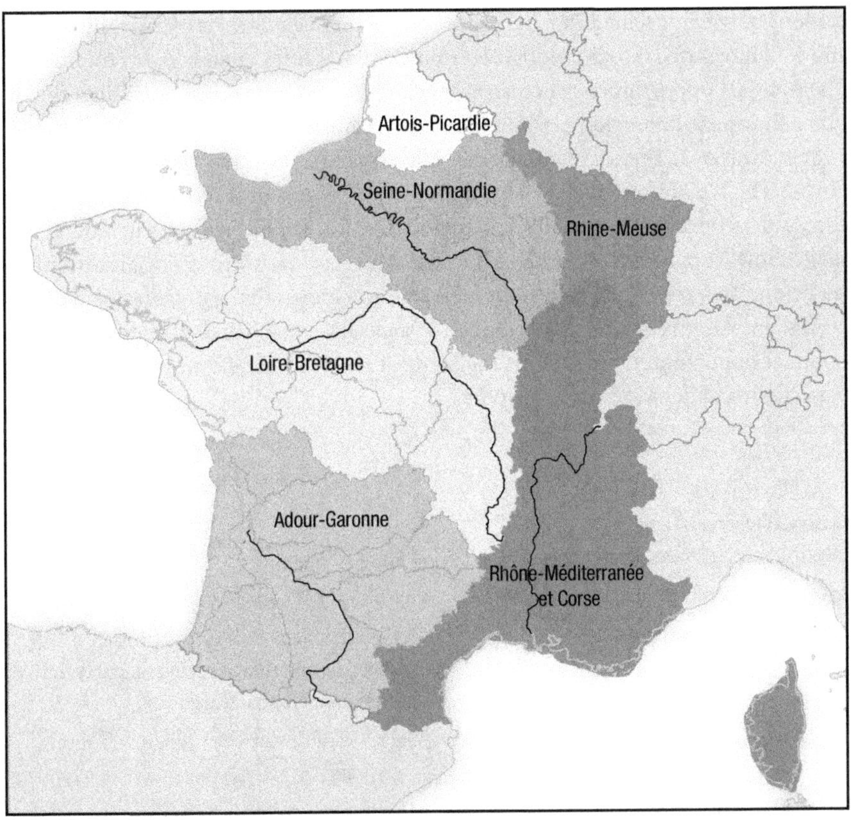

Figure 7.1 Major River Basin and Water Agency Boundaries in Metropolitan France (Adapted from European Environment Agency, 2011)

been stunted by comparison. This contrast is explained by the vertically integrated structure of French political institutions, which have facilitated coalition-building between third parties, especially nongovernmental environmental organizations, and national political elites. Even more important, the active encouragement of nongovernmental participation has led to a genuinely participatory forum for water resource decision-making. France's successful water agency system has been decisively shaped by the active inclusion of third-party advocacy groups, which have in turn forestalled the emergence of anything like the "reclamation nexus" between subnational water-user groups and federal officials which proved so inimical to collaborative and participatory basin management in the United States.

Table 7.1 depicts the salient features of water politics in France. Unlike in several of the other case studies, the dominant management issue is water quality rather than quantity. The primary institutional collective action dilemma, meanwhile, is vertical coordination between central and local levels of government, rather than the horizontal dilemmas which characterize the more decentralized political systems examined in this study. The French system provides several mechanisms less common in more decentralized countries, including strong regional governance institutions and closely integrated political party and bureaucratic structures that tie local officials closely to their counterparts at the center. Sectional cleavages, while less marked than in federal systems like the United States and India, are nonetheless evident between urban and rural areas. While France's cities are generally diverse and well integrated into the global economy, its rural areas remain highly traditional. The political opportunity structure in France is in some ways less open than that in other European countries, particularly in its tradition of political elitism (Kitschelt 1986). Moreover, France's strongly centralized political system means that political power which is decentralized to local levels can be recentralized with comparative ease, unlike the constitutionally guaranteed division of powers found in federal systems. However, on the other hand, France exhibits a very strong social movement tradition, and civil society groups like trade associations are powerful political actors. In water resource management, France has established a highly favorable political opportunity structure in the form of reserved seats for nongovernmental participants in decision-making bodies.

Table 7.1 Features of Water Politics in France

Case Characteristic	Issues
Driving management challenges	Water pollution
Institutional collective action dilemmas	Central–local coordination
Integration mechanisms	Regional governance institutions, vertically integrated political party and bureaucratic structure
Sectional cleavages	Marked divide between urban and rural regions
Political opportunity structure	Strong social movement tradition, reserved seats for civil society in water parliaments

Centralized Control and Civil Society in Water Resource Management

For most of French history, water resource management has been dominated by a succession of powerful centralized bureaucracies. In the seventeenth century, Louis XIV expanded the remit of a state agency called the Administration des Eaux et Forêts to include all of the country's navigable rivers, setting a precedent for unified central government control over navigation, water infrastructure development, and water resource allocation (Pritchard 2011, 30). Throughout France's subsequent history, control over water became a symbol of the centralizing French state in the form of the *grands corps*, the elite engineering bureaucracies which were responsible for building much of France's water-related infrastructure (Ghiotti 2006). One of these, the Génie rural, was responsible for the construction of irrigation works, while the better-known Corps des ponts et chaussées was responsible for constructing bridges, dams, and other hydrological works. In one major river basin, the Rhone, a national enterprise, the Compagnie Nationale du Rhône (CNR), was created to develop the basin for both hydropower generation and industrial production. Indeed, this novel integration of functions under a single state-owned organization later became an inspiration for the American Tennessee Valley Authority. Together, these agencies represented a formidable basis for the involvement of the French state in control of the nation's rivers. The *grands corps* continued to play a dominant role in the administration of most of France's major rivers until the twentieth century (Pritchard 2011). The influence of these agencies has historically been buttressed by a water rights regime in which most water uses were subject to state authorization (Sangarre and Larrue 2004). As a result, compared with the subnational actors more salient in the American and Indian cases, central government officials have historically been the dominant actors in French water resource management.

Yet while sectionalism has played a less significant role than in the other cases explored in this book, regional institution-building in France has remained plagued by considerable tensions between central authority and local autonomy. The French state has historically relied on a nested hierarchy of several levels of local government, including *communes* responsible for municipal services such as trash collection and *départments* responsible for local education, social services, and infrastructure, including water supply. Despite the centralization of the French state, these jurisdictions have over time proven capable of thwarting reforms instituted at the central level: several attempts on the part of the central government to merge or reorganize the *communes* before 1945 were defeated by the opposition of local officials. Instead of realizing a hoped-for improvement in the efficiency of local government, the effort served, in the estimation of two political commentators, only to give "evidence of the capacity of the French territorial system to resist any rationalization imposed by the government" (Le Gales and Pinson 2009, 214). Indeed, as these same observers later remarked, despite historical centralization, French politics "remains profoundly territorial, rooted in regions, rural communes, small towns and cities" (Le Gales and Pinson 2009, 223). As the French state has become more decentralized in recent decades, the tension between central control and local autonomy has grown more acute.

Centralization remains fundamental to French politics and political culture, but the institutions of government themselves have become significantly more decentralized than when the water agencies were first established in the 1960s. The landmark event in France's decentralization process was passage of the Decentralization Law in 1982, which formally created governments at the level of the *région*, in addition to the existing *départements* and *communes* at local and municipal levels, respectively. Unlike similar decentralization reforms elsewhere, the French effort principally attempted to strengthen the regional rather than the municipal level. Under the 1982 law, the regional governments were vested with budgetary authority and policymaking powers in the fields of economic planning, environmental protection, and regional infrastructure. Subsequent institutional reforms have strengthened regional governance even further. Constitutional amendments in 2003 codified the status of the regions and permitted them for the first time to pass laws and regulations that differ from those of the center (Le Gales and Pinson 2009). Moreover, regional urban governance institutions have proliferated in recent years and have come to be viewed as a model for other European states (Smith and Heywood 2000). Finally, in order to exercise their rapidly growing powers, the regions have established locally elected mayors and legislative councils. On the other hand, political autonomy at the regional level remains curtailed through the institution of the *préfet*, who functions as a representative of the center, enjoys broad powers of oversight, and can govern through administrative order in key areas such as public safety and security.

As the unique institution of the *préfet* suggests, the centralized French system provides considerably more integration mechanisms between central and local levels of government than do federal systems like that of the United States. French political leaders at both central and local levels are closely linked to one another through both formal and informal relationships. In contrast to federal systems, where political parties are usually highly decentralized, in France political parties are highly centralized; and even local officials are typically reliant on the support of partisans at the national level. Moreover, because local elected officials serve as conduits for policy implementation, they often develop reciprocal relationships with central officials, facilitating negotiation and interest-bargaining rather than the political conflict and litigation that prevails in federal systems like that of the United States, where subnational autonomy is frequently a salient political issue (Smith and Heywood 2000). Unlike in the United States, this hierarchy and centralization means that national elites control many of the levers of power, and they can more easily exert influence over local authorities. Historically, however, the vertically integrated nature of central–local relations in France has sustained a symbiosis between national and local political elites, which has in turn enabled them to bridge profound differences between urban and rural constituencies in order to secure support for major reforms (Le Gales and Pinson 2009). This capacity to create reform coalitions that include national elites proved crucial to the development of the water agencies against the concerted opposition of local officials.

The political opportunity structure in France reflects the country's vertically integrated politics. Points of access for nongovernmental organizations are numerous, but the prevalence of centralized, elite decision-making has historically guided French social movements toward more radical positions and more confrontational approaches to gaining political influence (Arzheimer and Carter 2006). In comparison to neighboring Switzerland, for example, France features a lower level of overall civil society participation but a higher level of activism and a greater propensity to employ more confrontational means to pressure political parties and political leaders (Meyer 2004). Civil society participation in French politics therefore takes a less collaborative form than in other European countries, but it can also exert greater and more targeted influence on specific policy issues. The tradition of confrontation also provides many French advocacy organizations with experience in exerting effective pressure on political leaders in pursuit of individual policy objectives. This proved to be the case in water resource management, in which concentrated third-party pressure persuaded the French government to create new institutions to govern the country's river basins.

Vertically integrated relationships between national and subnational political actors also enabled environmental organizations to bring water pollution control to the top of the political agenda in the 1960s and 1970s. In much the same way that the Izaak Walton League helped to catalyze a greater federal role in American water resource management, French anglers' associations (*les clubs de pêche*) and the National Union of Fishing Federations (Union nationale des fédérations de pêche) played a key role in the creation of the water agencies. Founded during the same period as the Izaak Walton League, in the early twentieth century anglers' associations began to lobby officials to regulate water pollution and to prosecute polluting enterprises for despoiling favorite fishing grounds. As in the United States, this localized advocacy eventually led to the creation of regional river basin management organizations (Bouleau 2009). Unlike in the case of the United States, however, the anglers' associations quickly formed a coalition with national political elites, who shared the basic objective of tightly supervising the actions of local officials. The regional approach embodied by the water agencies in fact represents a compromise between two forms of strategic behavior: the interest of nongovernmental organizations in increasing their voice in governance institutions, on the one hand, and the central state in enhancing its leverage over recalcitrant local authorities, on the other. Nevertheless, the water agencies successfully gained the support of all of these constituencies by incorporating the interests of both national political elites and civil society into planning and development within the various river basins (Organization for Economic Cooperation and Development 2005).

The water agencies are in fact only part of a sophisticated water governance infrastructure arranged around river basin boundaries. Each of the six water agencies in metropolitan France covers a major river basin and exercises both executive and legislative functions. Each basin includes a river basin committee (sometimes called a "water parliament") with representatives from central and local government, industry, farmers, municipal water users, and nongovernmental organizations. The Seine-Normandy Basin Committee, for instance, has 185 members, with 40% of the seats allocated to water-user group representatives, including agricultural and industrial users, 40% to local government, and 20% to central government representatives. Stakeholder representatives include delegates from the Seine-Maritime Agricultural Association, the Fishing Federation of the Aube, the Peugeot-Citroen motor company, water and hydropower utilities, a consumer advocacy association, and three environmental nongovernmental organizations. The agencies themselves effectively act as the administrative secretariat to support implementation of the committee's decisions, including preparation of basin management plans, levying fees on water users, and financing water infrastructure projects within each basin (Noel 1990).

The participation of nongovernmental stakeholders in agency decision-making has become institutionalized in the form of the Water Development and Management Master Plan, which must be periodically developed for each basin, along with subsidiary plans for each subbasin. Plans are formulated and reviewed every six to nine years through a series of public consultations between government and civil society organizations. Nongovernmental participation is further reinforced through a series of contracts which are awarded to environmental advocacy organizations to implement specific features of each plan, such as water quality monitoring. Coordination between central and local levels of government in each basin is facilitated by regional offices of the Ministry of the Environment, Sustainable Development, Transport, and Housing, which cooperates closely with each water agency (Organization for Economic Cooperation and Development 2005). The National Water Committee convenes water resource management stakeholders at the national level to make periodic decisions with applicability to the whole of France. This robust institutional architecture has proven to be remarkably conducive to collaborative and participatory decision-making. However, its development has been the product of sustained political bargaining and coalition-building between bureaucratic agencies, central and local levels of government, and civil society organizations, all of which have been decisively shaped by France's centralized unitary political system and attendant political opportunity structure.

Evolution of the Regional Tier in French Politics and Governance

For most of the prewar period, water resource management in France was dominated by technobureaucratic organizations like the Corps des ponts et chaussées and, in the Rhone River Valley, the state-owned CNR (Pritchard 2011). After 1945, however, bureaucratic officials and political leaders at the central government level began to fear the loss of influence over local officials, particularly the *préfets* who were responsible for overseeing administration in the various *départements*. The rapid economic growth of the postwar period created new economic interest groups, including industrial enterprises that, as in the more recent case of postreform China, developed close relationships with local officials. As reports emerged of corruption and collusion, the French national elite became convinced that the *préfets* had been captured by these provincial interests. By the 1960s, *préfets* were seen as not content merely with "representing the national interest but have [also] assumed in very substantial measure that of representing the regional interest" (Hayward 1969, 49). In order to combat this perceived collusion, French officials and political leaders became increasingly attracted to the idea of establishing new regional governance institutions to effectively sidestep local officials and re-establish control over local politics and administration.

The attraction of regional governance institutions drew on a broader sense among urban elites in the immediate postwar period that France was becoming increasingly polarized between a dynamic, prosperous Parisian urban region and a backward hinterland in need of comprehensive economic redevelopment. This idea was popularized by the geographer Jean-François Gravier in a book, published in 1947, proclaiming that France had become bifurcated into "Paris and the French desert" (*Paris et le désert français*) (Gravier 1947). To bridge this perceived chasm, the French government divided the country into regional economic planning zones, each under the jurisdiction of a new bureaucracy called the Interministerial Committee for Spatial Planning and Regional Development (Délégation interministérielle à l'aménagement du territoire et à l'attractivité régionale, or DATAR). As its name implies, DATAR was intended to facilitate the establishment of a new type of governance institution at the regional scale, which it was hoped would facilitate both interministerial coordination within government as well intersectoral collaboration across elements of the industrial economy (Ghiotti 2006, 11). Soon, however, this more limited objective was expanded into a broad vision for supplanting older forms of territorial administration that had become unsuited to the demands of modern economic development.

Regional institution-building increasingly attracted the support of the central bureaucracy, and the creation of the basin agencies became part of a broader regionalization of governance in postwar France through which Paris sought to tackle new policy issues, such as pollution, while simultaneously reasserting power over local authorities. The center's ultimate objective during this period was to "restore the authority of the regional agents of government" (Hayward 1969, 51). Yet in order to do so, the French government found that it needed to accommodate and form alliances with new economic interest and advocacy organizations, including environmentally oriented organizations like the anglers' associations. An unwieldy coalition therefore emerged between civil society groups and certain central bureaucracies, particularly DATAR, in support of creating an entirely new regional tier of government. This coalition had to overcome concentrated local opposition: "The Gaullist regime faced the resistance of local interests. It therefore aimed . . . to promote new groups of interests and new territorial levels to implement and support the modernization strategy" (Le Gales and Pinson 2009, 214). This coalition possessed a strong shared interest in subverting entrenched local elites, who were perceived to have become unresponsive to both central direction and the demands of addressing emerging issues like water pollution (Ashford 1982). Unlike in the Chinese case, then, national political elites and environmental advocacy groups forged a strong, durable alliance in opposition to geographically concentrated water user groups and their allies among local officials.

Unsurprisingly, subnational officials and their allies in Parliament were initially opposed to the expansion of regional governance institutions, especially the basin agencies, which they viewed as a threat to local autonomy. Yet unlike in other cases, notably that of the United States, this subnational opposition ultimately foundered when arrayed against a powerful bureaucratic–civil society coalition that favored the creation of a new regional tier of governance. This coalition was held together by a shared, if somewhat paradoxical, interest in weakening the power of local officials by creating a new tier of governance, which environmental groups believed would strengthen enforcement of water quality regulations and national elites believed would create greater leverage over local officials. This shared interest nonetheless was slow to emerge. In the early twentieth century, the anglers' associations began to lobby for more stringent water quality legislation against the wishes of national elites, who favored adopting a regulatory framework more favorable to industry. In the late 1950s, central officials instead proposed to adopt a flexible system of pollution standards which would vary across regions. The anglers' associations objected, asserting that pollution was a "moral issue" that could be solved only by means of uniform national pollution legislation. The impasse was resolved only when a senator sympathetic to the fishermen introduced an amendment providing that water quality assessments be periodically conducted in each region, with regulations tightened as necessary (Bouleau 2009). This mundane compromise, though seemingly trivial, would come to form the foundation of an enduring alliance in favor of a regional approach to water resource management in France.

As in the United States, French environmental advocacy organizations, led by the anglers' associations, employed the available political opportunity structure to form coalitions with national political elites. But unlike in the American case, the French reform coalition proved more durable and effective in overcoming local opposition and subnational politics. Eventually, this coalition proved able to persuade even the *préfets* that regionalization could enhance rather than diminish their powers by facilitating the expansion of government into new policy and issue areas (Ashford 1982). Unencumbered by the resistance of state officials which so frustrated the Roosevelt administration's attempts to replicate the Tennessee Valley Authority model, regional governance has thrived in France, even as it has withered in the United States. Following passage of the Water Law in 1964, the water agencies have accumulated considerable and widespread legitimacy throughout France's political system. The success of regional-scale governance in France thus stands in stark contrast to that of the United States, where it has proved fragmentary in comparison. The crucial difference, as the following case illustrates, has been the ability of a third-party reform coalition to negotiate interests at both central and subnational levels of government and thereby create a new, regional tier of governance that fosters both collaboration and participation.

CIVIL SOCIETY, COOPERATION, AND THE EVOLUTION OF THE WATER AGENCIES

Efforts to build institutions for cooperative river basin governance in France, as in other countries, have been shaped by a contest between the interests of actors at central and local levels. This contest is most aptly illustrated in the Rhone River basin in southern France, which, given its potential for hydropower generation, was the first region to attract the attention of the state for water resource management and development. As early as 1878, the Rhone was designated as a "region of national interest," and the state was given authority to "approve or reject development proposals as well as to undertake projects in the name of the public good" within it (Pritchard 2011, 35). The state's prominent role in the region, and its promised financing for construction projects, soon created considerable tension between the various *départements* in the basin. By the end of the nineteenth century, these had grown sufficiently acute that a special intergovernmental institution, the General Council (Conseil général), had been established to negotiate the competing interests of the Rhone basin *départements*. Even in France's highly centralized political system, these localized interests posed a challenge to Paris's aspirations for comprehensive river basin development. As the country's economy developed throughout the twentieth century, a combination of infrastructural and qualitative conflict would nonetheless create substantial pressure for the creation of more collaborative and participatory water management institutions.

The highly centralized French political system ensured that the development and management of river basins like that of the Rhone were shaped more by central–local than interjurisdictional conflict. Despite the center's more or less absolute legal and political authority, its efforts to build river basin management organizations were stymied by the opposition of local officials who feared the loss of age-old privileges and prerogatives. This dynamic played out in the Rhone, where in the early twentieth century local officials increased their pressure for state financing for hydropower and industrial development. The influence of the reclamation nexus ethos was betrayed by these officials' accompanying demands that such financing be more or less unconditional and directed to local enterprises. National leaders, on the other hand, unsurprisingly favored a strong role for central government agencies like the Corps des ponts et chaussées in making use of centrally provided financing. This debate was resolved through the creation of a unique institutional entity that would pursue local riparian development priorities in close cooperation with the center. The resulting CNR, created by a 1921 law, was effectively set up as a state-owned enterprise but one that had the novel remit of attempting the coordinated development of hydropower, navigation, and irrigation throughout the Rhone River basin (Pritchard 2011). The CNR therefore established an important institutional precedent for state-led management and governance at the river basin scale.

Despite the early precedent of regional development in the Rhone, the governance of France's river basins lay largely dormant as a political issue until the postwar period, when it re-emerged in the context of growing concern over pervasive water pollution. As in the case of the Izaak Walton League in the United States, the National Union of Fishing Federations played a crucial role in enhancing the political profile of water resource issues. The union demonstrated its political power in 1958, when its advocacy for stronger pollution-control regulations resulted in passage of legislation making pollution a crime under the Rural Code. But this early exercise of the union's influence proved insufficient to address the full scope of its water quality concerns, and by the mid-1960s the pace of economic development had increased water pollution to the point that many questioned the capacity of the French state to control it. Water quality became an issue of frequent mention and debate in Parliament, and prodded by the continued advocacy of civil society organizations like the union, members of Parliament began to criticize the state's perceived indifference to environmental concerns. In a 1964 session of the National Assembly (Assemblée nationale), for example, a deputy questioned whether the French state had become a "slave to industry," either unwilling or unable to control pollution on the country's rivers (Journal Officiel 1963a, 3146).

Political debate over the causes of water pollution soon focused on the alleged failure of local officials to enforce regulations on polluting enterprises within their jurisdictions. The government's official rapporteur introduced the 1964 Water Law to Parliament by highlighting this problem, charging that "prefects don't apply pollution regulations, and have not for some time" (Journal Officiel 1963b, 5181). A deputy from the Rhone region offered a similar example of prefects who refused to fine companies for water pollution because they had become persuaded of the companies' economic importance to their jurisdictions (Journal Officiel 1963c, 7227). Local pollution regulations, in fact, were claimed by members of Parliament to "reflect more the local balance of power than pollution realities" (Brun 2009). Yet efforts to construct a larger direct state role in pollution control quickly foundered on questions of financial and economic capacity. Deputies of the National Assembly fretted at estimates that wastewater treatment alone was likely to cost 8–10 billion francs, far more than could be provided for by the state budget at the time (Journal Officiel 1963b, 5180). These financial exigencies drew legislators toward an alternative mode of financing and institutional control—the creation of a new tier of government empowered to levy taxes on water users and to employ the proceeds for constructing water-treatment projects. In order to legitimize what was effectively a new mechanism of taxation for the French state, the government also determined that the new regional governments should feature councils or committees more broadly representative than the existing *départments* and *communes*.

This crucial emphasis on participatory decision-making informed the radical provisions of the French government's ultimate response to the problem of water pollution, the 1964 Water Law. In a decisive shift away from the traditional central–local hierarchy, the Water Law envisioned a regional approach to governance and proposed to create six *agences des bassins* (basin agencies) on river basin lines, covering the entire territory of metropolitan France. These basin agencies were to be granted a broad mandate, including the control of pollution, the construction and maintenance of hydrological public works, and crucially, given the government's acute financial constraints, the ability to levy taxes and collect fees from water users. In addition to these powers, the agencies were from the outset intended to directly incorporate civil society organizations into their decision-making structures, in recognition of their importance as crucial stakeholders. "The 1964 Law," notes a French water scholar, "can be understood as a way to pool the collective effort of different constituencies for pollution control" (Bouleau 2009). Governance of the new basin agencies was accordingly to be vested in basin committees composed of representatives from central and local governments, as well as different water-user groups (LegiFrance 2004). But while the inclusion of nongovernmental representation in the agency structure was novel, the Water Law's most far-reaching implication was to effectively supplant the traditional institutions of local government, the *départments* and the *conseils généraux* within the sphere of water resource management with the new basin agencies.

Accordingly, political debate over the law centered on powers and autonomy between central and local levels of government. The law's assertion of greater state control over water resources was complicated by the fact that in 1884 municipalities had been granted control of water supply and sanitation activities. Municipal governments, which predictably viewed the proposed creation of new regional river basin governance institutions as a threat to these historic prerogatives, therefore had a legal basis to contest the creation of the basin agencies (Fradin 2007, 91). Initially, national politicians appeared accommodating of these municipal concerns. In introducing the 1964 legislation, the government framed its provisions as merely an attempt to "complete and modernize" the 1884 legislation, rather than to create new authorities for higher levels of government (Journal Officiel 1963b, 5181). However, as with attempts to consolidate the *communes*, the government's apparent deference to local authorities failed to persuade local officials that their prerogatives would be adequately protected. Consequently, early drafts of the 1964 Water Law were strongly resisted by local governments, local elites, and sympathetic members of the Assemblée nationale (Pinson and Le Gales 2000, 7).

These debates, taking place before France's National Assembly, unfolded primarily on one side or the other of a sectional divide between urban and rural constituencies.. Rural farmers especially feared that the Water Law would threaten their existing water rights, which traditionally arose from ownership of riparian lands. The cause of preserving local autonomy and agricultural interests was most enthusiastically taken up by left-leaning members of Parliament. Indeed, opposition to the Water Law and the basin agencies became something of a pet cause for the leftist opposition parties of the 1960s. Edmond Garcin, a Communist deputy from the Rhone Valley constituency of Aubagne, for instance, charged that the Water Law threatened to abrogate provisions of the long-standing Napoleonic *Code rural*, which protected the riparian rights of small-scale agriculturalists. This threat to a long-standing pillar of French rural society, Garcin proclaimed, could not be tolerated (Journal Officiel 1963b, 5182). Marcel Massot, a Socialist deputy from Provence, similarly argued that existing laws were already sufficient to combat pollution and that revision threatened only to impose additional economic burdens on farmers seeking water diversion for irrigation (Journal Officiel 1963b, 5195). In the Rhone River basin, which because of the CNR had long been subject to a high degree of state investment and control, public opinion was particularly skeptical of the greater state role in water resources management promised by the Water Law (Pritchard 2011).

Apart from the concerns of these individual local interests, members of Parliament also voiced the objections of local officials that the proposed basin agencies would supplant the traditional duties of local governments. One deputy warned that the Law Commission, which vets proposed legislation to ensure it conforms with existing constitutional provisions, would need to "temper state power over water" before the Water Law could be accepted (Journal Officiel 1963b, 5183). A Communist Party deputy likewise remarked that while it was clear that pollution needed to be tackled, such regulation could be accomplished at a local level rather than through the creation of new institutions that would increase the reach of the state, to the detriment of customary prerogatives in rural areas (Journal Officiel 1963b, 5186). On another occasion, the same deputy proclaimed that "We fear some state control of water at the expense of local communities, including the loss of rights that are sometimes centuries-old" (Journal Officiel 1963a, 3147). Communist Deputy Garcin, meanwhile, urged that local government departments and municipalities be empowered with greater financing capacities to undertake their own pollution-control initiatives, rather than to rely on an expansion of the central state role in water resource management (Journal Officiel 1963b, 5182).

Other deputies voiced the related concern that in creating a new set of decision-making structures, the basin agencies might undermine the democratic legitimacy of local governments. Deputy Barbet, a Communist representative from the Seine region, warned that the proposed basin agencies threatened to remove from local elected officials "the power of control and the exercise of [their] rightful power" (Journal Officiel 1963b, 5193). Michel de Grailly, a member of the Union for a New Republic party and who also represented the Seine region, likewise charged that by creating a new tier of government the 1964 law exceeded the government's constitutional authority. In a statement that would not have been out of place in a federal legislature, de Grailly argued that by exceeding this authority, the new basin agencies were "inconsistent with the separation of powers" (Journal Officiel 1963c, 7231). These concerns reflected the scope of the proposed changes to the French system of governance contained in the Water Law. However, unlike in the more decentralized political systems examined in this book, the dilemma of autonomy did not lead subnational officials to uniformly and relentlessly oppose the proposed interjurisdictional collective action institutions. Instead, the French state simply co-opted local officials, eventually including them in a reform coalition that it adroitly forged with newly ascendant environmental advocacy groups.

By the time the Water Law was proposed, environmental matters had begun to decisively influence French politics. The *départements*, however, proved too sclerotic to address the complaints of organizations like the National Union of Fishing Federations. For these organizations, the prospect of a new structure of regional governance promised a more open political opportunity structure (Pritchard 2011). At the same time, the regional river basin organizations established by the Water Law offered the French state the ability to pursue twin strategic goals, both enhancing its own control over local governments while also accommodating new political interest groups like the National Union. The political value of the new institutions was reflected by the "largely political" composition of the basin committees, which attempted to carefully balance representation from central and local government, industry, and other water-user groups (Barrque 1995, 443). In the eyes of national politicians, the chief attraction of the basin agencies lay in the fact that not only were they "not linked to a pre-existing tier of government" but they also "facilitated a broad approach to both water quality and water quantity management that went beyond abstraction and discharge controls." Finally, and critically from the vantage point of the central government, the new basin agencies were "sufficiently large to generate appropriate levels of fiscal revenue" (Buller 1996, 292). The basin agencies thus provided a new institutional structure through which to enhance both the political and financial powers of the central government relative to entrenched local governments.

The central government's advocacy for creation of the basin agencies, and the 1964 Water Law more broadly, relied on the position that only a new type of government could effectively address emerging environmental issues. The regional approach to river basin management was portrayed by the government at the time as an ideal, and indeed the only practical, means of combating pollution. In a 1963 session of the Assemblée nationale, Marc Jacquet, the minister of transport and public works, expressed the view that the government's economic development plans, which relied on stimulating regional economic clusters, could not be effectively accomplished without addressing water quality (Journal Officiel 1963b, 5180). Other government ministers expressed a similar desire to realize the "integrated management" of both water and land use under a single, comprehensive institutional framework (Journal Officiel 1963b, 5181). Increasingly, the government argued that this degree of integration could be accomplished only through the creation of a new scale of governance over that of the *départments* and *communes.* In response to the suggestion that strengthened national pollution-control regulations might be more effective than creating a new tier of governance, Jacquet insisted that "the fight against pollution must be handled at a regional level" (Journal Officiel 1963c, 7221–22).

Even as national politicians couched advocacy for regional institution-building in terms of the fight against pollution, they attempted to neutralize opposition from water users facing greater regulation and taxation by appealing to notions of justice and fairness. The basin agencies were in fact subtly portrayed by the government as constituting a "new deal" for local water users, ensuring a fairer distribution of water resources by bypassing existing local governance structures. In introducing the 1964 Water Law to the Assemblée nationale, Jacquet proclaimed that its twin objectives were to fight pollution as well as to ensure a better and fairer distribution of water resources through consolidation of government powers (Journal Officiel 1963b, 5180). Following implementation of the water agencies, this "language of distribution" remained a critical justification for the transfer of powers traditionally vested in local government to the new regional institutions (Richard and Rieu 2008). Although such obfuscation failed to quiet the opposition of left-wing members of Parliament to the new institutions, it did effectively blunt public criticism of the proposed reforms.

Given the extensive centralization of the French state, the most important element of the central government's legislative strategy involved a careful balancing act between bureaucratic interests, particularly those of the *grands corps.* The 1964 Water Law included provisions that limited the participation of civil society groups in decision-making within each river basin, with the intent of ensuring that the state maintained significant control over the actions of the basin agencies. In the Rhone River basin, which had a longer history of regional planning and partial autonomy than any comparable area of France, the *grands corps* nonetheless remained the dominant political influence toward the end of the twentieth century (Pritchard 2011). Against this background, sustained lobbying by and deference to the historical role of the *grands corps* persuaded the government to retain virtually all enforcement powers in the central state administration rather than in the basin agencies (Meublat and Le Lourd 2001). Moreover, land-use and economic planning authority remained vested in central government bureaucracies, particularly DATAR (Barrque 1995). Central government representatives in the individual departments were accorded ex officio seats on the basin committees (Notte and Keesen 2009). Perhaps most indicative of the central government's deference to bureaucratic interests, however, is the fact that de facto control over the six basin agencies was evenly apportioned among the *grands corps*, one of which was granted implicit authority over each agency (Barrque 1995, 442). On the strength of this promise to expand agency control, the Water Law and basin agencies eventually secured the backing of most elements of the central government bureaucracy.

Yet despite the degree to which bureaucratic interests shaped the original design of the basin agencies, since 1964 they have greatly expanded the voice of third parties, particularly civil society organizations and water-user groups. The increasing importance of these groups is reflected by institutional changes incorporated in revisions to the 1964 law passed in 1992 and 2006, both of which have enhanced the role of nongovernmental actors in decision-making at the basin level. This evolution was the result of two external forces, the increasing political salience of the environmental movement in France after the 1960s and the increasing trend toward decentralization of administrative authority, particularly after passage of the 1982 Decentralization Law (Richard, Bouleau, and Barone 2010). Both processes had the effect of creating substantial pressure on the customarily hierarchical French bureaucracy, including the agencies which initially controlled the basin agencies, to engage nongovernmental stakeholders. This political pressure is reflected in the complaints of members of the Assemblée nationale into the 1990s that the water agencies failed to consult with local officials regarding planning decisions (Journal Officiel 1994, 2207) and critiques that the basin commissions continued to fall prey to "a long tradition to let the [central] government decide" matters (Barraque, Maurel, and Raymond 2004). By making the basin agencies more inclusive and participatory, the French state could accommodate such pressure without the need to undertake far-reaching reforms of the bureaucracies under its direct control.

Over time, the basin agencies have successfully addressed these concerns and have steadily accumulated legitimacy among political actors at both central and local levels by increasing access to civil society interests. This successful evolution has in turn set the stage for its continual development and successful institutional performance. In 1992, France's environment minister boasted that France's river basin agencies were "an original concept that brings together civil servants, engineers, elected officials, and industry. It works so well that Eastern Europe wants to copy us" (*Paris L'Express International* 1992). Amid this widespread political support, a 1992 revision to the Water Law strengthened the powers of the basin agencies, which were renamed "water agencies," and increased the representation of civil society organizations (Barrque 1995). The 1992 legislation stipulated that both basin and subbasin management plans were to be formulated with input from different water-user groups and were to be approved by the basin committee, on which nongovernmental groups were represented. In addition to these existing basin committees, which had been established by the 1964 Water Law, the 1992 revision expanded the political opportunity structure for civil society groups by creating local water commissions (*commissions locale de l'eau*) empowered to review local and subbasin water-use plans (Richard, Bouleau, and Barone 2010). Furthermore, the National Water Council (Comité national de l'eau), made up of eighty-two governmental and nongovernmental representatives, was established to set national water policy, extending the influence of nongovernmental organizations to the national level (Ministry of Environment, Sustainable Development, Transport, and Housing 2014).

By the turn of the century, buoyed by these reforms, the water agencies had become a central fixture of the French political system. The agencies emerged as significant political actors in their own right and have helped to stimulate cooperation between local governments on a variety of other regional governance issues, including transport and urbanization (Dubois 2001; Richard and Rieu 2008, 11). The water agencies have furthermore been promoted as model institutions by the French government abroad and have emerged as something of a point of national pride, especially for the Ministry of Environment and Sustainable Development under whose authority the water agencies presently reside (Agence d l'Eau Seine-Normandie 2003). In 2004, reforms to the water agencies led to a further de-emphasis on central state control by increasing the representation of local water users on basin committees (Lanfranchi 2008). The 2006 Law on Water and Aquatic Environments (*Lois sur l'eau et les milieu aquatiques*) strengthened the water agency system by establishing explicit constitutional authority for it to tax water users and created a national secretariat, known as the National Office of Water and Aquatic Environments (Office national de l'eau et des milieu aquatiques, or ONEMA), to assist the water agencies in exercising administrative and managerial functions. The 2006 law also enhanced nongovernmental participation still further by reserving 25% of seats on each local water commission for civil society representatives. As a result of the 2006 law, half of each commission must now be elected by local water users, while central and local governments are restricted to 25% of the seats on each body (Richard, Bouleau, and Barone 2010).

The strength of the water agency system is illustrated by the Seine-Normandy Basin Committee, one of the country's most active. The local basin committee has four permanent committees that reflect the scope of its functions and ability to coordinate across sectors, including Plans and Programs, Aquatic Environment, Coastal and Maritime Affairs, and Land-Use Planning and Flood Management. The coastal committee (Commission de littoral et de la mer) is especially notable as an example of collaborative, participatory management that extends beyond the basin itself and engages coastal stakeholders, including fishermen and tourism authorities. Just as notable is the fact that Seine-Normandy Basin Committee decision-making is supported by a robust technical infrastructure, including twenty-one expert members of a scientific council. These members include agricultural engineers, hydrologists, and sociologists, who are charged with delivering recommendations to the basin committee as a whole (Eau Seine-Normandie 2017).[1]

This feature of the Seine-Normandy Basin Committee highlights a crucial and distinctive aspect of the French water agency system: it institutionalizes cooperation not only between jurisdictions but also between water users, political elites, and technical elites. Unlike the technocratic model of river basin management in China, which excludes subnational political and water-user group representatives, or the intergovernmental model of many American river basin institutions, which feature only representatives of state governments, the French approach is fully inclusive. The inclusion of elites and civil society representatives in the same institutional arrangements enables the water agency system to avoid the pitfalls of either autocratic or overly diffused institution-building.

On the strength of its legitimacy among most, if not all, political actors in the French system, the water agencies have evolved into one of the most collaborative and participatory regional water basin governance institutions in the world. Most assessments of the water agencies' effect on environmental outcomes are notably positive. In a report, the Organization for Economic Cooperation and Development (2005) concluded, for example, that "The role of the river basin authorities . . . has expanded from purely financial tasks to assessment of the state of aquatic environments and to planning. The integrated management at major basin level, which is partnership-based and multi-annual, has proved highly effective." The institutions are also largely self-sustaining: three-quarters of agency funding comes from fees on residential water users, which are used to finance stringent wastewater treatment as well as ecological protection. Overall, such environmental protection activities account for over two-thirds of agency expenditure (Les Agences de l'Eau 2014). This combination of political legitimacy and institutional sustainability distinguishes the French basin agencies among the cases examined in this book and presents a particular contrast to the American river basin commissions, which both failed to develop political support and were eventually defunded during the Reagan administration, as well as the Chinese water conservancy commissions, which have for the most part failed to improve environmental outcomes.

The most important achievement of the water agencies, however, is that through their broad-based participatory governance mechanisms they have proved adept at preventing conflict between water users within France's river basins (Sangarre and Larrue 2004). Though isolated political conflicts do occur between water users within these basins, little of the sustained conflict between geographically concentrated water-user groups or subnational jurisdictions evident in the other cases examined in this book has occurred in France. Even more significantly, the water agency model has proven to be an effective catalyst of international cooperation over water. Under a 2005 law, each water agency is empowered and encouraged to devote 1% of its budget to international aid and cooperation projects, and water agencies are twinned with a particular geographic region to encourage water-related technical cooperation. The Loire-Bretagne Water Agency, for example, is twinned with Sub-Saharan African and Southeast Asian countries, while the Artois-Picardie Water Agency takes the lead on cooperative projects with European and Central Asian countries. The agencies themselves served as an inspiration for the European Union's Water Framework Directive (WFD), which mandates all member states to create basin-wide cooperative governance institutions that include both governmental and civil society participants (Ministry of Environment, Sustainable Development, Transport, and Housing 2014). The internationalization of the water agency model has moreover proved to be self-reinforcing: implementation of the WFD has further enhanced the institutional architecture of the French water agencies and helped spur establishment of the ONEMA (Richard, Bouleau, and Barone 2010). By the mid-2010s, the French water agencies had developed into perhaps the world's foremost example of a cooperative, institutionalized approach to interjurisdictional collective action in shared river basins. Yet this success stems less from the institutional design of the water agencies themselves than from the coalition of political leaders and third-party civil society organizations that have propelled their creation and evolution over the past half-century.

COUNTRY CASE CONCLUSION

In comparative perspective, France is notable for both the quantity and the quality of its regional governance institutions. Its approach to the problem of governing interjurisdictional river basins has resulted in by far the highest degree of institutionalized collective action among the countries examined in this book. This highly institutionalized approach to managing interjurisdictional river basins is manifested in the water agencies, which unlike their counterparts in other countries span all the country's major river basins, include genuinely participatory governance structures, and have accumulated legitimacy and capacity over time. Yet, rather paradoxically, this institutional architecture emerged out of an effort on the part of French national political leaders to reassert control over protectionist local authorities. The French water agencies were in many ways conceived as an elite and technocratic project. The fact that they have evolved into cooperative, polycentric institutions is due to France's political opportunity structure and the way that third-party advocacy organizations, especially the anglers' associations, exploited it. France's highly integrated political opportunity structure helped to ensure that once third parties and political elites forged a reform coalition in support of regional governance, they could outflank and co-opt subnational opponents. Unlike many of the other cases examined in this book, nationwide institutional reform in France was not undermined, or even much informed, by the influence of subnational politics and interests.

The case of France, as summarized in Table 7.2, differs from the others explored in this book in terms of the role of noninstitutional factors like physical water scarcity and how sectionalism influenced conflict and cooperation over water. Water scarcity was not a significant issue; instead, water quality was the principal management issue and the original cause for conflict between water users. This did not, however, make these conflicts less acute, suggesting that scarcity itself does not necessarily drive conflict between water users. Instead, in the case of France water-use conflicts between economic sectors, especially industry and urban water users, catalyzed grievances that eventually led to civil society mobilization. These divisions had little to do with riparian geography; the strength of interjurisdictional collective action institutions ensured that there was little conflict between upstream and downstream water users, and tensions between water users pertained primarily to water uses rather than flows. Even so, it is notable that conflict between water users did take on a sectional character, in that it pitted largely rural, agrarian water-user groups against urban water users. But like so much in French politics, this divide was national in scale, not local. Consequently, the linkage between water conflict and sectional identity that typifies subnational hydropolitics in other countries is largely absent. Finally, and most vividly, the French case demonstrates the

Table 7.2 Summary of Subnational Hydropolitics in France

Factor	Characteristics	Effect on Interjurisdictional Collective Action
Decentralization	• Both formally and informally highly centralized; party structures, elites, and institutions like Corps des ponts et chaussées strongly hierarchical • Central government employed regional governance as a means to enhance power over local authorities	• Central government highly involved in attempts to address emergent interjurisdictional water issues through regional governance institutions like DATAR, Agences des bassins
Sectional identity	• Primary divide is between rural and urban identities	• Substantial resistance from rural constituencies to regional water management institutions
Political opportunity structure	• Centralization results in fewer points of access, but institutional and elite integration makes it easier for third parties like the anglers' associations to advocate once access achieved	• Institutionalized participation in river agencies promoted adaptive and inclusive management practices, supported by ONEMA

Abbreviations: DATAR, Délégation interministérielle à l'aménagement du territoire et à l'attractivité régionale; ONEMA, Office national de l'eau et des milieu aquatiques.

crucial role of third-party advocacy organizations, in alliance with national political elites, in lobbying for the creation of collaborative, participatory water resource management institutions. As in the case of China, regional river basin governance institutions in France were born of an attempt to enhance central control over local authorities and the failure of subnational actors to achieve satisfactory environmental outcomes. But unlike in the case of China, France provided a distinct role for third-party actors in regional governance. Instead, as in successful periods of regional institution-building in the United States, France's political opportunity structure enabled third parties to create a constituency for institutionalized cooperation between subnational water users.

Despite this similarity between the French and American cases, it is important to detail the ways in which the French political opportunity structure proved more effective in sustaining this constituency over time, particularly in the face of subnational opposition. In contrast to that of the United States, the French political system is hierarchical and densely integrated at both national and subnational levels, a fact which shaped the evolution of the water agencies in several important ways. First, the clear and long-standing central government role in water resource management and regional development provided a strong precedent for the involvement of national political leaders in river basin governance. Second, the strong tradition of social movement activism at the national level ensured that the environmental movement, and groups seeking to advance it, were rapidly accommodated in national political debates. Third and finally, the high level of integration between central and local political leaders encouraged negotiation and interest-bargaining between them. This combination of factors has shaped the evolution of the water agencies and basin committees, which effectively convene water users, government, and third parties to support collaborative and participatory basin management at the regional scale. These institutions reflect a distinctly French contribution to understanding subnational hydropolitics: institutionalized cooperation requires bridging national and parochial, as well as popular and elite, interests. Inclusion of third parties is the glue that holds this delicate structure together. The French case illustrates, more clearly than any other, that by not only permitting but encouraging the involvement of third-party civil society organizations, governments can prevent conflict between water users and promote cooperation in its stead.

NOTE

1. I am indebted to Regina Rossman for bringing the details of the Seine-Normandie example to my attention.

Conclusions and Policy Recommendations

Arizona governor B.B. Moeur's 1934 declaration of martial law to prevent construction of a dam to supply California with water now belongs to history. The open political conflict that characterized the early relationship between the states of the Colorado River basin has given way to extensive cooperation, even as water scarcity poses an ever-greater challenge for management of the basin's water resources. Yet halfway around the world, in India, history threatens to repeat itself. In April 2013, the Indian press reported a most unusual case of espionage: the police in Kerala, a southern state engaged in a long-running river basin dispute with the adjacent state of Tamil Nadu, announced that they had uncovered a spy working on behalf of their neighbor, who they charged had "developed a wide network with officials and ministers to source vital info regarding Kerala's stand on interstate water issues" (Express News Service 2013). Both the 1934 invocation of martial law and the more recent 2013 allegation of espionage are poignant examples of behavior common to countries but not to subnational states. Though fortunately rare, they attest to the gravity of conflict over water that occurs within countries, rather than between them.

These subnational hydropolitics are both surprisingly common and conceptually puzzling, for they occur despite the existence of relationships and institutions, ranging from water ministries to political party structures, that may be expected to prevent them. This book helps to explain why water so often proves to be a focal point for political conflict within nations as well as among them. This conclusion first reprises the argument of the book and its implications, then discusses directions for future research. Finally, it offers some policy recommendations for countries like India where sustainable water resource management is undermined by extensive subnational political conflict over water. In a world where water needs are growing rapidly even as water supply variability increases due to climate change, it is critical to understand the politics of water at subnational as well as international levels—and how this understanding can be leveraged to improve water resource management.

This book has sought to explain subnational hydropolitics in terms of three factors: decentralization, sectional identity, and political opportunity structure. While this understanding of conflict and cooperation over water is based on the specific cases examined in this book, it raises several broad questions for the study of environmental politics. First, and most fundamentally, these cases suggest that the dynamics of subnational and international conflict and cooperation over shared natural resources differ in degree more so than in kind. The problems of managing forests, fisheries, and other natural resources that are shared among states and provinces may more closely resemble those shared between countries than has previously been recognized. Second, these cases indicate that the causes of water conflict are as much ideational as institutional, suggesting the need to balance concern for institutional design with that of sectional politics. Third and finally, the cases examined in this book suggest a testable hypothesis concerning interjurisdictional collective action. These cases suggest that collective action institutions that are imposed on existing subnational jurisdictions are likely to fail, while at the same time institutionalized interjurisdictional collective action is unlikely to arise on its own. Instead, cooperation in shared subnational river basins is most likely to stem from the inducement of third parties, acting in concert with national political elites. Further research should attempt to validate this hypothesis across additional case studies.

UNDERSTANDING WATER POLITICS: SUMMARY AND IMPLICATIONS

This book began by posing a three-part puzzle of interjurisdictional water conflict: first, its intensity can rival that of international water disputes; second, existing institutions often fail to prevent it; and third, its intensity varies markedly across countries and across time. The cases of interjurisdictional water conflict examined in this book suggest that this puzzle can be explained by the interaction of three factors as laid out in Chapter 3. Decentralization explains why interjurisdictional water conflicts arise, namely because it tends to encourage weak forms of interjurisdictional cooperation and to impede the creation of strong regional institutions to promote collective action. Sectional identity, meanwhile, explains why such conflicts persist over time. In the absence of institutionalized cooperation, sectional identity is a powerful force for political mobilization and one that subnational politicians often embrace as a means of building support within their individual jurisdictions. Geographically concentrated water-user groups like irrigation districts in the United States and farmers' groups in India allied with subnational politicians to expand the base of support for their objectives by appealing to sectional identity. This strategy, though often initially effective in building local support for water infrastructure projects or in negotiations over water-sharing agreements, makes adopting cooperative strategies more difficult later on. When linked to shared waterways, sectional tensions give voice to deeply rooted grievances that make water conflicts difficult to resolve. But a third factor, political opportunity structure, explains how cooperation can emerge even amid such apparently intractable conflicts. When third parties enjoy adequate access to the political system, they can form coalitions to advocate for creation of strong interjurisdictional institutions, thereby creating a constituency for cooperation instead of conflict over shared water resources. These observations carry several specific implications for better understanding the causes of water conflict, as well as how to resolve and prevent it.

First, the cases examined in this book lend little credence to the long-held belief that immutable physical characteristics, including acute water scarcity and riparian geography, represent the root cause of water conflict. Instead, these physical factors represent simply a set of initial conditions. Subnational politics themselves are animated by key political actors, especially subnational politicians and third parties, operating within institutional frameworks. In the cases of both the Colorado and the Krishna, water scarcity did initially lead geographically concentrated water-user groups to advocate zero-sum approaches to water-sharing agreements that maximized control of water for their own regions at the expense of neighboring jurisdictions. But if anything, the cases revealed an inverse relationship between the intensity of water scarcity and conflict: in the later stages of the case of the Colorado, acute water scarcity actually helped to catalyze institutionalized cooperation. Geography played a more complicated role in conflict dynamics, but the parties to water conflicts were divided less by upstream–downstream riparian geographies, as traditional hydropolitical theory would suggest, than by deeper sectional cleavages. In cases like the Krishna, protracted conflict was the result of subnational politicians and geographically concentrated water users attempting to build support by appealing to sectional identity. In effect, shared water resources became the stage upon which sectional tensions were played out. On the other hand, the development of collaborative and participatory interjurisdictional institutions, catalyzed by third-party pressure, prevented conflict between water users, especially in the later phases of the Colorado case as well as in the French case.

These differential outcomes suggest a second major implication for understanding water conflict, namely that while collective action institutions are essential to preventing it, these institutions rarely arise on their own, and are much more often created as a result of pressure from alliances between third-party environmental organizations and national political elites. The extent to which they are able to form these coalitions in turn depends on political opportunity structure in water resource management. Table C.1 illustrates how the case studies together demonstrate the relationship between political opportunity structures at the national level and subnational water conflict. Those countries with closed structures, namely China and India, are subject to widespread subnational water conflict. The more open, but still exclusive, political opportunity structure of the United States is reflected in a combination of conflict and cooperation. Among the cases examined

Table C.1 Relationship Between National Political Opportunity Structure and Subnational Water Conflict

Political Opportunity Structure Type	Open	Closed
Inclusive	Cooperation *France*	Conflict *India*
Exclusive	Conflict/cooperation *USA*	Conflict *China*

in this book, only France, which features both an open political opportunity structure alongside measures to actively encourage third-party participation in water resource management, features generally cooperative water politics. This linkage proves to be robust in terms of not only cross-case comparison but within-case variation as well: third-party involvement is critical to explaining the relative success of the French basin agencies in comparison to their counterparts in other countries, for example, but also to explaining the improving performance of interjurisdictional cooperation in the Colorado River basin over time.

But the story of effective institution-building in these cases is not simply one of democratizing decision-making. Indeed, a key lesson of the cases examined in this book is that tackling complex-adaptive challenges like ecosystem protection depends on bridging both sectional and elite politics and building partnerships between political elites, water-user groups, and bureaucratic officials. Third parties are not so much an antidote to these disparate interests as a binding agent that can hold them together within a single institutional framework that enables collaborative, participatory, basin-wide management. In both the United States and France, coalitions between national political elites and environmental advocacy groups, often aided by the co-option of bureaucratic and some local officials, eventually led to the adoption of measures like streamflow modulation that benefit natural ecosystems. In the case of France, an unlikely alliance between the anglers' associations and national political elites to improve water quality and maintain fish stocks led directly to creation of the basin agencies and committees, and these entities have continually adapted to meet new management challenges, especially complex-adaptive ones like ecosystem protection. In the United States, patient advocacy on the part of environmental organizations in combination with sympathetic officials in the Department of the Interior led ultimately to Minute 319, which restored environmental flows to the lower Colorado River after the better part of a century. These examples illustrate that effective institution-building faces something of a Goldilocks problem: it must be neither too much representative of sectional or elite political interests nor too little. The really critical role played by third parties is to help central states, constrained as they are in tackling many aspects of water resource management, realize such a happy medium.

Indeed, the final main lesson of the cases examined in this book is that while the challenges of governing international transboundary water resources have been the subject of much previous scholarly attention, central governments can find it equally as difficult to govern waters that lie entirely within their own borders and sometimes even more so. Whether or not this is the case has much to do with decentralization. In decentralized countries, the combination of institutional constraints on central government action and the resonance of shared water resources to sectional constituencies can create interjurisdictional collective action dilemmas as marked as those of any international river basin. Under decentralization, the political units that share river basins are often large and powerful enough to resist efforts to impose institutionalized collective action. But, on the other hand, the dilemma of autonomy means that interjurisdictional collective action is also unlikely to emerge organically. Just as at the international level, interjurisdictional collective action under decentralization is therefore both costly and difficult. Subnational water politics are different in degree from international hydropolitics but not in kind. Instead, common dynamics of conflict and cooperation appear to link water politics at both international and subnational levels.

Though this book is focused on subnational hydropolitics, the idea of a continuum raises several implications for understanding international water conflict as well. The first such issue relates to decentralization. The cases examined in this book suggest that more decentralized countries are likely to be at higher risk of interjurisdictional water conflict, though much depends on the particular institutional arrangements of the country at issue. The natural implication is that more decentralized countries might also be more prone to conflict with their neighbors in international transboundary basins. Several researchers have raised this issue, but systematic study does not appear to have been conducted (Giordano, Giordano, and Wolf 2002; Wirsing 2007). The emphasis placed in this book on decentralization as an enabling condition for subnational hydropolitics nonetheless suggests that its significance for international water conflict warrants further study. Indeed, a plausible pathway linking decentralization and the probability of international conflict relates to the second factor emphasized in this theory of subnational hydropolitics, sectional identity. Because more decentralized countries tend to codify sectional cleavages, expressions of sectional identity may place greater pressure on national governments to adopt a more confrontational stance with respect to shared water resources, just as subnational governments often do. However, the implication that either more decentralized or more heterogeneous countries may be more likely to engage in conflict must be balanced with this book's conclusion that political opportunity structure mediates the prospect of interjurisdictional water conflict. Countries that feature more opportunities for nongovernmental organizations (NGOs) and other third parties to participate in the political process may therefore be more likely to cooperate with their neighbors in international as well as subnational shared river basins. Indeed, in the case of the Colorado examined in Chapter 4, environmental NGOs worked with state as well as national governments to improve ecosystem management on both sides of the U.S.–Mexican border. These potential implications for the study of international as well as subnational hydropolitics point to several directions for future research in the field of water politics.

DIRECTIONS FOR FUTURE RESEARCH: TOWARD A FIELD OF WATER POLITICS

The cases examined in this book show that interjurisdictional water conflict results not from scarcity, geography, or even a lack of collective action per se but rather from the interaction of identity and a particular set of institutional arrangements. Where these are too weak or too cumbersome to restrain the influence of sectional politics, conflict often results. This understanding of conflict and cooperation over water explains how institutions, including constitutional-level arrangements like central–local relations, interact with the ideational aspects of water and the issues of power, identity, and fairness that are inherent to it. Even as the management of shared water resources is shaped by institutions at multiple levels of government, it is also exposed to sectional grievances and tensions rooted in the uniquely poignant normative issues concerned in allocating, supplying, and controlling water. Linking these institutional and ideational aspects of water points the way toward developing water politics as a distinct field of study within that of environmental politics.

Of course, the study of water politics is hardly novel. But it has yet to congeal as a subdiscipline and remains largely bifurcated between scholars concerned with the international and subnational levels and between those focused on institutions versus issues of identity and power. Even so, given the increasing demands placed on water resource management institutions around the world, few areas of environmental politics are more deserving of concerted, additional scholarly effort. Accordingly, I suggest that scholars concerned with the political aspects of water work toward building a more coherent subfield of water politics based on three principles: a comparative methodological approach, a theoretical framework centered on institutional collective action, and a research agenda that aims to better understand relationships between economic interest groups, political elites, social groups, and social movements at different levels of governance.

Most previous studies of water politics have been concerned with either individual localities or single transboundary river basins like the Nile or the Mekong. The focus on anecdotal case studies is understandable: as this book has made clear, the institutional and ideational aspects of shared water resources vary considerably across countries and regions. However, the promise of a comparative approach is that it allows scholars to identify previously unrecognized or underappreciated factors, like the structure of central–local relations and political opportunity structure, that influence the dynamics of conflict and cooperation over shared water resources. Comparative approaches also open new lines of conceptual engagement between the study of water politics and well-developed literatures like those concerned with federalism and decentralization. Such engagement in turn promises to yield new insights on lingering questions like how political system and regime type influence the course of collective action over time (Hargrave and Van de Ven 2006; Gerlak, Heikkila, and Lubell 2008).

The future study of water politics would likewise be well served by coalescing around collective action as a conceptual foundation. Of all the theoretical frameworks for understanding water politics, collective action is the most mature and sophisticated; and with some modification, it can provide the field of water politics with a coherent theory to support sustained scholarly effort. The principal weakness of collective action theory as it currently stands is its reluctance to address issues of power, sectional identity, and distributional politics that this book shows are inherent to shared water resources. However, an important element of collective action is the role of trust and reciprocity in creating and sustaining cooperation between resource users (Adger 2003; Pretty 2003; Mannemar Sonderskov 2009). The issues of identity that this book highlights map readily on to a trust-centered understanding of collective action: where sectional cleavages and grievances exist, they undermine the trust needed to build collaborative and participatory water resource management institutions. With this conceptual modification, collective action theory presents a compelling framework that explains why, where power asymmetries and brittle institutions coexist, conflict is likely but, where higher levels of social capital between water users and stronger, more flexible institutions coincide, cooperation is likely to prevail instead.

Within this framework, future research in water politics can refine theories of conflict and cooperation by focusing on several key questions. The first centers on the role of subnational political units. As this study has demonstrated, subnational units and the politicians who lead them play an unexpectedly critical role in fostering conflict and cooperation over shared water resources. Subnational units occupy a distinct meso-scale in water politics: they are usually powerful enough to challenge both central states and their neighbors when their autonomy is threatened yet also too small and, their powers too circumscribed, to manage shared water resources on their own. This scalar position warrants further attention from researchers. Indeed, the case studies examined in this book show that subnational officials frequently determine the success or failure of national water resource policy reforms.[1]

Subnational officials are often the most hesitant to adopt such reforms for fear of surrendering their autonomy or prerogatives to the central government or nongovernmental actors or of limiting their freedom of action relative to counterparts in other jurisdictions. This phenomenon is striking in its commonality across countries, and it may go some way toward explaining the observation that while "There is broad consensus on what constitutes good water resources management . . . all countries are far from managing water according to these principles" (World Bank 2004, 28). The role of subnational units and actors outlined in this book moreover forms a testable hypothesis that could be validated through additional cross-national study: that institutionalized collective action will be both quantitatively less common and qualitatively less robust between subnational jurisdictions than within them. Put another way, collective action institutions are easier to create when they are contained within single subnational political units.

However, answering this question will require a major effort in cross-national data collection. At present, few if any data exist on the global incidence of interjurisdictional water conflict or on the strength of water resource management institutions across countries. Constructing data sets on these phenomena would therefore be a significant contribution to the study of water politics and would enable a robust, cross-national test of the factors that contribute to the success and failure of water resource management policies. Quantitative, cross-national study would also allow future research to address a question raised but not answered by the case studies examined in this book: whether certain forms of decentralization reliably explain water resource management outcomes. The model of subnational hydropolitics presented in this book suggests, in particular, that federal political systems face greater challenges in securing interjurisdictional cooperation over shared water resources for two reasons. First, they typically feature institutions like powerful judiciaries, which permit subnational states to pursue interjurisdictional water disputes. Second, and more importantly, federal systems tend to institutionalize the very sectional cleavages that I have argued are the animating factor in water politics. However, forms of federalism vary widely, and answering this question requires analysis of many more political systems than has been possible in this book (Garrick et al. 2014).[2]

Second, a greater focus on substate actors would help to better contextualize the role of central states in water resource management. Suspicion of the central state runs throughout the literature on water politics. For many neo-Malthusians and scholars writing in the political ecology tradition, central states often seek to control populations and enhance their power by controlling sources of water. For institutionalists and collective action theorists, meanwhile, the central state's inclination toward command-and-control is inimical to complex-adaptive environmental management issues, such as those entailed in governing interjurisdictional river basins. This book adds nuance to these perspectives: the common tendency of central states to pursue a technobureaucratic approach to water resource management does leave them generally ill-equipped to meet complex-adaptive environmental challenges. However, in a few of the cases examined in this book, national political leaders and central states provided an important impetus for water resource management reform and interjurisdictional cooperation. In the Colorado River case, some federal officials enthusiastically supported third-party pressure for more collaborative and participatory approaches to water governance, while in the case of France, the state reacted by mandating that a certain number of seats in the country's water parliaments be reserved for civil society. My account therefore suggests that the ability of central states to meet such challenges is contingent on the political opportunity structure for nongovernmental involvement that obtains in each country. Even so, further research is again needed to validate this proposition.

Third and finally, future research in water politics should aim to better understand factors that both promote and inhibit third-party involvement in natural resource governance. While previous studies have highlighted the critical role of third party–induced cooperation in international transboundary river basins (Lowi 1993; Weinthal 2002), less attention has been paid to the different political opportunity structures offered to nongovernmental actors within countries. In particular, social movement theory raises the intriguing question, ripe for further empirical testing, of how informal institutions for civil society mobilization might promote better water resource management. Democracies like the United States adopt a "pluralistic" attitude toward social movements and organizations, offering access but no particular encouragement. Corporatist democracies like Norway, on the other hand, adopt an inclusive approach that actively encourages and solicits civil society involvement in virtually all important political decisions and processes (Dryzek et al. 2003). In addition, some peculiar features of water resource management may present special issues for third-party involvement. The "policy transitions" identified by Dustin Garrick (2015) that characterize the evolution of river basin management across countries might, for example, present more or less favorable opportunities for third-party involvement. A comparative study that sheds light on how these less formal institutional features influence third-party involvement in water resource management would greatly contribute to strengthening the field of water politics.

In addition to simply understanding the role of civil society, future work in water politics should develop a proscriptive strategy to support its involvement in water resource decision-making over time. The cases examined in this book suggest that participatory decision-making results from not simply a set of initial reforms but rather a continual process in which relationships between central and subnational states, water-user groups, and civil society are enhanced over time. The case of the United States suggests that framework legislation like the National Environmental Policy Act offers critical, initial encouragement to nongovernmental actors to engage in the political process. However, other cases examined in this book also indicate that the formal right to access the political process should be complemented by the kind of statutory access that NGOs enjoy to France's water parliaments. Perhaps the most important lesson of these case studies is that participatory and collaborative water resource management is most effectively promoted by inclusive states that actively encourage civil society participation in water resource decision-making. They therefore both bolster the case for civil society involvement in addressing pressing environmental challenges and offer several lessons for how to best support this engagement.

POLICY RECOMMENDATIONS

At the time of writing, several of the countries examined in this book were facing at least one form of interjurisdictional water conflict. In the United States, a major interstate dispute is brewing between Florida and Georgia over allocation of the Apalachicola River, while Texas and New Mexico, Wyoming and Colorado, and Mississippi and Tennessee have either filed or threatened to file lawsuits related to water-sharing agreements. In India, meanwhile, long-running disputes in peninsular river basins like the Krishna and Cauvery have been joined by new and re-emergent conflicts over the Ravi–Beas and Sutlej–Yamuna Canal Link that delivers water to New Delhi. This book has explained these different forms of subnational conflict in the context of a more general theory of water politics, stressing both institutional and ideational factors. However, it also offers several practical recommendations for preventing subnational political conflict over water and for advancing more sustainable water resource management in the United States, India, and other countries.

At the broadest level, the cases examined in this book reinforce the need for political leadership to prevent and resolve water conflicts. In most countries and contexts, national political leaders are reluctant to play this role. Electoral incentives rarely favor their involvement in political disputes over water, and its distinctive characteristics, including its complex value, its fungibility, and its multiple uses, require substantial investments of time and effort to understand—a real challenge for preoccupied politicians. Perhaps just as important, many water resource management issues unfold over multiple electoral cycles. While occasional flooding and drought crises can seize national political agendas, other causes of disputes between water users, like infrastructure construction and water pollution, represent slow-moving crises; and national political leaders must usually contend with far more pressing ones.

Yet as the case studies examined in this book attest, the politics of water respond well to constructive, high-level intervention; indeed, subnational water conflict festers most when national leaders are distracted and disengaged. Leaders in pursuit of a legacy are well advised to take on the task of bringing water users to the table. The best piece of advice for leaders who do so is to exercise a form of strategic patience. Water conflicts, as this book has demonstrated, are often rooted in sectional tensions that go far beyond water. Navigating these politics requires maintaining a certain neutrality but also a determination to employ both carrots and sticks to negotiate agreements. Political leaders engaged in these negotiations must understand that they will take time but that consistency and credibility are key to overcoming what is often a legacy of mistrust.

This book also provides several specific tools to help policymakers prevent water conflicts. First, the U.S. case indicates that countries should adopt framework legislation like the National Environmental Policy Act to codify civil society involvement in decision-making and create specific points of access, such as an environmental impact assessment process. Second, the case of France demonstrates that the involvement of nongovernmental representatives and organizations should be not merely permitted but, where possible, actively encouraged, including by specifically designating a certain number of seats in decision-making bodies for civil society. Third, representation in decision-making bodies should balance both national and subnational interests, as well as those of resource users and third parties. The case of India suggests the value of establishing specialized water resource management institutions in order to avoid some of the political manipulation prevalent in the Krishna basin, for example. Third, the cases of the Colorado, Krishna, and Yellow River basins suggest that water resource management institutions must make a deliberate attempt to engage subnational officials and seek consensus between them on basic management strategies. These case studies also suggest several specific reforms for the United States, India, and China.

In the case of the United States, the president and Congress should work together to restore funding to institutions created under the Water Resource Planning Act of 1965, especially the regional river basin commissions and the National Water Resources Council. In keeping with the provisions of the original legislation, the creation of river basin commissions should not be imposed by mandate but rather should be undertaken by agreement between states that share particular river basins. However, Congress should also take several steps to ensure that, unlike the original legislation, rejuvenated river basin governance institutions adequately incorporate nongovernmental constituencies. First, it should stipulate that at least half the seats available on the newly reconstituted commissions be reserved for nongovernmental representatives, ideally including some ordinary citizens elected by other water users. Second, the rejuvenated river basin commissions should be encouraged to establish subbasin advisory councils to convene water-user groups in particular regions, in order to further enhance participation by all relevant stakeholders and water users. Third and finally, the National Water Resources Council should be reconstituted to provide high-level strategic vision and guidance with respect to water resource policy and should include representatives from agriculture, industry, water utilities, and environmental groups.

Similar reforms would also likely enhance the performance of China's river basin commissions, which at present include only representatives of the central government. Although political sensitivities make the direct inclusion of nongovernmental actors in the commissions impractical, the commissions might instead establish subbasin-scale committees with nongovernmental advisory councils. This approach follows the government's recent model of including civil society at lower levels of administration in certain policy areas, including environmental protection. Though not an ideal solution, the advisory committees could still help to ameliorate the myriad, localized disputes that hobble water resource management in China. In addition, the river basin commissions themselves should be reformed to include representatives not only from the Ministry of Water Resources but also provincial governments and other ministries, including those of Agriculture, Environmental Protection, Land and Resources, and Housing and Urban–Rural Development. Although incumbents may well resist such reforms as a threat to their influence, expanding the commission architecture would do much to enhance the legitimacy of water resource management policies among the subnational governments that must implement them.

India, meanwhile, faces perhaps the greatest challenge in establishing effective basin-scale governance institutions, even as it possesses a strong legal foundation to do so. The government of India should immediately invoke its authority under the Seventh Schedule to the Constitution to establish river boards in all basins subject to an interstate dispute. These river boards should be constituted with representatives from central and state water resource departments, as well as farmers' associations and environmental organizations. Finally, the reconstituted river boards should be chaired by a respected leader of national stature. Composed in this manner, river boards offer an alternative forum to the highly charged political arena, in which existing sectional tensions and cleavages have been projected onto water-sharing agreements. In addition to reconstituting the river boards, the Union government should adopt a proposed suggestion from the Ministry of Water Resources to abolish the five current interstate water tribunals and replace them with a single, dedicated tribunal whose members specialize in interstate water dispute resolution (*Economic Times* 2015). Less conventionally, the government should make it a priority to embed ministry officials within state water resource departments to build trust, communication, and coordination at the state and local levels.

At the same time that governments must undertake reforms to prevent subnational water conflict, third parties must also enhance their efforts to ensure that cooperative institutions are created in river basins shared between subnational jurisdictions. A variety of foundations, academic institutions, and NGOs have focused on encouraging institution-building in international transboundary river basins; but much less effort has been invested in creating such institutions at the subnational level.

Third parties who aspire to help prevent conflict over water must therefore look to strengthen not only international institutions but interjurisdictional ones as well. This new focus is especially important for multilateral institutions, which have to date focused primarily on international transboundary rivers. In decentralized and federal country contexts, interjurisdictional conflict nonetheless represents a substantial risk factor for water-related development projects. Moreover, for countries like India, these interjurisdictional conflicts present an opportunity to multilateral institutions to deploy their experience in addressing international transboundary water conflict. Multilateral institutions should accordingly develop new protocols and procedures for assessing political economy risks at the subnational as well as national levels and should expand their capacity to engage subnational political actors. With the aid of these reforms, third parties can enhance their ability to induce cooperation between subnational jurisdictions, thereby promoting sustainable management of waters shared within countries as well as between them.

Such sustainable water resource management is likely to remain a work in progress for some time to come. The struggle to cooperatively govern interjurisdictional river basins is a common, and often neglected, feature of state-building and political development in many countries, especially those described in this book. The study of how states approach the problem of interjurisdictional cooperation in shared river basins is significant not only because of its relevance to water conflict and pressing water resource management challenges but also for what it reveals about the nature of political development across countries. Shared water resources are the classic institutional collective action problem, and how different political systems respond to it represents a kind of stress test for how flexibly countries can adapt to emerging environmental challenges like water quality and ecosystem protection. Precisely because river basins cut across such sectional borders and boundaries, they open a window onto these animating political forces. The politics of water are shaped as much by the underlying institutional and ideational as by the physical terrain.

This realization highlights the promise and the peril of sharing water between groups and regions as well as countries. It is tempting to conclude that the distinctive properties of water as a resource, including its unique value to every aspect of human life, its relative scarcity in many parts of the world, and the emotive politics of deciding who gets how much water, make conflict over water a tragically predictable outcome. The conceit underlying most thinking on water politics has therefore long been that conflict between those who share it is probable, if not inevitable. But the case studies explored in this book belie this belief: subnational conflict over water is the symptom of pre-existing sectional tensions, not its cause. When water resource managers and the institutions of which they are part embrace the role of third parties, cooperation is not only possible but likely. Even as the world enters an age marked by growing demand for water, shifting rainfall and streamflow patterns, and an epidemic of sectional tensions in some parts of the world, conflict is far from inevitable. The politics of water are animated by tensions and grievances among those who use it, and their contests can become reflected in the rivers, lakes, and streams that they share. But with the aid of third parties these shared subnational waters can, like their international counterparts, become catalysts of cooperation instead of focal points for sectional discord.

NOTES

1. This role has not gone entirely unnoticed by scholars, particularly in writing about Latin America. See Lemos and de Oliveira (2004).
2. The relationship between federal systems and water resource management has been explored at some length by Garrick et al. (2014).

GLOSSARY

Decentralization A form of government and a process that grants or transfers some combination of political power, fiscal–economic resources, and administrative responsibility to subnational levels

Institution-building The process of establishing, maintaining, adapting, and reforming formal rules, mechanisms, processes, and organizations to support interjurisdictional collective action in shared river basins

Interjurisdictional collective action Effort undertaken by several subnational political jurisdictions, like states, provinces, or counties, in pursuit of common objectives, in this case the management of shared river basins

Political opportunity structure The set of formal and informal institutions, including laws, norms, and the prevailing views of political office-holders, that determines access for individuals and nongovernmental organizations to the political process

Reclamation nexus A political coalition composed of geographically concentrated water-user groups, especially irrigators, subnational politicians, and members of national legislatures who seek central government financing for water infrastructure projects that principally benefit particular regions within individual subnational jurisdictions

Reform coalition A political coalition composed of third parties, especially national political elites and environmental nongovernmental organizations, that advocates for collaborative and participatory institutions for basin governance

Sectionalism The tendency to identify and politically organize around a particular region or geographically concentrated social group, united by a common identity as distinct from that of neighboring regions or groups

Subnational state A subnational political jurisdiction invested with significant political power, fiscal–economic resources, and administrative responsibility, such that it is sufficiently powerful to act independently from the central government

Third parties Entities other than subnational governments or economic water users in a given river basin which nonetheless have an interest in the management of shared water resources, and have reason to favour creation of collaborative, participatory governance institutions

BIBLIOGRAPHY

Abers, Rebecca, and Margaret Keck. "Mobilizing the State: The Erratic Partner in Brazil's Participatory Water Policy." *Politics & Society* 37, no. 2 (2009): 289–314.

———. "Muddy Waters: The Political Construction of Deliberative River Basin Governance in Brazil." *International Journal of Urban and Regional Research* 30, no. 3 (2006): 601–22.

———. *Practical Authority: Agency and Institutional Change in Brazilian Water Politics.* New York: Oxford University Press, 2013.

Adger, W. Neil. "Social Capital, Collective Action, and Adaptation to Climate Change." *Economic Geography* 79, no. 4 (October 2003): 387–404.

Adger, W. Neil, Katrina Myrvang Brown, and Emma L. Tompkins. "The Political Economy of Cross-Scale Networks in Resource Co-Management." *Ecology and Society* 10, no. 2 (2005): 9–23.

Administrative Reforms Commission. *Report of the Study Team on Centre–State Relationships*. Vol. 1. Simla: Government of India Press, 1968.

Advisory Commission on Intergovernmental Relations. *Multistate Regionalism.* Commission Report. Washington, DC: US Government Printing Office, 1972.

Agashe, Anand. "State's Dilemma over Krishna Projects." *Times of India,* September 8, 1992a, 19.

———. "Tussle over Krishna Water Projects." *Times of India,* November 2, 1992b, 3.

———. "Will Krishna Valley Development Project Inundate the Best Efforts to Complete It?" *Times of India,* July 18, 1996, 13.

Agashe, Anand, and Milind Kokje. "Sena-BJP Govt. Plans to Float Bonds to Tap Krishna Waters." *Times of India,* July 4, 1996, 1.

Agence d l'Eau Seine-Normandie. *Seine-Normandy Basin, France*. Pilot Case Study, United Nations World Water Assessment Programme, Paris, New York, and Oxford: United Nations Economic, Social, and Cultural Organization and Berghahn Books, 2003.

Ahmad, Shahid, Khalid Aziz, and Mujib Khan. "Inter-Jurisdictional Water Management in Pakistan's Indus Basin." In *Federal Rivers: Managing Water in Multi-Layered Political Systems,* edited by Dustin Garrick, George Anderson, Daniel Connell, and Jamie Pittock, 243–58. Cheltenham, UK: Edward Elgar, 2014.

Ainsworth, Ed. "Indians Enter Suit on Colorado River Water." *Los Angeles Times,* March 27, 1953, A1.

———. "Setup Behind Dispute over Water Outlined." *Los Angeles Times,* August 30, 1946, A2.

———. "Water Report Sets off Controversies." *Los Angeles Times,* December 18, 1950, 1.

Alam, Undala, Ousmane Dione, and Paul Jeffrey. "The Benefit-Sharing Principle: Implementing Sovereignty Bargains on Water." *Political Geography* 28 (2009): 90–100.

Albaic, Jose, Elena Calvo, and Encarna Esteban. "River Basin Governance and Water Policies in Spain." In *River Basin Governance and Water Policies in Spain*, edited by Dustin Garrick, George Anderson, Daniel Connell, and Jamie Pittock, 141–57. Cheltenham, UK: Edward Elgar, 2014.

Alcamo, Joseph, Martina Florke, and Michael Marker. "Future Long-Term Changes in Global Water Resources Driven by Socio-Economic and Climatic Changes." *Hydrological Sciences Journal* 52, no. 2 (2007): 247–75.

Amen, Mark, Noah Toly, Patricia McCarney, and Klaus Segbers. "Sighting or Slighting Cities in International Relations." In *Cities and Global Governance: New Sites for International Relations*, edited by Mark Amen, Noah Toly, Patricia McCarney, and Klaus Segbers, 13–32. Farnham, UK: Ashgate, 2011.

Amenta, Edwin. "Making the Most of a Historical Case Study: Configuration, Sequence, Casing, and the US Old-Age Pension Movement." In *The Sage Handbook of Case-Based Methods*, edited by David Byrne and Charles Ragin, 351–66. London: Sage, 2009.

Andersson, Krister, Clark Gibson, and Fabrice Lehoucq. "Municipal Politics and Forest Governance: Comparative Analysis of Decentralization in Bolivia and Guatemala." *World Development* 34, no. 3 (2006): 576–95.

Andersson, Krister, and Elinor Ostrom. "Analyzing Decentralized Resource Regimes from a Polycentric Perspective." *Policy Science* 41, no. 71 (2008): 71–93.

Archibold, Randal. "Western States Agree to Water-Sharing Pact." *New York Times*, December 10, 2007, A18.

Arizona Colorado River Commission. *Second Report of the Colorado River Commission of Arizona*. Phoenix: Colorado River Commission of Arizona, 1927.

Arizona Republican. "Division of the Waters of the Colorado River Is Maxwell's Theme." Proquest Historical Newspapers. December 17, 1922a. Accessed December 2011.

———. "Kiwanis Hear Judge Sloan Explain Benefits to State of Colorado River Compact." Proquest Historical Newspapers. December 27, 1922b. Accessed December 2011.

———. "Millions of Homes Built by Compact Says Hoover." Proquest Historical Newspapers. November 26, 1922c. Accessed December 2011.

———. "Only a Basis for Co-operation." Proquest Historical Newspapers. November 27, 1922d. Accessed December 2011.

———. "When Mines Have Passed New Arizona Will Arise Says Governor Campbell." Proquest Historical Newspapers. November 6, 1922e. Accessed December 2011.

Armstrong, Robert. "Coolidge to Aid Compact." *Los Angeles Times*, November 18, 1923, I1.

Arnell, Nigel. "Climate Change and Global Water Resources: SRES Emissions and Socio-Economic Scenarios." *Global Environmental Change* 14, no. 1 (2004): 31–52.

Arulampalam, Wiji, Sugato Dasgupta, Amrita Dhillon, and Bhaskar Dutta. "Electoral Goals and Center–State Transfers: A Theoretical Model and Empirical Evidence from India." *Journal of Development Economics* 88, no. 1 (2009): 103–19.

Arzheimer, Kai, and Elisabeth Carter. "Political Opportunity Structures and Right-Wing Extremist Party Success." *European Journal of Political Research* 45 (2006): 419–43.

Ashford, Douglas. *British Dogmatism and French Pragmatism: Central–Local Policymaking in the Welfare State*. London: George Allen and Unwin, 1982.

Assal, Munzoil A. M. "Sudan: Identity and Conflict over Natural Resources." *Development* 49, no. 3 (2006): 101–105.

Associated Press. "Artificial Flood Created to Rejuvenate the Grand Canyon." *New York Times,* March 27, 1996, B8.

———. "Species Act Battle; Law Protecting Endangered Animals, Plants Faces Review in Congress." *Boston Globe,* December 6, 1981, 1.

August, Jack. *Dividing Western Waters: Mark Wilmer and Arizona v California.* Fort Worth: Texas Christian University Press, 2007.

Awkesasne Notes. "Dry Death in Arizona." *Akwesasne Notes,* September 30, 1975, 32.

Bakker, Karen. *An Uncooperative Commodity: Privatizing Water in England and Wales.* Oxford: Oxford University Press, 2003.

———. *Privatizing Water: Governance Failure and the World's Urban Water Crisis.* Ithaca, NY: Cornell University Press, 2010.

———. "The Politics of Hydropower: Developing the Mekong." *Political Geography* 18 (1999): 209–232.

Barraque, Bernard, Pierre Maurel, and Richard Raymond. *Participation and Social Learning in the Dordogne River Basin, France: Stakeholder Workshop Report.* HarmoniCOP Project Workpackage 5. Castelnaud la Chapelle, France: EPIDOR, 2004.

Barrett, Scott. *Environment and Statecraft: The Strategy of Environmental Treaty-Making.* New York: Oxford University Press, 2003.

Barrios, Eugenio. "Water Management and Ecosystems: A New Framework in Mexico." In *Federal Rivers: Managing Water in Multi-Layered Political Systems,* edited by Dustin Garrick, George Anderson, Daniel Connell, and Jamie Pittock, 105–19. Cheltenham, UK: Edward Elgar, 2014.

Barrque, Bernard. "Les politiques de l'eau en Europe." *Revue francaise de science politique* 45, no. 3 (1995): 420–53.

Basu, Sudip Ranjan. "Comparing China and India: Is the Dividend of Economic Reforms Polarized?" *European Journal of Comparative Economics* 6, no. 1 (2009): 57–99.

Batterbury, Simon, and Jude Ferdando. "Rescaling Governance and the Impacts of Political and Environmental Decentralization: An Introduction." *World Development* 34, no. 11 (2006): 1851–63.

Beland, Daniel, and Andre Lecours. "Sub-State Nationalism and the Welfare State: Quebec and Canadian Federalism." *Nations and Nationalism* 12, no. 1 (2006): 77–96.

Benedict, Laurence. "Arizona Files Dam Suit." *Los Angeles Times,* October 7, 1930, 1.

Benford, Robert, and David Snow. "Framing Processes and Social Movements: An Overview and Assessment." *Annual Review of Sociology* 26 (2000): 611–39.

Bennett, John. "Anthropological Contributions to the Cultural Ecology and Management of Water Resources." In *Man and Water: The Social Sciences in Management of Water Resources,* edited by L. Douglas James, 34–81. Lexington: University of Kentucky Press, 1974.

Bennett, Lynne. "The Integration of Water Quality into Transboundary Allocation Agreements: Lessons from the Southwestern United States." *Agricultural Economics* 24 (2000): 113–25.

Berardo, Ramiro, and John Scholz. "Self-Organizing Policy Networks: Risk, Partner Selection, and Cooperation in Estuaries." *American Journal of Political Science* 54, no. 3 (2010): 632–49.

Berber, F. J. *Rivers in International Law.* London: Stevens and Son, 1959.

Berger, Suzanne. *The French Political System.* New York: Random House, 1974.

Berkes, Fikret. "Cross-Scale Institutional Linkages: Perspectives from the Bottom Up." In *The Drama of the Commons,* edited by Susan Stonich, Paul Stern, Nives Dolsak, Thomas Dietz, Elinor Ostrom, and Elke Weber, 293–321. Washington, DC: National Academies Press, 2002.

———. "Evolution of Co-Management: Role of Knowledge Generation, Bridging Organizations and Social Learning." *Journal of Environmental Management* 90, no. 5 (2009): 1692–1702.

———. "From Community-Based Resource Management to Complex Systems. The Scale Issue and Marine Commons." *Ecology and Society* 11, no. 1 (2006): 45–60.

Bernauer, Thomas, and Tobias Bohmelt. "Basins at Risk: Predicting International River Basin Conflict and Cooperation." *Global Environmental Politics* 14, no. 4 (2014): 116–38.

Bernauer, Thomas, Tobias Bohmelt, Halvard Buhaug, Petter Gleditsch, Theresa Tribaldos, Eivind Weibust, and Gerdis Winschnath. "Water-Related Intrastate Conflict and Cooperation (WARRIC): A New Event Dataset." *International Interactions: Empirical and Theoretical Research in International Relations* 38, no. 4 (2012): 529–45.

Bernauer, Thomas, Tobias Bohmelt, and Vally Koubi. "Environmental Changes and Violent Conflict." *Environmental Research Letters* 7 (2012): 1–8.

Bernstein, Leonard. "San Diego Pushes Ambitious System to Reclaim Water." *Los Angeles Times*, May 28, 1989, 1.

Besley, Timothy, and Stephen Coate. "Centralized Versus Decentralized Provision of Local Public Goods: A Political Economy Approach." *Journal of Public Economics* 87, no. 12 (2003): 2611–37.

Bhattacharyya, Harihar. "UPA (2004–) and Indian Federalism." In *New Dimensions of Politics in India: The United Progressive Alliance in Power*, edited by Lawrence Saez and Gurharpal Singh, 26–38. New York: Routledge, 2012.

Bickers, Kenneth, Stephanie Post, and Robert Stein. "The Political Market for Intergovernmental Cooperation." In *Self-Organizing Federalism: Collaborative Mechanisms to Mitigate Institutional Collective Action Dilemmas*, edited by Richard Feiock and John Scholz, 161–78. New York: Cambridge University Press, 2010.

Bouleau, Gabrielle. "La contribution des pêcheurs à la loi sur l'eau de 1964." *Economie rurale* 309 (2009): 9–21.

Biswas, Asit. "Indus Water Treaty: The Negotiating Process." *Water International* 17 (1992): 201–209.

Blackbourn, David. *The Conquest of Nature: Water, Landscape, and the Making of Modern Germany*. New York: Random House, 2011.

Blair, William. "3 Justices Strongly Oppose Provision Allowing US to Apportion Supplies." *New York Times*, June 3, 1963.

———. "States Act to Bar 'TVA' in Midwest." *New York Times*, January 24, 1952, 30.

———. "West's Governors Seek to Bar MVA." *New York Times*, December 1, 1949, 36.

Blakeslee, Sandra. "Restoring an Ecosystem Torn Asunder by a Dam." *New York Times*, June 11, 2002a, F1.

———. "U.S. Panel Backs a Risky Effort to Save a Grand Canyon Fish." *New York Times*, April 26, 2002b, A16.

Blanchard, Robert. "Knight's Charges Denied by Engle." *Los Angeles Times*, May 10, 1958, B1.

Blatter, Joachim, and Helen Ingram. *Reflections on Water: New Approaches to Transboundary Conflicts and Cooperation*. Cambridge, MA: MIT Press, 2001.

Blomquist, William, and Peter deLeon. "The Design and Promise of the Institutional Analysis and Development Framework." *Policy Studies Journal* 39, no. 1 (2011): 1–7.

Bobekova, Elvira. "Third Party Conflict Management of Transboundary River Disputes." *International Journal of Conflict Management* 26, no. 4 (2015): 402–26.

Boberg, Jill. *Liquid Assets: How Demographic Changes and Water Management Policies Affect Freshwater Resources*. Santa Monica, CA: RAND Corporation, 2005.

Boelens, Rutgerd. *Water, Power, and Identity: The Cultural Politics of Water in the Andes*. Abingdon, UK: Routledge, 2015.

Boelens, Rutgerd, David Getches, and Jorge Armando Guevara Gil. *Out of the Mainstream: Water Rights, Politics and Identity*. London: Earthscan, 2010.

Bohmelt, Tobias, Thomas Bernauer, Halvard Buhaug, Nils Gleditsch, Theresa Tribaldos, and Gerdis Wischnath. "Demand, Supply, and Restraint: Determinants of Domestic Water Conflict and Cooperation." *Global Environmental Change* 29 (2014): 337–48.

Bouleau, Gabrielle, Sylvain Barone, Pierre Maurel, Audrey Richard, and Geraldine Abrami. *French Case Study Report: Implementing the WFD on the Thau Basin*. Research Report No. 2.3. Limoges, France: Office International de l'Eau, 2009.

Bowman, Ann. "Social Forces, Space and Boundaries." In *The Dynamics of Federalism in National and Supranational Political Systems*, edited by Michael Pagano and Robert Leonardi, 86–100. New York: Palgrave Macmillan, 2007.

Boyarsky, Bill. "Sharing the Colorado River's Water: West Braces for a Change." *Los Angeles Times*, September 23, 1985a, 3.

———. "Imperial District Delays Vote on Water Sharing." *Los Angeles Times*, July 10, 1985b, 6.

Braun, Dietmar. "The Territorial Division of Power in Comparative Public Policy Research: An Assessment." In *Public Policy and Federalism*, edited by Dietmar Braun, 27–56. Aldershot, UK: Ashgate, 2000.

Brean, Henry. "Cooperation Keys Latest Pact to Protect Lake Mead." *Las Vegas Review-Journal* (December 15, 2014). Accessed January 1, 2015. http://www.reviewjournal.com/news/cooperation-keys-latest-pact-to-protect-lake-mead.

Briscoe, John, and R. P. S. Malik. *India's Water Economy: Bracing for a Turbulent Future*. Washington, DC: World Bank, 2006.

Brochmann, Marit, and Nils Petter Gleditsch. "Shared Rivers and Conflict—A Reconsideration." *Political Geography* 31 (2012): 519–27.

Brodsgarrd, Kjeld. "Institutional Reform and the Bianzhi System in China." *China Quarterly* 170 (2002): 361–86.

Brown, L. David. "Bridging Organizations and Sustainable Development." *Human Relations* 44, no. 8 (1991): 807–31.

Brun, Alexandre. "Gestion de l'eau en France." *Economie rurale* 309 (2009): 4–8.

Brunn, Robert. "Far West Bids for Cohesion." *Christian Science Monitor*, November 7, 1950, 3.

Bulkeley, Harriet. "Reconfiguring Environmental Governance: Towards a Politics of Scales and Networks." *Political Geography* 24 (2005): 875–902.

Buller, Henry. "Towards Sustainable Water Management: Catchment Planning in France and Britain." *Land Use Policy* 13, no. 4 (1996): 289–302.

Burgess, Robin, Matthew Hansen, Benjamin Olken, Peter Potapov, and Stefanie Sieber. *The Political Economy of Deforestation in the Tropics*. Centre for Climate Change Economics and Policy. London: Grantham Research Institute on Climate Change and the Environment, 2012.

Burns, John. "Strengthening Central CCP Control of Leadership Selection: The 1990 Nomenklatura." *China Quarterly* 138 (1994): 458–91.

Cai, Hongbin, and Daniel Treisman. "Does Competition for Capital Discipline Governments? Decentralization, Globalization, and Public Policy." *American Economic Review* 95, no. 3 (2005): 817–30.

California Colorado River Commission. *The Boulder Canyon Project: To Convert a Natural Menace into a National Resource*. San Diego: California Colorado River Commission, September 10, 1930.

Camillo, Charles, and Matthew Pearcy. *Upon Their Shoulders: A History of the Mississippi River Commission from Its Inception Through the Advent of the Modern Mississippi River and Tributaries Project*. Vicksburg: Mississippi River Commission, 2004.

Cammisa, Anne Marie. *Governments as Interest Groups: Intergovernmental Lobbying and the Federal System*. Westport, CT: Praeger, 1995.

Carpenter, Daniel. *The Forging of Bureaucratic Autonomy: Reputations, Networks, and Policy Innovation in Executive Agencies, 1862–1928*. Princeton, NJ: Princeton University Press, 2001.

Carpenter, Delph. *Report and Supplemental Report of Delph Carpenter, Commissioner for Colorado on the Colorado River Commission*. Vol. 1, *Pamphlets on Water Rights and the Use of Water Power*. Denver: Colorado River Commission, 1923.

Cash, David, W. Neil Adger, Fikret Berkes, Po Garden, Louis Lebel, Per Olsson, Lowell Pritchard, and Oran Young. "Scale and Cross-Scale Dynamics: Governance and Information in a Multi-Level World." *Ecology and Society* 11, no. 2 (2006): 1–12.

Central Waterpower, Irrigation and Navigation Commission. *River Valley Projects in India: Seven Year Plan*. New Delhi: Government of India Press, 1949.

Chakrabarty, Bidyut. *Indian Politics and Society Since Independence: Events, Processes and Ideology*. New York: Routledge, 2008.

Chandler, William. *The Myth of TVA: Conservation and Development in the Tennessee Valley, 1933–1983*. Cambridge, MA: Ballinger, 1984.

Chandrashekhar, B. "Telangana, Maharashtra CMs Ink Pact on Godavari Water Projects." *The Hindu*, March 8, 2016. Accessed September 20, 2016. http://www.thehinud.com/news/national/telangana-maharashtra-cms-ink-pact-on-godavari-water-projects/article8327039.ece.

Cheikbossian, Guillaume. "Lobbying and Rent-Seeking for Public Goods in a Fiscally Decentralized System." *Annals of Economics and Statistics* 75–76 (2004): 331–51.

———. "Rent-Seeking, Spillovers, and the Benefits of Decentralization." *Journal of Urban Economics* 63, no. 1 (2008): 217–28.

Chellaney, Brahma. *Water: Asia's New Battleground*. Washington, .D.C: Georgetown University Press, 2013.

Chene, Jean-Michel. "Integrated Water Resources Management: Theory Versus Practice." *Natural Resources Forum* 33, no. 1 (2009): 2–5.

Cheng, Youwei. "Huanghe de lishi yanbian yu Henan de xingshuai" [Historical evolution of the Yellow River and the rise and fall of Henan]. In *Huanghe yu Henan luntan wenji* [Collected works of the forum on Henan and the Yellow River], edited by Yuguo Niu, 189–96. Zhengzhou, China: Huanghe Shuili Chubanshe [Yellow River Conservancy Press], 2008.

Chi, Chao-ting. *Key Economic Areas in Chinese History as Revealed in the Development of Public Works for Water Control*. New York: Paragon Books, 1963.

Chicago Daily Tribune. "California Hard Hit by Colorado River Ruling." May 9, 1960, 1.

———. "Regional Pacts Avert US Curbs on Local Duties." May 23, 1948, 13.

———. "Senate Settles State Row over Water Division." Proquest Historical Newspapers. December 12, 1928. Accessed December 2011.

Childs, Marquis. "Cleaning up the Colorado." *Washington Post*, September 11, 1973, A21.

Chong, Howard, and David Sunding. "Water Markets and Trading." *Annual Review of Environmental Resource Economics* 31 (2006): 239–64.

Chowdhury, Arnab Roy, and Ngamjahao Kipgen. "Deluge Amidst Conflict: Hydropower Development and Displacement in the North-East Region of India." *Progress in Development Studies* 13, no. 3 (2013): 195–208.

Christian Science Monitor. "Far West States Line Up in Fight on Water Rights." December 12, 1947, 14.

———. "No Action in Sight on Colorado River." September 17, 1965, 3.

———. "Regional Base Outlined for US Planning." June 3, 1937a, 1.

———. "States See Help in Borah Plan." February 27, 1937b, 4.

Clark, Walter Gordon. *The Colorado River: History, Seven States Compact and Future Development*. Phoenix, AZ, 1924.

Clarke, Susan. "'Thinking Federally' from a Governance Perspective." In *The Dynamics of Federalism in National and Supranational Political Systems*, edited by Michael Pagano and Robert Leonardi, 55–85. New York: Palgrave Macmillan, 2007.

Coates, James. "U.S. Pulls Plug on New Dams." *Chicago Tribune*, November 15, 1987, 25.

———. "Water Woes Put Denver in a Bind." *Chicago Tribune*, October 8, 1989, 3.

Coelho, Ana Carolina, Benedito Braga, Rodrigo Flecha, Nelson Freitas, and Osman da Silva. "Main Challenges and Responses to Federalism and Water Security in Brazil." In *Federal Rivers: Managing Water in Multi-Layered Political Systems*, edited by Dustin Garrick, George Anderson, Daniel Connell, and Jamie Pittock, 120–38. Cheltenham, UK: Edward Elgar, 2014.

Colorado River Commission. *South Platte River Compact: Colorado Proceedings*. Denver, CO: Bradford-Robinson Printing, 1925.

Colorado River Commission of Arizona. *Report of the Colorado River Commission of Arizona. Annual Report, February 2, 1933–May 3, 1935*. Phoenix: Colorado River Commission of Arizona, 1935.

———. *Annual Report of the Colorado River Commission of Arizona, November 1, 1939–November 1, 1940*. Phoenix: Colorado River Commission of Arizona, 1940.

Commission on Centre–State Relations. *Report of the Commission on Centre–State Relations*. Vol. 1 New Delhi: Government of India Press, 1988.

Committee for Consultations on the Situation in Andhra Pradesh. *Report*. New Delhi: Government of India, 2010.

Conca, Ken. "Decoupling Water and Violent Conflict." *Issues in Science and Technology* 29, no. 1 (2012): 39–48.

Congressional Quarterly. "Water Resources Planning Act Passed by Congress." In *CQ Almanac 1965*, 21st ed., 759–61. Washington, DC: Congressional Quarterly, 1965. Accessed January 7, 2014, http://library.cqpress.com/cqalmanac/document.php?id=cqal65-1257794.

Connell, Daniel. *Water Politics in the Murray-Darling Basin*. Annandale, NSW, Australia: Federation Press, 2007.

Connell, Rich. "Bradley Accused of 'Inciting Water Wars.'" *Los Angeles Times*, October 31, 1985, 3.

Cooper, Bethany, Lin Crase, and Nicholas Pawsey. "Best Practice Pricing Principles and the Politics of Water Pricing." *Agricultural Water Management* 145 (2014): 92–97.

Crohurst, H. R. "Water Pollution Abatement in the United States." *American Journal of Public Health* 26 (1936): 176–80.

Damodar Valley Corporation. *Annual Report of the Damodar Valley Corporation, 1948–1949*. Calcutta, India: Santi Ghose, 1949.

———. "Damodar Valley Corporation Act. Act No. XIV of 1948." 2012. Accessed December 16, 2017. http://indiacode.nic.in/amendmentacts2012/The%20Damodar%20Valley%20Corporation%20(Amendment)%20Act.pdf.

———. *Eight Years of DVC*. Calcutta, India: Damodar Valley Corporation, 1956.

———. *Why the Damodar Scheme Deserves Top Priority Among India's Development Projects*. Calcutta, India: SC Ghose, 1948.

Date, Vidyadhar. "Krishna Waters: State's Problem of Plenty." *Times of India*, July 31, 1991, 13.

Davis, Tony. "How California Got on Board for the Water Deal." *Arizona Daily Star* December 21, 2014. Accessed January 2, 2015. http://tucson.com/news/blogs/desertblog/how-california-got-on-board-for-the-water-deal/article_b55d3960-88e4-11e4-ae69-67118550eb7f.html.

Debaere, Peter, Brian Richter, Kyle Frankel Davis, Melissa Duvall, Jessica Ann Gephart, Clark E. O'Bannon, Carolyn Pelnik, Emily Maynard Powell, and Tyler William Smith. "Water Markets as a Response to Scarcity." *Water Policy* 16, no. 4 (2014): 625–49.

Deccan Chronicle. "Water Brings Telangana State and Andhra Pradesh Close." September 9, 2014. Accessed November 22, 2014. http://www.deccanchronicle.com/140909/nation-politics/article/water-brings-telangana-state-and-andhra-pradesh-close.

Delaware River Basin Commission. *Delaware River Basin 2006–2007 Report*. Trenton, NJ: Delaware River Basin Commission, 2007.

Delli Priscoli, Jerome. "River Basin Organizations." Program in Water Conflict Management and Transformation, Oregon State University. 2007. Accessed August 14, 2012. http://www.transboundarywaters.orst.edu/research/case_studies/Documents/RiverBasinOrganizations.pdf.

Delli Priscoli, Jerome, and Aaron T. Wolf. *Managing and Transforming Water Conflicts*. New York: Cambridge University Press, 2009.

Dern, George. "The Colorado River." Address of Governor George H. Dern of Utah, Before the Western Division of the Chamber of Commerce of the United States. Colorado Springs, CO, December 7, 1926.

Derthick, Martha. *Between State and Nation: Regional Organizations of the United States*. Washington, DC: Brookings Institution, 1974.

Desai, Raj, Lev Freinkman, and Itzhak Goldberg. "Fiscal Federalism in Rentier Regions: Evidence from Russia." *Journal of Comparative Economics* 33, no. 4 (2005): 814–34.

De Stefano, Lucia, Paris Edwards, Lynette de Silva, and Aaron Wolf. "Tracking Cooperation and Conflict in International Basins: Historic and Recent Trends." *Water Policy* 12 (2010): 871–84.

Diksit, Ramesh. *The Political Geography of Federalism: An Inquiry into Origins and Stability*. New Delhi, India: Macmillan, 1975.

Dimitrov, Radoslav. "Water, Conflict, and Security: A Conceptual Minefield." *Society and Natural Resources: An International Journal* 15, no. 8 (2002): 677–91.

Dinar, Ariel, Mark Rosegrant, and Ruth Meinzen-Dick. "Water Allocation Mechanisms—Principles and Examples." Policy Research Working Paper 1779. World Bank, Washington, DC, 1997.

Dominy, Floyd. *Pacific Southwest Water Plan*. Bureau of Reclamation Report. Washington, DC: US Department of the Interior, 1964.

Dorfman, Robert. "The Technical Basis for Decision-Making." In *The Governance of Common Property Resources: Papers Presented at a Forum Conducted by Resources for the Future, Inc. in Washington, DC, January 21–22, 1974*, edited by Edwin Haefele, 5–25. Washington, DC: Resources for the Future, 1974.

Downs, Anthony. *Inside Bureaucracy*. Boston: Little, Brown, and Company, 1967.

Dryzek, John, Downes Downes, Christian Hunold, David Schlosberg, and Hans-Kristian Hernes. *Green States and Social Movements: Environmentalism in the United States, United Kingdom, Germany, and Norway*. New York: Oxford University Press, 2003.

D'Souza, Rohan. *Drowned and Dammed: Colonial Capitalism and Flood Control in Eastern India*. New Delhi, India: Oxford University Press, 2006a.

———. *Interstate Disputes over Krishna Waters: Law, Science, and Imperialism*. New Delhi, India: Orient Longman, 2006b.

Dubochet, Lucy. "Civil Society in a Middle-Income Country: Evolutions and Challenges in India." *Journal of International Development* 24 (2012): 714–27.

Dubois, Dominique. "Mission d'etude et de reflexion sur l'organisation des pouviors publics dans le domiane de la protection de l'environnement." La documentation francaise. October 2001. Accessed September 19, 2011. http://www.ladocumentationfrancaise.fr/rapports-publics/014000798/index.shtml.

Duchacek, Ivo. *Comparative Federalism: The Territorial Dimension of Politics*. New York: Holt, Rinehart and Winston, 1970.

Duffus, R. L. "'Little Capitals' Proposed." *New York Times*, January 5, 1936, E10.

Dwivedi, Ranjit. *Conflict and Collective Action: The Sardar Sarovar Project in India*. New Delhi, India: Routledge, 2006.

Dziobek, Claudia, Carlos Gutierrez Mangas, and Phebby Kufa. "Measuring Fiscal Decentralization—Exploring the IMF's Databases." IMF Working Paper, International Monetary Fund, Washington, DC, 2011.

E, Jingping. "Zhi Huanghe guihua xiuding de zongti shexiang" [General consideration on revision of the Yellow River control plan]. Zhongguo Shuili [China Water Conservancy]. February 12, 1999. Accessed November 29, 2012. http://www.chinawater.net.cn/cwr_journal/199902/990207.html.

Easter, K. William, Mark Rosegrant, and Ariel Dinar. "Formal and Informal Markets for Water: Institutions, Performance, and Constraints." *World Bank Research Observer* 14, no. 1 (1999): 99–116.

Eau Seine-Normandie. Le comite de bassins. Accessed August 8, 2017. http://www.eau-seine-normandie.fr/le-comite-de-bassin/presentation.

Economic Times. "PM Narendra Modi critical of Tribunals, asks if they are a barrier to justice." April 5, 2015. Accessed April 7, 2015. http://economictimes.indiatimes.com/news/politics-and-nation/pm-narendra-modi-critical-of-tribunals-asks-if-they-are-barrier-for-justice/articleshow/46813862.cms.

Economic Weekly. "Inter-State River Waters Controversy." May 19, 1962, 805–806. Accessed November 14, 2014. http://www.epw.in/system/files/pdf/1962_14/20/interstate_river_waters_controversy.pdf.

Editorial Board. "Florida's Water Wars." *Herald-Tribune*, October 11, 2013, A6.

Elazar, Daniel. *American Federalism: A View from the States*. New York: Harper and Row, 1984.

———. *Exploring Federalism*. Tuscaloosa: University of Alabama Press, 1987.

Espy, Willard. "More TVAs? The Issues Weighed." *New York Times*, May 7, 1950, 182.

European Environment Agency. "WISE River Basin Districts (RBDs)." July 1, 2011. Accessed December 27, 2017. https://www.eea.europa.eu/data-and-maps/data/wise-river-basin-districts-rbds-1.

Express News Service. "TN Mole Taps Information on Water Disputes." *New Indian Express*, April 27, 2013. Accessed February 9, 2015. http://www.newindianexpress.com/states/kerala/TN-mole-taps-information-on-water-disputes/2013/04/27/article1563925.ece.

Faith, Roger. "Rent-Seeking Aspects of Bureaucratic Competition." In *Toward a Theory of the Rent-Seeking Society*, edited by James Buchanan and Gordon Tullock, 332–43. College Station: Texas A&M University Press, 1980.

Falletti, Tulia. "A Sequential Theory of Decentralization: Latin American Cases in Comparative Perspective." *American Political Science Review* 99, no. 3 (2005): 327–46.

Feiock, Richard. "The Institutional Collective Action Framework." *Policy Studies Journal* 41, no. 3 (2013): 397–425.

Feiock, Richard, and John Scholz. "Self-Organizing Governance of Institutional Collective Action Dilemmas: An Overview." In *Self-Organizing Federalism: Collaborative Mechanisms to Mitigate Institutional Collective Action Dilemmas*, edited by Richard Feiock and John Scholz, 3–32. New York: Cambridge University Press, 2010.

Feitelson, Eran. "Impediments to the Management of Shared Aquifers: A Political Economy Perspective." *Hydrogeology Journal* 14, no. 3 (2006): 319–29.

Filippov, Mikhail, Peter Ordeshook, and Olga Shvetsova. *Designing Federalism: A Theory of Self-Sustainable Federal Institutions*. New York: Cambridge University Press, 2004.

Fisher, Franklin, and Annette Huber-Lee. "Economics, Water Management, and Conflict Resolution in the Middle East and Beyond." *Environment* 48, no. 3 (2006): 26–41.

Fleming, Roscoe. "Eisenhower Pushes Upper Colorado Basin Project." *Christian Science Monitor*, September 24, 1954, 3.

———. "Upper Colorado Basin Projects Eyed." *Christian Science Monitor*, April 14, 1951, 7.

———. "Upper Colorado States Concerned at Water Plans." *Christian Science Monitor*, October 15, 1963, 5.

———. "Water Dispute Swirls in West." *Christian Science Monitor*, April 11, 1955, 3.

Floress, Kristin, Linda Stalker Prokopy, and Shorna Broussard. "It's Who You Know: Social Capital, Social Networks, and Watershed Groups." *Society and Natural Resources: An International Journal* 24, no. 9 (2011): 871–86.

Foster, Charles. *Experiments in Bioregionalism: The New England River Basins Story*. Hanover, NH: University Press of New England, 1984.

Fradin, G. "Integrated Water Resource Management at River Basin Level in France: A Participatory Way to Finance and Monitor Water Investments in a Sustinable Manner." In *International Congress on River Basin Management*, 88–98. Ankara, Turkey: General Directorate of State Hydraulic Works, 2007.

Fradkin, Philip. "Rep. Aspinall, Boss of the West: Coloradan Backs Land and Water Uses." *Washington Post*, December 5, 1971, E1.

France v. Spain, Lake Lanoux Arbitration. 12 RIAA 281 (International Court of Arbitration, November 16, 1957).

Francis, Lorania. "Sea Water Distillation Plans Hit." *Los Angeles Times*, February 3, 1949a, 9.

———. "Truman Urges Halt in Water Fight; Opposes Central Arizona Project." *Los Angeles Times*, February 8, 1949b, 1.

Francis, Warren. "Arizona's Water Suit Challenged." *Los Angeles Times*, May 20, 1953a, 24.

———. "Chapman Aids Arizona Water Plan." *Los Angeles Times*, July 29, 1952a, 14.

———. "More Funds Asked for State Projects." *Los Angeles Times*, January 22, 1952b, 7.

———. "Official Urges Congress Settle Colorado Rights." *Los Angeles Times*, March 19, 1949a, 6.

———. "Outlays on California Public Works Debated at Colorado River Hearing." *Los Angeles Times*, April 29, 1949b, 13.

———. "Senate Group Swamped by River Flow Estimates." *Los Angeles Times*, March 26, 1949c, 6.

———. "Shelve All Water Plans, Board Urges." *Los Angeles Times*, December 18, 1950, 21.

———. "US asks Supreme Court edict on Colorado Water." *Los Angeles Times*, November 3, 1953b, 18.

———. "US Drops Plea on Colorado River." *Los Angeles Times*, November 10, 1953c, 15.

Franda, Marcus. *West Bengal and the Federalizing Process in India*. Princeton, NJ: Princeton University Press, 1968.

Frey, Frederick. "The Political Context of Conflict and Cooperation over International River Basins." *Water International* 18, no. 1 (1993): 54–68.

Fry, Earl H. *States in the International Economy: An American Overview*. Vol. 2, *States and Provinces in the International Economy*, edited by Douglas M. Brown and Earl H. Fry, 23–44. Berkeley, CA: Institute of Governmental Studies Press, 1993.

Furlong, K., N. P. Gleditsch, and H. Hegre. "Geographic Opportunity and Neo-Malthusain Willingness: Boundaries, Shared Rivers and Conflict." *International Interactions* 32, no. 1 (2006): 1–30.

Gansu Dizhi Kuangchan Bao [Gansu Geological and Minerals Report]. "Shei zai wuran Huanghe?" [Who's polluting the Yellow River?]. *Xibu Kaifa* [Western Opening and Development] (2004): 37–38.

Gansusheng Heishanxia shuidian kaifa gongcheng lingdao xiaozu bangongshi [Office of the Leading Group for Power Development in Heishanxia, Gansu Province]. "Huanghe Heishanxia heduan diaoti guihua qingkuang jianjie" [Short introduction to the adjustment of the planning situation for the Heishanxia stretch of the Yellow River]. In *Huanghe Gansu Heishanxia heduan duoji kaifa* [Multistage development of the Heishanxia stretch of the Yellow River]. March 27, 2006. Accessed December 12, 2017. http://www.china5e.com/news/news-942516-1.html.

Gansu Shui Han Zai Hai Bianweihui [Gansu Flooding and Drought Disasters Publications Committee]. *Gansu Shui Han Zaihai* [Gansu flooding and drought disasters]. Zhengzhou, China: Huanghe Shuili Chubanshe [Yellow River Conservancy Press], 1996.

Gao, Bowen. "Jianyi zai Huanghe zhongyou shixing xiao liuyu zonghe zhili" [Advice on enacting comprehensive sub-basin control in the middle reaches of the Yellow River]. *Renmin Ribao* [People's Daily], January 15, 1981, 3.

Garg, Santosh. *International and Interstate River Disputes*. New Delhi, India: Laxmi Publications, 1999.

Garrick, Dustin. *Water Allocation in Rivers Under Pressure: Water Trading, Transaction Costs and Transboundary Governance in the Western US and Australia*. Cheltenham, UK: Edward Elgar, 2015.

Garrick, Dustin, George Anderson, Daniel Connell, and Jamie Pittock. "Federal Rivers: A Critical Overview of Water Governance Challenges in Federal Systems." In *Federal Rivers: Managing Water in Multi-Layered Political Systems*, edited by Dustin Garrick, George Anderson, Daniel Connell, and Jamie Pittock, 3–19. Cheltenham, UK: Edward Elgar, 2014.

Garrick, Dustin, Lucia De Stefano, Fai Fung, Jamie Pittock, Edella Schlager, Mark New, and Daniel Connell. "Managing Hydroclimactic Risks in Federal Rivers: A Diagnostic Assessment." *Philosophical Transactions of the Royal Society A* 371 (2013): 1–16.

Gautam, Sid. "Inter-State Water Disputes: A Case Study of India." *Water Resources Bulletin* 12, no. 5 (1976): 1061–70.

Gebert, Rita. "The Cauvery River Dispute: Hydrological Politics in Indian Federalism." MA thesis, University of British Columbia, 1983.

Geltzler, Joshua. *A History of Water Rights at Common Law*. Oxford: Oxford University Press, 2004.

Gentzler, Keith, and John Hines. "Delaware, Susquehanna, and Potomac: Three Federal Compact Commissions of the Northeast United States." In *Interstate Water Allocation in Alabama, Florida, and Georgia*, edited by Jeffrey Jordan and Aaron Wolf, 78–101. Gainesville: University Press of Florida, 2006.

Gerlak, Andrea, Tanya Heikkila, and Mark Lubell. "The Promise and Performance of Collaborative Governance." In *The Oxford Handbook of U.S. Environmental Policy*, edited by Sheldon Kamieniecki and Michael Kraft, 413–36. New York: Oxford University Press, 2008.

Gerring, John, Daniel Ziblatt, Johan Van Gorp, and Julian Arevalo. "An Institutional Theory of Direct and Indirect Rule." *World Politics* 63, no. 3 (July 2011): 377–433.

Gervasoni, Carlos. "A Rentier Theory of Subnational Regimes: Fiscal Federalism, Democracy, and Authoritarianism in the Argentine Provinces." *World Politics* 62, no. 2 (2010): 302–40.

Getches, David. "Competing Demands for the Colorado River." *University of Colorado Law Review* 56 (1984–1985): 413–80.

Ghiotti, Stephane. "Les territoires de l'eau et la decentralisation. La gouvernance de bassin versant ou les limites d'une evidence." *Developpement durable et territoires* 6 (2006): 2–26.

Gibbons, Roger. "Federal Societies, Institutions, and Politics." In *Federalism and the Role of the State*, edited by Herman Bakvis and William Chandler, 15–31. Toronto: University of Toronto Press, 1987.

Gibson, Edward. "Boundary Control: Subnational Authoritarianism in Democratic Countries." *World Politics* 58, no. 1 (2005): 101–32.

———. "Politics of the Periphery: An Introduction to Sub-National Authoritarianism and Democratization in Latin America." *Journal of Politics in Latin America* 2 (2010): 3–12.

Gibson, Edward, and Ernesto Calvo. "Federalism and Low-Maintenance Constituencies: Territorial Dimensions of Economic Reform in Argentina." *Studies in Comparative International Development* 35 (2000): 32–55.

Gimpel, James, and Jason Schnuknecht. *Patchwork Nation: Sectionalism and Political Change in American Politics*. Ann Arbor: University of Michigan Press, 2004.

Giordano, Mark, Zhongping Zhu, Ximing Cai, Shangqi Hong, Xuecheng Zhang, and Yunpeng Xue. *Water Management in the Yellow River Basin: Background, Current Critical Issues and Future Research Needs.* Comprehensive Assessment Research Report, Comprehensive Assessment Secretariat, International Water Management Institute. Colombo, Sri Lanka: Comprehensive Assessment of Water Management in Agriculture, 2004.

Giordano, Meredith, Mark Giordano, and Aaron Wolf. "The Geography of Water Conflict and Cooperation: Internal Pressures and International Manifestations." *Geographical Journal* 168, no. 4 (2002): 293–312.

Gizelis, Theodora-Ismene, and Amanda E. Wooden. "Water Resources, Institutions, & Intrastate Conflict." *Political Geography* 29 (2010): 444–53.

Gleditsch, Nils Petter, Kathryn Furlong, Havard Hegre, Bethany Lacina, and Taylor Owen. "Conflicts Over Shared Rivers: Resource Scarcity or Fuzzy Boundaries?" *Political Geography* 25 (2006): 361–82.

Gleick, Peter. "The Human Right to Water." *Water Policy* 1, no. 5 (1998): 487–503.

———. "Water and Conflict: Fresh Water Resources and International Security." *International Security* 18, no. 1 (1993): 79–112.

Gleick, Peter, and Meena Palaniappan. "Peak Water Limits to Freshwater Withdrawal and Use." *Proceedings of the National Academy of Sciences USA* 107, no. 25 (2010): 11155–62.

Goodman, Jack. "Protecting the Colorado's Banks." *New York Times,* February 2, 1969, XX15.

Government of India Act (1935. 26 Geo. 5. Ch. 2).

Government of Madras (Public Works Department). "Agreement Dates 18th February 1892 Between the Governments of Mysore and Madras Regarding New Irrigation Works." In *Agreements on Development of Inter-State and International Rivers,* edited by Central Water Commission, 169–71. New Delhi: Government of India, 1979.

Government of Maharasthra. "Maharasthra Krishna Valley Development Corporation Act of 1996." In *Maharasthra Government Gazette.* Bombay: Government of Maharasthra, 1996.

Gravier, Jean-Francois. *Paris et le desert francais.* Paris: Flammarion, 1947.

Green, Nat. "Positive Spillover? Impact of the Songhua River Benzene Incident on China's Environmental Policy." *China Environment Forum.* March 2009. https://www.wilsoncenter.org/publication/positive-spillover-impact-the-songhua-river-benzene-incident-china-s-environmental.

Greenhouse, Linda. "Court Refuses to Increase Water Rights of 5 Tribes." *New York Times,* March 31, 1983, A19.

Greer, Charles. *Water Management in the Yellow River Basin of China.* Austin: University of Texas Press, 1979.

Greider, William. "A Tribal Water Fight." *Washington Post,* June 29, 1969, 47.

Gulhati, N. D. *Development of Inter-State Rivers: Law and Practice in India.* New Delhi, India: Allied Publishers, 1972.

Guojia Fazhan Jihua Weiyuanhui [National Development Planning Commission]/Shuilibu [Ministry of Water Resources]. "Huanghe kegong shuiliang niandu fenpei ji ganliu shuiliang diaodu fangan" [Yellow River annual water quantity useage allocation and drought water quantity adjustment plan]. Shanxi Hewu Ju [Shanxi Province Water Affairs Bureau]. 1998. Accessed December 16, 2017. http://www.gov.cn/flfg/2006-07/28/content_367584.htm.

Guojia Renda [National People's Congress]. "Kuaxingzhengquyu de shuiwuran jiufen yingdang ruhe chuli?" [How should intrans-jurisdictional water pollution disputes be handled?]. *Zhongguo Renda Wang* [China National People's Congress Net]. April 17, 2012. Accessed April 17, 2012. http://www.yellowriver.gov.cn/xwzx/sszl/201202/t20120213_113788.html.

Guowuyuan [State Council]. *Guowuyuan Bangongting Zhuanfa Guojiajiwei he Shuidianbu guanyu Huanghe Kegong shuiliang fenpei fangan baogao de tongzhi [Notice by the State Council Office to the State Planning Commission and the Ministry of Water and Power regarding the Report on the Ye.* Guowuyuan Wenjian [State Council Document]. Beijing: State Council, 1987.

Gustafsson, Jan-Erik. *Water Resources Development in the People's Republic of China.* Stockholm, Sweden: Royal Institute of Technology, 1984.

Gyawali, Dipak. "Institutional Forces Behind Water Conflict in the Ganga Plains." *GeoJournal* 47, no. 3 (1999): 443–52.

Hammer, Chris. *The River: A Journey Through the Murray-Darling Basin.* Melbourne, Australia: Melbourne University Press, 2010.

Hardy, Scott, and Tomas Koontz. "Rules for Collaboration: Institutional Analysis of Group Membership and Levels of Action in Watershed Partnerships." *Policy Studies Journal* 37, no. 3 (2009): 393–414.

Hargrave, Timothy, and Andrew Van de Ven. "A Collective Action Model of Institutional Innovation." *Academy of Management Review* 31, no. 4 (2006): 864–88.

Harris, Leila, and Samer Alatout. "Negotiating hydro-scales, forging states: Comparison of the upper Tigris/Euphrates and Jordan River basins." *Political Geography* 29 (2010): 148–56.

Harrison, Kathryn. *Passing the Buck: Federalism and Canadian Environmental Policy.* Vancouver, Canada: University of British Columbia Press, 1996.

Harriss, John. "Antinomies of Empowerment: Observations on Civil Society, Politics and Urban Governance in India." *Economic and Political Weekly* 42, no. 26 (2007): 2716–24.

Hart, Henry. "Toward a Political Science of Water Resources Decisions." In *Man and Water: The Social Sciences in Management of Water Resources,* edited by L. Douglas James, 122–63. Lexington: University of Kentucky Press, 1974.

Hayward, J. E. S. "From Functional Regionalism to Functional Representation in France: The Battle of Brittany." *Political Studies* 27, no. 1 (1969): 48–75.

Hebert, Ray. "Unity Urged in Colorado River States." *Los Angeles Times,* December 2, 1960, A16.

Heikkila, Tanya, Edella Schlager, and Mark Davis. "The Role of Cross-Scale Institutional Linkages in Common Pool Resource Management: Assessing Interstate River Compacts." *Policy Studies Journal* 39, no. 1 (2011): 121–45.

Henansheng Huanghe Hewuju [Henan Provincial Yellow River Water Resource Bureau]. "Jiantan Huanghe shuiziyuan guanli yu baohu" [Talking about Yellow River Water Resource Management and Protection]. *Shuilibu* [MWR]. September 15, 2011. Accessed April 5, 2012. http://slfjq.mwr.gov.cn/ssthj/201109/t20110915_305094.html.

Hidalgo, F. Daniel, Suresh Naidu, Simeon Nichter, and Neal Richardson. "Economic Determinants of Land Invasions." *Review of Economics and Statistics* 92, no. 3 (2010): 505–23.

Hill, Gladwin. "Critics Scrutinize Coast Water Board." *New York Times,* December 31, 1978, 10.

Homer-Dixon, Thomas. *Environment, Scarcity, and Violence.* Princeton, NJ: Princeton University Press, 2000.

Hooghe, Liesbet and Gary Marks. "Types of Multi-Level Governance." In *Handbook on Multi-level Governance,* edited by Henrik Enderlein, Sonja Walti, and Michael Zurn, 17–31. Cheltenham, UK: Edward Elgar, 2010.

———. "Unraveling the Central State, but How? Types of Multi-Level Governance." *American Political Science Review* 97, no. 2 (2003): 233–43.

Hou, Liqiang. "River Chief System Established to Push Anti-Pollution Efforts." *China Daily,* January 23, 2017. Accessed May 10, 2017. http://usa.chinadaily.com.cn/epaper/2017-01/23/content_28035727.htm.

Howe, Charles. "The Return to the River Basin; The Increasing Costs of 'Jurisdictional Externalities.'" *Journal of Contemporary Water Research and Education* 131, no. 1 (2005): 26–32.

Hsiang, Solomon, and Marshall Burke. "Climate, Conflict, and Social Stability: What Does the Evidence Say?" *Climatic Change* 123 (2014): 39–55.

Huanghe Sanmenxia Shuili Shuniu Zhibian [Yellow River Sanmenxia Water Conservancy Project Editorial Committee]. *Huanghe Sanmenxia Shuili Shuniu zhi* [History of the

Yellow River Sanmenxia Water Conservancy Project]. Beijing: Zhongguo Dabaike Quanshu Chubanshe [China Dabaike Quanshu Press], 1993.

Huanghe Shuili Weiyuanhui [Yellow River Conservancy Commission]. *Dongying Shi Huanghe Zhi* [History of the Yellow River at Dongying City]. Shandong, China: Qilu Shushe [Qilu Press], 1995.

———. *Huanghe Guihuazhi* [History of Yellow River Planning]. Vol. 6. Zhengzhou, China: Henan Renmin Chubanshe [Henan People's Press], 1991.

Huanghe Shuili Weiyuanhui Huanghezhi Zongbianjishi [Yellow River Conservancy Commission, Yellow River History Editorial Office]. *Henan Huanghezhi* [Yellow River History of Henan]. Zhengzhou, China: Huangheshuiliweiyuanhui, 1986.

Huangshuihui [Yellow River Conservancy Commission]. *"Huanghe shuiliang tiaodu tiaoli" zhiding yu shijian* [Formulation and implementation of the "Yellow River Water Quantity Dispatching Regulations"]. August 14, 2011. Accessed April 5, 2012. http://www.yellowriver.gov.cn/zlcp/kjcg/kjcg07/201108/t20110814_103291.html.

Hubei Sheng Huanggang Diqu Shuili Dianli Ju [Huanggang Prefecture, Hubei Bureau of Water Conservancy and Electric Power]. "Management of the Meichuan Reservoir Irrigation Project." In *Water Management Organization in the People's Republic of China*, edited by James Nickum, 77–102. Armonk, NY: ME Sharpe, 1977.

Huffman, James. "Comprehensive River Basin Management: The Limits of Collaborative, Stakeholder-Based, Water Governance." *Natural Resources Journal* 49 (2009): 117–49.

Hundley, Norris. *Water and the West: The Colorado River Compact and the Politics of Water in the American West.* Berkeley: University of California Press, 1975.

IANS. "Goa CM Urges 'Big Brother' Karnataka to Check Violence over Water Dispute." *Times of India*, July 30, 2016. Accessed December 16, 2017. https://www.oneindia.com/india/goa-cm-urges-big-brother-karnataka-check-violence-over-water-dispute-2167877.html

Imperial, Mark. "Institutional Analysis and Ecosystem-Based Management: The Institutional Analysis and Development Framework." *Environmental Management* 24, no. 4 (1999): 449–65.

———. "Using Collaboration as a Governance Strategy." *Administration and Society* 37, no. 3 (2005): 281–320.

Imperial, Mark, and Tracy Yandle. "Taking Institutions Seriously: Using the IAD Framework to Analyze Fisheries Policy." *Society & Natural Resources: An International Journal* 18, no. 6 (2005): 493–509.

Ingram, Helen. "Patterns of Politics in Water Resources Development." *Natural Resources Journal* 11 (1971): 102–18.

Ingram, Helen, John Whiteley, and Richard Perry. "The Importance of Equity and the Limits of Efficiency in Water Resources." In *Water, Place, and Equity*, edited by John Whiteley, Helen Ingram, and Richard Perry, 1–32. Cambridge, MA: MIT Press, 2008.

International Conference on Water and the Environment. The Dublin Statement on Water and Sustainable Development. January 31, 1992. Accessed January 3, 2010. http://www.un-documents.net/h2o-dub.htm.

Irwin, Don. "Water Measure's Sponsors Gloomy." *Los Angeles Times*, September 8, 1966, 3.

Iyer, Ramaswamy. "Indian Federalism and Water Resources." *International Journal of Water Resources Development* 10, no. 2 (1994): 191–202.

Jahiel, Abigail. "The Contradictory Impact of Reform on Environmental Protection in China." *China Quarterly* 149 (1997): 81–103.

Jain, Purnendra. *Japan's Subnational Governments in International Affairs.* Abingdon, UK: Routledge, 2005.

Jamwal, Nidhi. "Mahadayi Dispute: Karnataka Villagers Say They Wanted Water, but Got Lathis." Scroll.in. August 26, 2016. Accessed September 3, 2016. http://scroll.in/article/814801/mahadayi-dispute-karnataka-villagers-say-they-wanted-water-but-got-laathis.

Jiang, Hong. "Anhui daibiaotuan: jinkuai jianli lianfang liankong jizhi, zhili kuajie shuiwuran" [Anhui delegation: Establish joint prevention and control mechanisms to manage transboundary water pollution]. *Phoenix News*, March 5, 2013. http://news.ifeng.com/mainland/special/2013lianghui/detail_2013_03/05/22769706_0.shtml.

Johnson, Rich. *The Central Arizona Project, 1918–1968*. Tucson: University of Arizona Press, 1977.

Jones, C. A. "River Compact Rift Analyzed." *Los Angeles Times*, September 3, 1927, 2.

Jothi, Prakash. "Gowda Assures State of Fiscal Aid for Krishna Valley Irrigation Projects." *Times of India*, August 13, 1996, 3.

Journal Officiel. "1re Seance de Mercredi 16 Octobre 1963." *Debats Parlementaires—Assemblée Nationale—Compte Rendu Integral—10e Seance*. No. Annee 1963–1964. Paris: Assemblée Nationale, October 17, 1963a.

———. "1re Seance du Jeudi 14 Novembre 1963." *Debats Parlementaires—Assemblée Nationale—Compte Rendu Integral—8e Seance*. Mo. Annee 1963–1964. Paris: Assemblée Nationale, November 15, 1963b.

———. "Seance de 2 Mai 1994." *Debats Parlementaires—Assemblée Nationale—10e Legislature—Questions Ecrites—Remises à la presidence de l'assemblée nationale et reponses des ministres*. No. Annee 1994. Paris: Assemblée Nationale, May 2, 1994.

———. "Seance du Mercredi 16 Octobre 1963." *Debats Parlementaires—Assemblée Nationale—Compte Rendu Integral—10e Seance*. No. Annee 1963–1964. Paris: Assemblée Nationale, October 17, 1963c.

Jung, Ali Nawaz. "Supplement to 1944 Agreement Executed in December 1945 by the Representatives of Governments of Hyderabad, Mysore, Madras, and India." In *Agreements on Development of Inter-State and International Rivers*, edited by Central Water Commission, 191. New Delhi: Government of India, 1979.

Kahl, Colin. *States, Scarcity, and Civil Strife in the Developing World*. Princeton, NJ: Princeton University Press, 2006.

Kamalesh, Ray. "The DVC: Need for Re-thinking." *Times of India*, February 15, 1963, 8.

Keating, Michael. *The New Regionalism in Western Europe: Territorial Restructuring and Political Change*. Cheltenham, UK: Edward Elgar, 1998.

Keck, Margaret, and Kathryn Sikkink. *Activists Beyond Borders: Advocacy Networks in International Politics*. Ithaca, NY: Cornell University Press, 1998.

Kennedy, Paul. "US Report Charts Vast Development of Missouri Basin." *New York Times*, February 21, 1953, 1.

Kenworthy, E. W. "Environmental Defense Group Suing Federal Agency to Prevent the Use of Irrigation Water for Industry." *New York Times*, October 17, 1973, 13.

Kenworthy, Tom. "Arid State Gets a Taste of Imported Water, and It's Bitter." *Washington Post*, January 1, 1994a, A14.

———. "Dam Limits Proposed to Aid Grand Canyon." *Washington Post*, January 7, 1994b, A3.

Kincaid, John. "Foreign Relations of Sub-national Units: Constituent Diplomacy in Federal Systems." In *Federalims in a Changing World—Learning from Each Other*, edited by Raoul Blindenbacher and Arnold Koller, 74–96. Montreal: McGill-Queen's University Press, 2003.

King, Gary, Robert Keohane, and Sidney Verba. *Designing Social Inquiry: Scientific Inference in Qualitative Research*. Princeton, NJ: Princeton University Press, 1994.

Kitschelt, Herbert. "Political Opportunity Structures and Political Protest: Anti-Nuclear Movements in Four Democracies." *British Journal of Political Science* 16, no. 1 (1986): 57–85.

Knudson, Thomas. "Colorado Farmers Threaten to Sue U.S. to Stop Water Delivery." *New York Times*, June 20, 1987, 15.

Koontz, Tomas, Toddi Stelman, JoAnn Carmin, Katrina Korfmacher, Cassandra Moseley, and Craig Thomas. *Collaborative Environmental Management*. Washington, DC: Resources for the Future, 2004.

Kramer, Annika, Aaron T. Wolf, Alexander Carius, and Geoffrey D. Dabelko. "The Key to Managing Conflict and Cooperation over Water." A World of Science. 11, no. 1 (2013): 4-12.

Kraul, Chris. "San Diegans to Be Asked to Reduce Water Use." *Los Angeles Times*, May 19, 1988, 1.

Kriesi, Hanspeter. "The Political Opportunity Structure of New Social Movements: Its Impact on Their Mobilization." In *The Politics of Social Protest: Comparative Perspectives on States and Social Movements*, edited by J. Craig Jenkins and Bert Klandermans, 167–98. Minneapolis: University of Minnesota Press, 1995.

Kriesi, Hanspeter, Ruud Koopmans, Jan Willem Duyvendak, and Marco Giugni. "New Social Movements and Political Opportunities in Western Europe." *European Journal of Political Research* 22, no. 2 (1992): 219–44.

Krishna Water Disputes Tribunal. *Further Report of the Krishna Water Disputes Tribunal*. Vol. 2. New Delhi: Government of India, 1976.

———. *The Report of the Krishna Water Disputes Tribunal*. Vol. 1. New Delhi: Government of India, 1973a.

———. *The Report of the Krishna Water Disputes Tribunal with the Decision*. Vol. 2. New Delhi: Government of India, 1973b.

Krishnakumar, P. K. "Telangana Raises Concerns of Inter-State Sharing of Water." *Economic Times*, August 2, 2013. Accessed November 22, 2014. http://articles.economictimes.indiatimes.com/2013-08-02/news/41008271_1_andhra-pradesh-rice-production-seemandhra.

Kruckman, Arnold. "Vast Irrigation Project: Southwest Pushing Plan to Reclaim 3,000,000 Acres of Desert Land." *New York Times*, April 3, 1921.

Lampton, David. "Water: Challenge to a Fragmented Political System." In *Policy Implementation in Post-Mao China*, edited by David Lampton, 157–89. Berkeley: University of California Press, 1987.

Lancaster, John. "Lujan Bars Testimony by 2 Officials." *Washington Post*, April 27, 1990, A17.

Landry, Pierre. *Decentralized Authoritarianism in China: The Communist Party's Control of Local Elites in the Post-Mao Era*. New York: Cambridge University Press, 2008.

Lanfranchi, Guillaume. *Evolution of the Water Governance in the French River Basin*. Technical Synthesis. Limoges, France: Office International de l'Eau, 2008.

Lankford, Bruce, and Nick Hepworth. "The Cathedral and the Bazaar: Monocentric and Polycentric River Basin Management." *Water Alternatives* 3, no. 1 (2010): 82–101.

Larkins, Jeremy. *From Hierarchy to Anarchy: Territory and Politics Before Westphalia*. New York: Palgrave Macmillan, 2010.

Leach, William, and Paul Sabatier. "Are Trust and Social Capital the Keys to Success? Watershed Partnerships in California and Washington." In *Swimming Upstream: Collaborative Approaches to Watershed Management*, by Paul Sabatier, Will Focht, Mark Lubell, Zev Trachtenberg, Arnold Vedlitz, and Marty Matlock, 233–49. Cambridge, MA: MIT Press, 2005.

Le Gales, Patrick, and Gilles Pinson. "Local/Regional Governments and Centre–Periphery Relations in the Fifth Republic." In *The French Fifth Republic at Fifty: Beyond Stereotypes*, edited by Sylvain Brouard, Andrew Appleton, and Amy Mazur, 208–25. New York: Palgrave Macmillan, 2009.

LegiFrance. "Act No. 64-1245 of 16 December 1964 on the Rules and the Distribution of Water and the Fight Against Pollution." Civil Code. February 21, 2004. Accessed December 16, 2017. http://eur-lex.europa.eu/legal-content/EN/TXT/?uri=CELEX%3A32007D0051.

LeMarquand, David. "Politics of International River Basin Cooperation and Management." *Natural Resources Journal* 16 (1976): 883–902.

Lemos, Maria, and Joao Lucio Farias de Oliveira. "Can Water Reform Survive Politics? Institutional Change and River Basin Management in Ceara, Northeast Brazil." *World Development* 32, no. 12 (2004): 2121–37.

Les Agences de l'Eau. *The Water Agencies and French Water Policy*. Paris: Ministere de l'Ecologie, du Developpement durable, des Transports et du Logement, 2014.

Li, Chengsi, Tong Kenan, and Xingfei Long. "Kuajie shuiwuran jiufen weihe duonian nanjue?" [Why are transboundary water pollution disputes intractable?]. *Zhongguo huanjing bao* [China Environmental Report]. March 12, 2015. http://www.chinanews.com/ny/2015/03-12/7122747.shtml.

Li, Dong. "Guowuyuan Huanghe fenshui yu shuidiao fangan" [State Council Yellow River Water Division and Water Transfer Plan]. Shui Xinxi Wang [Water Information Net]. March 2, 2010. Accessed December 17, 2017. http://www.hydropower.org.cn/showNewsDetail.asp?nsId=2355 Li, Gang. "Haunghe zhili mingnian quanmian chengbao" [Comprehensive contracting for Yellow River control from next year]. *Renmin Ribao* [People's Daily], December 26, 1987, 1.

Li, Guangchun. "Jiaqiang jinggong guanli gongzuo ba Huanghe de shiye banhao" [Strengthenining operation and administration work to improve Yellow River undertakings]. *Gongzuo Yanjiu* [Operations Research] (1999): 54–56.

Li, Hao, Tao Liu, and Wei Huang. "Kuajie shuiziyuan chongtu dongyin yu xietiao moshi yanjiu" [Research on the drivers and resolution modes of transbounday water resource disputes]. *Ziran Ziyuan Xuebao* [Journal of Natural Resources] 25, no. 5 (2010).

Li, Lianjiang, and Kevin O'Brien. "Villagers and Popular Resistance in Contemporary China." *Modern China* 22, no. 1 (1996): 28–61.

Lichtenstein, Grace. "Drought Perils 13 Western States." *New York Times*, February 2, 1977a, 1.

———. "West's Water Fight Intensifies as Carter Plans Project Cuts." *New York Times*, March 7, 1977b, 1.

Lichtenthaler, Gerhard. *Political Ecology and the Role of Water: Environment, Society, and Economy in Northern Yemen*. Aldershot, UK: Ashgate, 2003.

———. "Power, Politics, and Patronage: Adaptation of Water Rights Among Yemen's Northern Highland Tribes." *Etudes rurales* 155–156 (2000): 143–66.

Lieberthal, Kenneth. "China: The Three Gorges Project." In *Culture and Negotiation: The Resolution of Water Disputes*, edited by Guy Olivier Fuare and Jeffrey Zubin, 176–95. London: Sage, 1993.

———. "Introduction: The 'Fragmented Authoritarianism' Model and Its Limitations." In *Bureaucracy, Politics, and Decision-Making in Post-Mao China*, edited by Kenneth Lieberthal and David Lampton, 1–30. Berkeley: University of California Press, 1992.

Lieberthal, Kenneth, and Michel Oksenberg. *Policy Making in China: Leaders, Structures, and Processes*. Princeton, NJ: Princeton University Press, 1988.

Lijphart, Arend. *Patterns of Democracy: Government Forms and Performance in Thirty-Six Countries*. New Haven, CT: Yale University Press, 1999.

Lilienthal, David. "The TVA: An Experiment in the "Grass Roots" Administration of Federal Functions." *Australian Journal of Public Administration* 4, no. 4 (1942): 155–172.

Lipscomb, Molly, and A. M. Mobarak. "Decentralization and Pollution Spillovers: Evidence from the Re-Drawing of County Borders in Brazil." *Review of Economic Studies* 84 (2017): 464-502.

Liu, Bin, and Robert Speed. "Water Resources Management in the People's Republic of China." In *Water Resources Management in the People's Republic of China*, edited by Xuetao Sun, Robert Speed, and Dajun Shen, 5–20. New York: Routledge, 2010.

Liu, Hong, and Peng An. "Jiji xietiao shengji shuishi jiufen, cujin Jing-Jin-Ji dushiquan hexie fazhan" [Actively resolve inter-provincial water disputes, encourage harmonious development in the Beijing–Tianjin–Hebei Capital Region]. *Haihe shuili* [Water Resources] 2 (2007): 14–16.

Lok Sabha. "Synopsis of Debates: Inter-State Water Disputes (Amendment) Bill, 2001–Contd." Parliament of India. August 3, 2001. Accessed October 30, 2011. http://parliamentofindia.nic.in/lsdeb/ls13/ses7/030801.html.

Los Angeles Times. "Arizona Swats Compact Again." Proquest Historical Newspapers. January 30, 1923a. Accessed December 2011.

———. "Arizona Water Treaty Urged." November 5, 1931. Accessed December 2011.

———. "Arizonans Give Assurance on Accord." Proquest Historical Newspapers. October 4, 1934. Accessed December 2011.

———. "Dam Chairman Will Mediate." Proquest Historical Newspapers. February 25, 1929a. Accessed December 2011.

———. "Democrats of Arizona Upset." Proquest Historical Newspapers. February 24, 1924. Accessed December 2011.

———. "Gov. Hunt Firm on Pact Stand." Proquest Historical Newspapers. October 2, 1923b. Accessed December 2011.

———. "Governors Give Warning on River Compact Issue." Proquest Historical Newspapers. October 21, 1926. Accessed December 2011.

———. "Agreement on Colorado Water Fails." April 12, 1956, 25.

———. "Arizona Disputes California Claim." August 29, 1953a, 10.

———. "Arizona Project Seen Staying in Committee." October 12, 1950, 10.

———. "Arizona Water Plan Held Peril to California." April 2, 1948a, 10.

———. "Arizona Will Oppose California Water Plan." June 24, 1967, A12.

———. "Arizonans Speak Out Against Water Project." May 12, 1953b, 25.

———. "Colorado River Project Given House Approval." September 6, 1968a, 4.

———. "Compromise on the Colorado." August 9, 1968b, A4.

———. "Far-Reaching Letter-Writing Drive Urged in Fight over Water Rights." May 8, 1948b, 4.

———. "Ickes Approves Water Accord with Arizona." February 10, 1944, 3.

———. "Krug Deplores Water Quarrels." July 30, 1947a, 8.

———. "River Water Row Starts." April 18, 1937, 9.

———. "State Set to Fight Colorado Plans." February 9, 1955, 11.

———. "State Water Resources to Be Topic." October 13, 1947b, A3.

———. "States Push Water Plans: Organization Formed at Peaceful Session on Colorado River Supply." June 24, 1938, 17.

———. "Upper Colorado Panel Opposes Water Proposal." March 6, 1965, 5.

———. "Utah Approves River Pact Entry by Arizona." December 19, 1943, 13.

———. "New River-Dam Bill Presented." Proquest Historical Newspapers. February 14, 1928. Accessed December 2011.

———. "Notable Guests Cheer Hoover at Luncheon." Proquest Historical Newspapers. December 6, 1922. Accessed December 2011.

———. "Pact Urged by Hoover." Proquest Historical Newspapers. March 7, 1925. Accessed December 2011.

———. "President Confers on Colorado." Proquest Historical Newspapers. March 20, 1929b. Accessed December 2011.

———. "River Pact in Force." Proquest Historical Newspapers. June 26, 1929c. Accessed December 2011.

Lowi, Miriam. *Water and Power: The Politics of a Scarce Resource in the Jordan River Basin*. New York: Cambridge University Press, 1993.

Lubell, Mark. "Collaborative Watershed Management: A View from the Grassroots." *Policy Studies Journal* 32, no. 3 (2004): 341–61.

Lubell, Mark, Mark Schneider, John Scholz, and Mihriye Mete. "Watershed Partnerships and the Emergence of Collective Action Institutions." *American Journal of Political Science* 46, no. 1 (2002): 148–63.

Maass, Arthur. *Muddy Waters: The Army Engineers and the Nation's Rivers*. New York: Da Capo Press, 1974.

Macey, Jonathan. "Federal Deference to Local Regulators and the Economic Theory of Regulation: Toward a Public-Choice Explanation of Federalism." *Virginia Law Review* 76 (1990): 265–292.

Madden, Richard. "Carter Considers Plan to Continue Portions of Some Water Projects." *New York Times*, April 14, 1977, 18.

Maddock, Thomas. "The Control and Division of Interstate Water." *Journal of the American Water Works Association* 24, no. 11 (1932): 1659–75.

Mandarano, Lynn A., Jeffrey P. Featherstone, and Kurt Paulsen. "Institutions for Interstate Water Resources Management." *Journal of the American Water Resources Association* 44, no. 1 (2008): 136–47.

Mannemar Sonderskov, Kim. "Different Goods, Different Effects: Exploring the Effects of Generalized Social Trust in Large-N Collective Action." *Public Choice* 140 (July 2009): 145–60.

Marks, Gary, and Doug McAdam. "Social Movements and the Changing Structure of Political Opportunity in the European Union." *West European Politics* 19, no. 2 (1996): 249–78.

Marshall, Graham. "Nesting, Subsidiarity, and Community-Based Environmental Governance Beyond the Local Level." *International Journal of the Commons* 2, no. 1 (2008): 75–97.

———. "Transaction Costs, Collective Action and Adaptation in Managing Complex Social-Ecological Systems." *Ecological Economics* 88 (2013): 185–94.

Matthews, Nathanial, and Jeremy Schmidt. "False Promises: The Contours, Contexts and Contestation of Good Water Governance in Lao PDR and Alberta, Canada." *International Journal of Water Governance* 2 (2014): 21–40.

Matthews, Olen Paul, and Dan St. Germain. "Boundaries and Transboundary Water Conflicts." *Journal of Water Resources Planning and Development* 133, no. 5 (2007): 386–396.

Mawdsley, Emma. "Redrawing the Body Politic: Federalism, Regionalism, and the Creation of New States in India." *Commonwealth and Comparative Politics* 40, no. 3 (2002): 34–54.

McCaffrey, Stephen. "A Human Right to Water: Domestic and International Implications." *Georgetown International Environmental Law Review* 5, no. 1 (1992–1993): 1–24.

———. "Convention on the Law of the Non-Navigational Uses of International Watercourses." Audiovisual Library of International Law. May 21, 1997. Accessed October 21, 2014. http://legal.un.org/avl/ha/clnuiw/clnuiw.html.

McCool, Daniel. *River Republic: The Fall and Rise of America's Rivers*. New York: Columbia University Press, 2012.

McElheny, Victor. "Desalting Plant Faces Year Delay." *New York Times*, June 27, 1974, 7.

McLaughlin, Paul, and Marwan Khawaja. "The Organizational Dynamics of the U.S. Environmental Movement: Legitimation, Resource Mobilization, and Political Opportunity." *Rural Sociology* 65, no. 3 (2000): 422–39.

McLean, Iain. *Public Choice: An Introduction*. Oxford: Basil Blackwell, 1987.

McMillan, Samuel Lucas. *The Involvement of State Governments in US Foreign Relations*. New York: Palgrave Macmillan, 2012.

Meadowcroft, James. "Greening Leviathan: The Rise of the Environmental State?" *Environmental Politics* 25, no. 1 (2016): 1–23.

Menon, M. S. "Lessons Learnt from Cauvery: Nationalise Interstate Rivers." *The Hindu*, September 2, 2003. Accessed February 5, 2013. http://www.hindu.com/thehindu/op/2003/09/02/stories/2003090200120300.htm.

Mertha, Andrew. "China's 'Soft' Centralization: Shifting Tiao/Kuai Authority Relations." *China Quarterly* 184 (2005): 791–810.

———. *China's Water Warriors: Citizen Action and Policy Change*. Ithaca, NY: Cornell University Press, 2008.

Mertha, Andrew, and William Lowry. "Unbuilt Dams: Seminal Events and Policy Change in China, Australia, and the United States." *Comparative Politics* 39 (2006): 1–20.

Meublat, Guy, and Philippe Le Lourd. "Les agences de bassin: Un modele francais de decentralisation pour les pays emergents? La renovation des institutions de l'eau en Indonesie, au Bresil et au Mexique." *Tiers-Monde* 42, no. 166 (2001): 375–401.

Meyer, David. "Institutionalizing Dissent: The United States Structure of Political Opportunity and the End of the Nuclear Freeze Movement." *Sociological Forum* 8, no. 2 (1993): 157–79.

———. "Protest and Political Opportunities." *Annual Review of Sociology* 30 (2004): 125–45.

Miller, Kathleen, Steven Rhodes, and Lawrence Macdonnell. "Water Allocation in a Changing Climate: Institutions and Adaptation." *Climatic Change* 35, no. 2 (1997): 157–77.

Ministry of Environment, Sustainable Development, Transport, and Housing. "Water Policy in France: Experience at the Aervice of the International Community." 2014. Accessed December 17, 2017. http://www.pfvt.fr/index.php?option=com_easyfolderlistingpro&view=download&format=raw&data=eNpFkF9rgzAUxb-K5L34p6Mbt09rTaGwWFnt-ihRr21AoyQROsa--5JG2VNyzz3nx-FyiCP40bAB0g5dg4psNbzEQETPb6jD9LS_MJoV5zC_7D6O-_fieMpWjLLdXCc9r_LDV-ECsQ1MGpVH6KcEJFpWtwm1-ce_AinLp-amNx-VvEc3RkDcs_GqaMhWQORDCruRm7tbXCdrIGE-VZ2ouRGD1Az7Stm-NJuZrehwYa6tm9E0pWUSxUl55QZVMA42_B0IGRwUl7U3AsGHmX9j085t8TEKC1-q2IsRbgyv7z1K666s-vsHZWhmcw.

Ministry of Environmental Protection. "Guanyu yufang yu chuzhi kuashengjie shuiwuran jiufen de zhidao yijian" [Guidance on Prevention and Disposal of Water Pollution Disputes Across Provincial Borders]. Ministry of Environmental Protection. July 7, 2008. Accessed December 16, 2017. http://www.mep.gov.cn/gkml/hbb/bwj/200910/t20091022_174603.htm.

Ministry of Irrigation and Power (India). *Report of the Irrigation Commission,* Vol. 1. Faridabad: Government of India, 1972.

Ministry of Law and Justice (India). Constitution of India. 2011. Accessed February 6, 2013. http://indiacode.nic.in/coiweb/welcome.html.

Ministry of Water Resources. "About MWR." Ministry of Water Resources of the People's Republic of China. 2012. Accessed December 5, 2012. http://www.mwr.gov.cn/english/aboutmwr.html.

Mollinga, Peter. "Water, Politics, and Development: Framing a Political Sociology of Water Resources Management." *Water Alternatives* 1, no. 1 (2008): 7–23.

Montero, Alfred. "The Politics of Decentralization in a Centralized Party System: The Case of Spain." *Comparative Politics* 38, no. 1 (2005): 63–82.

Moore, Michael, Aimee Mulville, and Marcia Weinberg. "Water Allocation in the American West: Endangered Fish Versus Irrigated Agriculture." *Natural Resources Journal* 36, no. 2 (1996): 319–58.

Moore, Scott. "China's Domestic Hydropolitics: An Assessment and Implications for International Transboundary Dynamics." *International Journal of Water Resources Development* (2017): doi: 10.1080/07900627.2017.1313157.

———. "Hydropolitics and Inter-Jurisdictional Relationships in China: The Pursuit of Localized Preferences in a Centralized System." *China Quarterly* 219 (2014a): 760–80.

———. "Modernization, Authoritarianism, and the Environment: The Politics of China's South–North Water Transfer Project." *Environmental Politics* 23, no. 6 (2014b): 947–64.

Moore, Scott, and Joshua Fisher. "Challenges and Opportunities in GRACE-Based Groundwater Storage Assessment and Management: An Example from Yemen." *Water Resources Management* 26, no. 6 (2012): 1425–53.

Muller, Mike. "Allocating Powers and Functions in a Federal Design: The Experience of South Africa." In *Federal Rivers: Managing Water in Mult-Layered Political Systems,* edited by Dustin Garrick, George Anderson, Daniel Connell, and Jamie Pittock, 179–94. Cheltenham, UK: Edward Elgar, 2014.

Mullin, Megan. *Governing the Tap: Special District Governance and the New Local Politics of Water.* Cambridge, MA: MIT Press, 2009.

Murali, Janaki. "Mahadayi Protests: Karnataka's Long History of Disputes over Water-Sharing." *Firstpost,* August 2, 2016. Accessed September 3, 2016. http://www.firstpost.com/india/mahadayi-protests-karnatakas-long-history-of-disputes-over-water-sharing-2929170.html.

Murphy, Dean. "California Water Users Miss Deadline on Pact for Sharing." *New York Times,* January 1, 2003a, A11.

———. "County Water Board Sues U.S. in Dispute over Colorado River." *New York Times,* January 12, 2003b, L16.

———. "In a First, U.S. Officials Put Limits on California's Thirst." *New York Times,* January 5, 2003c, L1.

Musemwa, Muchparara. "Disciplining a 'Dissident' City: Hydropolitics in the City of Bulawayo, Matabeleland, Zimbabwe, 1980–1994." *Journal of Southern African Studies* 32, no. 2 (2006): 239–54.

Muthien, Yvonne, and Meshack Khosa. " 'The Kingdom, the Volkstaat and the New South Africa': Drawing South Africa's New Regional Boundaries." *Journal of Southern African Studies* 21, no. 2 (1995): 303–22.

Myint, Tun. *Governing International Rivers: Polycentric Politics in the Mekong and the Rhine.* Cheltenham, UK: Edward Elgar, 2012.

Nadeau, Remi. *The Water Seekers*. Santa Barbara, CA: Peregrine Smith, 1974.

New York Times. "3 Endangered Fish Delay Utah's Construction of Dam and Reservoir." December 7, 1980, A2.

———. "Carl Hayden Day." October 1, 1968, 42.

———. "Carter Aide Assails Water Projects." March 27, 1977, 24.

———. "Feud over Use of Colorado River Water Still Goes On." September 24, 1950, E12.

———. "US Eases Its Stand on Water for Indians." December 15, 1953, 41.

———. "Water Diversion Tunnel Is Dedicated in Colorado as Its Opponents Protest." July 5, 1972, 78.

NewsX Bureau. "Congress Calls for Odisha Bandh over Mahanadi River Issue." August 16, 2016. Accessed September 3, 2016. http://www.newsx.com/national/38199-congress-calls-for-odisha-bandh-over-mahanadi-river-issue.

Ningxia Ribao [Ningxia Daily]. "Quanguo rendachangweihui guanxin Ningxia zhongdian duban jianyi yishi" [NPC Standing Committee gives an on-the-spot report on key supervision work in Ningxia]. Quanguo Renmin Daibiao Dahui [National People's Congress]. October 3, 2010. Accessed November 8, 2012. http://www.npc.gov.cn/npc/xinwen/2010-10/08/content_1597926.htm.

Ningxia Xinwen Wang [Ningxia News Network]. "Zonghe kaifa Huanghe Heishanxia heduan, jiasu Daxibei guotu kaifa zhengzhi" [Comprehensive development of the Heishanxia stretch of the Yellow River, acceleration of the great development and renovation of national territory in the northwest]. Huanghe Daliushu shuili shuniu gongcheng [Yellow River Daliushu Water Conservancy Project]. September 8, 2008. Accessed April 3, 2012. http://www.nxein.com/sszt/dls/zjlz/200809/20080908170401.html.

Noel, Jean-Claude. "Water Policy in France: The River Basin Agencies." *Water International* 15, no. 2 (1990): 93–100.

Notte, O., and A. M. Keesen. *Quick Scan: WFD Implementation in France*. Utrecht, The Netherlands: Center for Environmental Law and Policy, University of Utrecht. 2009.

Oakerson, Ronald, and Roger Parks. "The Study of Local Public Economies: Multi-Organizational, Multi-Level Institutional Analysis and Development." *Policy Studies Journal* 39, no. 1 (2011): 147–67.

Oates, Wallace. "An Essay on Fiscal Federalism." *Journal on Economic Literature* 37, no. 3 (1999): 1120–49.

O'Brien, Kevin, and Lianjiang Li. "Selective Policy Implementation in Rural China." *Comparative Politics* 31, no. 2 (1991): 167–86.

Times of India. "Centre's Duty in Krishna Dispute." *Times of India*, October 6, 1960, 8.

Office of the Chief Engineer. "Agreement Made Between the Government of HEH The Nizam and the Government of Madras for the Partial Utilization of the Tungabhadra Waters, June 1944." In *Agreements on Development of Inter-State and International Rivers*, edited by Central Water Commission, 185–86. New Delhi: Government of India, 1979.

Ohlsson, L. "Water Conflicts and Social Resource Scarcity." *Physical Chemistry of the Earth* 25, no. 3 (2000): 213–20.

Oi, Jean. "The Role of the Local State in China's Transitional Economy." *China Quarterly* 144 (1995): 1132–49.

Oksenberg, Michel, and James Tong. "The Evolution of Central–Provincial Fiscal Relations in China, 1971–1984: The Formal System." *China Quarterly* 125 (1991): 9–10.

Olivera, Oscar. *Cochabamba! Water Rebellion in Bolivia*. Cambridge, MA: South End Press, 2004.

Organization for Economic Cooperation and Development. *OECD Environmental Performance Reviews: France*. Paris: OECD Publishing, 2005.

Organization for Economic Cooperation and Development. *Stakeholder Engagement for Inclusive Water Governance*. OECD Studies on Water. Paris: OECD Publishing, 2015.

Orissa Diary. "Mahanadi Peace Jatha to Be Organised in Both Odisha, Chhattisgarh." *Orissa Diary*, August 24, 2016. Accessed December 16, 2017. http://www.cansouthasia.net/a/cansa_news_information.php?id=23&title=mahanadi-peace-jatha-to-be-organised-in-both-odisha-chhattisgarh.

Ostrom, Elinor. "Background on the Institutional Analysis and Development Framework." *Policy Studies Journal* 39, no. 1 (2011): 7–27.

———. "Beyond Markets and States: Polycentric Governane of Complex Economic Systems." *American Economic Review* 100 (2010): 641–72.

———. "Collective Action and the Evolution of Social Norms." *Journal of Economic Perspectives* 14, no. 3 (2000): 137–58.

———. "Constituting Social Capital and Collective Action." *Journal of Theoretical Politics* 6 (1994a): 527–62.

———. *Governing the Commons: The Evolution of Institutions for Collective Action*. Cambridge: Cambridge University Press, 1990.

———. "Rules, Games, and Common Pool Resource Problems." In *Rules, Games, and Common Pool Resources*, edited by Elinor Ostrom, Roy Gardner, and James Walker, 3–22. Ann Arbor: University of Michigan Press, 1994b.

———. *Understanding Institutional Diversity*. Princeton, NJ: Princeton University Press, 2005.

Ostrom, Elinor, Joanna Burger, Christopher Field, Richard Norgaard, and David Policansky. "Revisiting the Commons: Local Lessons, Global Challenges." *Science* 284, no. 5412 (1999): 278–82.

Ostrom, Elinor, and Roy Gardner. "Coping with Asymmetries in the Commons: Self-Governing Irrigation Systems Can Work." *Journal of Economic Perspectives* 7, no. 4 (1993): 93–112.

Ostrom, Elinor, Larry Schroeder, and Susan Wynne. *Institutional Incentives and Sustainable Development: Infrastructure Policies in Perspective*. Boulder, CO: Westview Press, 1993.

Ostrom, Elinor, James Walker, and Roy Gardner. "Covenants with and without a Sword: Self-Governance Is Possible." *American Political Science Review* 86, no. 2 (1992): 404–17.

Ostrom, Vincent, Charles Tiebout, and Robert Warren. "The Organization of Government in Metropolitan Areas: A Theoretical Inquiry." *American Political Science Review* 55 (1961): 831–42.

Otero, Iago, Giorgos Kallis, Raul Aguilar, and Ruiz Vicenc. "Water Scarcity, Social Power and the Production of an Elite Suburb: The Political Ecology of Water in Matadepera, Catalonia." *Ecological Economics* 70, no. 7 (2011): 1297–1308.

Paavola, Jouni. "Interstate Water Pollution Problems and Elusive Federal Water Pollution Policy in the United States, 1900–1948." *Environment and History* 12, no. 4 (2006): 435–65.

Pagano, Michael. "In the Eye of the Beholder: The Dynamics of Federalism in National and Supernational Political Systems." In *The Dynamics of Federalism in National and Supernational Political Systems*, edited by Michael Pagano and Robert Leonardi, 1–18. New York: Palgrave Macmillan, 2007.

Pahl-Wostl, Claudia, Louis Lebel, Christian Knieper, and Elena Nikitina. "From Applying Panaceas to Mastering Complexity: Toward Adaptive Water Governance in River Basins." *Environmental Science and Policy* 23 (2012): 24–34.

Palmer, Kyle. "Arizona Claims Owership of River." *Los Angeles Times*, August 18, 1925a, 1.

———. "Legislature Refuses to Drop River Pact Terms." *Los Angeles Times*, October 23, 1926, 1.

———. "Passage of River Pact Advocated." *Los Angeles Times*, January 25, 1923, 11.

———. "River Legislation Faces New Danger of Defeat." *Los Angeles Times*, January 16, 1927, 1.

———. "Willing to Kill Pact: Swing Bill Group Out in Open." *Los Angeles Times*, March 14, 1925b, 1.

Pani, Narendar. "By the Rivers of Telangana." The Hindu Business Line. December 6, 2013. Accessed November 22, 2014. http://www.thehindubusinessline.com/opinion/columns/narendar-pani/by-the-rivers-of-telangana/article5430494.ece.

Paris L'Express International. "Background, Structure of Environmental Agency." April 17, 1992, West Europe edition, 8.

Paul, Darel E. *Rescaling International Political Economy: Subnational States and the Regulation of the Global Political Economy*. New York: Routledge, 2005.

Peng, Huan. "Kuaxingzhengqu liuyu shuiwuran zhili" [Control of trans-jurisdictional river basin water pollution]. *Zhongguo Chengshi Yanjiu (Dianzi qikan)* [E-Journal of China Urban Studies] 4, no. 1 (2009): 36–40.

Perry, Tony. "Babbitt Warns California District on Colorado River Usage." *Washington Post*, December 19, 1997, A41.

Persha, Lauren, and Krister Andersson. "Elite Capture Risk and Mitigation in Decentralized Forest Governance Regimes." *Global Environmental Change* 24 (2014): 265–76.

Peterson, Iver. "Arizona Cheers as Canal Carries Colorado River Water to Phoenix." *New York Times*, November 17, 1985, A1.

Pinson, Gilles, and Patrick Le Gales. *State Restructuring and Decentralization Dynamics in France: Politics Is the Driving Force*. Cahiers du pole Ville 7. Paris: Centre d'Etudes Europeennes de Sciences Po Paris, 2000.

Pisani, Donald. *Water and American Government*. Berkeley: University of California Press, 2002.

Planning Commission. "Summary Record of Discussions at the Inter-State Conference on the Utilisation of the Krishna and Godavari Waters Held in the Committee Room of the Planning Commission, New Delhi, on 27th and 28th July, 1951." Water Resources Information System of India. 1951. Accessed November 13, 2014. http://india-wris.nrsc.gov.in/wrpinfo/images/7/7f/67.pdf.

Planning Commission. "Summary Record of Proceedings of the Inter-State Conference on the Tungabhadra High Level Canal Held at Residency Building, Bangalore on 18th June 1956." In *Agreements on the Development of Inter-State and International Rivers*, edited by Central Water Commission, 197–98. New Delhi: Government of India, 1979.

Post, Stephanie. "Metropolitan Area Governance and Institutional Collective Action." In *Metropolitan Governance: Conflict, Competition, and Cooperation*, edited by Richard Feiock, 67–92. Washington, DC: Georgetown University Press, 2004.

Postel, Sandra, and Aaron Wolf. "Dehydrating Conflict." *Foreign Policy* 126 (2001): 60–67.

Prasad, Srinivasa. "Krishna Water Dispute Is Back to Square One, Almost." *Times of India*, May 1, 2000, 9.

Press Trust of India. "Protests in Karnataka Against Supreme Court Direction to Release Cauvery Water." NDTV. September 6, 2016. Accessed September 6, 2016.

http://www.ndtv.com/karnataka-news/protests-in-karnataka-against-supreme-court-direction-to-release-water-1454862.

Pretty, Jules. "Social Capital and the Collective Management of Resources." *Science* 302, no. 5652 (2003): 1912–14.

Price, Monroe, and Gary Weatherford. "Indian Water Rights in Theory and Practice: Navajo Experience in the Colorado River Basin." *Law and Contemporary Problems* 40, no. 97 (1976): 97–131.

Pritchard, Sara. *Confluence: The Nature of Technology and the Remaking of the Rhone*. Cambridge, MA: Harvard University Press, 2011.

PRS Legislative Research. Damodar Vaelly Corporation (Amendment) Bill. 2011. Accessed February 18, 2013. http://www.prsindia.org/billtrack/the-damodar-valley-corporation-amendment-bill-2011--1884/.

Prud'homme, Remy. "The Dangers of Decentralization." *World Bank Research Observer*, 1995: 201-220.

PTI. "Telangana Bill Gives Key Role to Centre in Water Management." *Times of India*, December 18, 2013. Accessed November 22, 2014. https://timesofindia.indiatimes.com/india/Telangana-bill-gives-key-role-to-Centre-in-water-management/articleshow/27572459.cms.

Publications Division, Ministry of Information and Broadcasting (India). *Our River Valley Projects*. Faridabad: Government of India Press, 1961.

Purdum, Todd. "7 States in West Reach Outline." *New York Times*, July 28, 2000, A12.

———. "U.S. Acts to Meet Water Needs in the West." *New York Times*, December 19, 1997, A36.

Qi, Guizhen. *Lun Zhongguo de dazhengfu yu qiangzhengfu* [On China's powerful and strong government]. Beijing: Shehui Kexue Wenxian Chubanshe [Social Sciences Academic Press], 2011.

Rahman, H. M. Tuihedur, Gordon Hickey, and Swapan Kumar Sarker. "A Framework for Evaluating Collective Action and Informal Institutional Dynamics Under a Resource Management Policy of Decentralization." *Ecological Economics* 83 (2012): 32–41.

Raleigh, Clionadh, and Henrik Urdal. "Climate Change, Environmental Degradation and Armed Conflict." *Political Geography* 26, no. 6 (2007): 674–94.

Ramana, M. V. V. *Inter-State River Water Disputes in India*. Madras, India: Orient Longman, 1992.

Rao, C. V. M. Krishna. "Telangana to File Suit in Supreme Court for Water Share." *Deccan Chronicle*, July 23, 2014. Accessed November 22, 2014. http://www.deccanchronicle.com/140723/nation-current-affairs/article/telangana-file-suit-supreme-court-water-share.

Rao, Dodda Srinivasa. *Inter-State Water Disputes in India: Constitutional and Statutory Provisions and Settlement Machinery*. New Delhi, India: Deep and Deep Publications, 1998.

Rao, G. Venkataramana. "Silence of Centre on Apex Council Meet Worries A.P." *The Hindu*, July 31, 2016. Accessed September 7, 2016. http://www.thehindu.com/news/cities/Vijayawada/silence-of-centre-on-apex-council-meet-worries-ap/article8923461.ece.

Rao, M. Govinda, and Nirvikar Singh. *The Political Economy of Federalism in India*. New Delhi, India: Oxford University Press, 2005.

Rao, V. G. Prasad. "Krishna Pact Augurs Well." *Times of India*, May 2, 1983, 9.

Ratner, Blake, Ruth Meinzen-Dick, Candace May, and Eric Haglund. "Resource Conflict, Collective Action, and Resilience: An Analytical Framework." *International Journal of the Commons* 7, no. 1 (2013): 183–208.

Rawls, Wendell. "Arizonans Criticize Big Water Project." *New York Times*, March 22, 1977, 27.

Rector, Chad. *Federations: The Political Dynamics of Cooperation*. Ithaca, NY: Cornell University Press, 2009.

Reinhold, Robert. "Farmers in West May Sell Something More Valuable Than Any Crop: Water." *New York Times*, April 6, 1992, B9.

Renmin Ribao [People's Daily]. "Kongzhi shuitiliushi zujin shengchan fazhan baohu Sanmenxia Shuiku" [Control soil erosion and promote production, development, and protection of Sanmenxia Reservoir]. November 18, 1959.

———. "Shuili shuniu gongcheng gaijian huode chubu gongcheng" [Water hub project renovation receives early success]. December 20, 1974.

Ribot, J. C. *Democratic Decentralization of Natural Resources: Institutionalizing Popular Participation*. Washington, DC: World Resources Institute, 2002.

Richard, Sophie, Gabrielle Bouleau, and Sylvain Barone. "Water Governance in France: Institutional Framework, Stakeholders, Arrangements and Process." In *Water Governance and Public Policies in Latin America and Europe*, edited by P. Jacobi and P. Sinisgali, 137–78. Sao Paulo, Brazil: Anna Blume, 2010.

Richard, Sophie, and Thierry Rieu. "Une approche historique de la gouvernance pour eclairer la gestion concertee de l'eau en France." Paper presented at the 13th IWRA World Water Congress 2008. Montpellier, France, September 1–4, 2008.

Richards, Alan, and Nirvikar Singh. "Inter-State Water Disputes in India: Institutions and Policies." *Water Resources Development* 18, no. 4 (2002): 611–25.

Riker, William. *Federalism: Origin, Operation, Significance*. Boston: Little, Brown and Company, 1964.

Riker, William. "Party Organization." In *The Development of American Federalism*, edited by William Riker, 217–21. Boston: Kluwer Academic, 1987.

Ripley, Anthony. "Study by Water Commission Is Critical of Large Federal Projects." *New York Times*, December 10, 1972a, 81.

———. "West's Thirst for Water Is Questioned." *New York Times*, January 17, 1972b, 1.

RMIT International. *Local Legislation to Support Transjurisdictional Water Pollution Management*. ADB Technical Assistance Report TA 3588-PRC. Melbourne, Australia: RMIT International, 2003.

Robertson, David. *Federalism and the Making of America*. New York: Routledge, 2012.

Rodden, Jonathan. *Hamilton's Paradox: The Promise and Peril of Fiscal Federalism*. New York: Cambridge University Press, 2006.

Rodriguez-Pose, Andres, and Nicholas Gill. "The Global Trend Towards Devolution and Its Implications." *Environment and Planning C: Government and Policy* 21, no. 3 (2003): 333–51.

Rogers, Peter, Radhika de Silva, and Ramesh Bhatia. "Water Is an Economic Good: How to Use Prices to Promote Equity, Efficiency, and Sustainability." *Water Policy* 4, no. 1 (2002): 1–17.

Roosevelt, Franklin. "Message to Congress Suggesting the Tennessee Valley Authority." FDR Presidential Library. April 10, 1933. Accessed September 19, 2011. http://docs.fdrlibrary.marist.edu/odtvacon.html.

Rucker, Randal, and Price Fishback. "The Federal Reclamation Program: An Analysis of Rent-Seeking Behavior." In *Water Rights: Scarce Water Allocation, Bureaucracy, and the Environment*, edited by Terry Anderson, 45–82. San Francisco: Pacific Institute for Public Policy Research, 1983.

Sabatier, Paul, Will Focht, Mark Lubell, Zev Trachtenberg, Arnold Vedlitz, and Mary Matlock. *Swimming Upstream: Collaborative Approaches to Watershed Management*. Cambridge, MA: MIT Press, 2005.

Sabhu, Asun. "Steps to Make Optimum Use of Krishna Waters." *Times of India*, June 4, 1976, 1.

Saez, Lawrence. *Federalism Without a Centre: The Impact of Political and Economic Reform on India's Federal System*. New Delhi, India: Sage Publications, 2002.

Salman, Salman, and Daniel Bradlow. *Regulatory Frameworks for Water Resources Management: A Comparative Study*. Washington, DC: World Bank, 2006.

Salman, Salman, and Siobhan McInerney-Lankford. *The Human Right to Water: Legal and Policy Dimensions*. Washington, DC: World Bank, 2004.

Salman, Salman M. A., and Kishor Uprety. *Conflict and Cooperation on South Asia's International Rivers: A Legal Perspective*. The Hague, The Netherlands: Kluwer Law International, 2002.

Samuels, David, and Richard Snyder. "The Value of a Vote: Malapportionment in Comparative Perspective." *British Journal of Political Science* 31 (2001): 651–71.

Sangarre, I., and C. Larrue. "The Evolution of the Water Regime in France." In *The Evolution of National Water Regimes in Europe*, edited by Ingrid Kissling-Naf and Stefan Kuks, 187–234. Dordrecht, The Netherlands: Kluwer Academic, 2004.

Sarsons, Heather. "Rainfall and Conflict: A Cautionary Tale." *Journal of Development Economics* 115 (2015): 62–72.

Saunders, J. Owen, and Michael Wenig. "Whose Water? Canadian Water Management and the Challenges of Jurisdictional Fragmentation." In *Eau Canada: The Future of Canada's Water*, edited by Karen Bakker, 119–42. Vancouver, Canada: UBC Press, 2007.

Scharpf, Fritz. "The Joint Decision-Trap: Lessons from German Federalism and European Integration." *Public Administration* 66, no. 3 (1988): 239–78.

Schiff, Jennifer. "Silencing the Opposition: The State v. Civil Society in India's Ganges River Basin." *International Studies Perspectives* 15 (2014): 229–42.

Schlager, Edella, and William Blomquist. *Embracing Watershed Politics*. Boulder: University Press of Colorado, 2008.

Schlager, Edella, and Tanya Heikkila. "Resolving Water Conflicts: A Comparative Analysis of Interstate River Compacts." *Policy Studies Journal* 37, no. 3 (2009): 367–92.

Schlager, Edella, Tanya Heikkila, and Carl Case. "The Costs of Compliance with Interstate Agreements: Lessons from Water Compacts in the Western United States." *Publius: The Journal of Federalism* 42, no. 3 (2012): 494–515.

Schmidt, William. "Floods Along Colorado River Set Off a Debate over Blame." *New York Times*, July 17, 1983a, A1.

———. "Watt Choice for Southwest Water Job Assailed." *New York Times*, April 30, 1983b, 1.24.

Schneider, Mark, John Scholz, Mark Lubell, Denisa Mindruta, and Matthew Edwardsen. "Building Consensual Institutions: Networks and the National Estuary Program." *American Journal of Political Science* 47, no. 1 (2003): 143–58.

Schrijver, Nico. *Sovereignty over Natural Resources: Balancing Rights and Duties*. New York: Cambridge University Press, 2008.

Schultz, M. "Turkey, Syria and Iraq: A Hydropolitical Security Complex." In *Hydropolitics: Conflicts over Water as a Development Constraint*, edited by L. Ohlsson. London: Zed Books, 1995.

Scott, Anthony. "Assigning Powers over the Canadian Environment." In *Competition and Structure: The Political Economy of Collective Decisions: Essays in Honor of Albert Breton*, edited by Gianluigi Geleotti, Pierre Salmon, and Ronald Wintrobe, 174–219. Cambridge: Cambridge University Press, 2000.

Scott, James. *Seeing Like a State: How Certain Schemes to Improve the Human Condition Have Failed*. New Haven, CT: Yale University Press, 1998.

Seetharaman, G., Hakeem Irfan, and Indulekha Aravind. "Cauvery, Indus Treaty and Brahmaputra: River Water Proving to Be a Source of Potential Conflict." *Economic Times*, October 2, 2016. Accessed October 4, 2016. http://economictimes.indiatimes.com/news/politics-and-nation/cauvery-indus-treaty-and-brahmaputra-river-water-proving-to-be-source-of-potential-conflict/articleshow/54632042.cms.

Selznick, Philip. *TVA and the Grass Roots: A Study in the Sociology of Formal Organization*. Berkeley: University of California Press, 1949.

Semple, Robert. "U.S.–Mexico Pact Details Boundary." *New York Times*, August 22, 1970, 1.

Severo, Richard. "Colorado River Salt Annoys Mexicans." *New York Times*, June 11, 1972, 6.

Shaanxi Zhengfu [Shaanxi Government]. *Shaanxisheng renmin zhengfu guanyu yinfa Huanghe jingji xiezuoqu zai Xibudakaifa zhong jinyibu jiaqiang lianhe yu xiezuo de shishi yijian de tongzhi* [Shaanxi Provincial People's Government notice]. Shaanxi Zhengbao [Shaanxi Government Report]. 16–20. Shaanxi, China: Shaanxi Provincial People's Government, November 27, 2000.

Shabecoff, Philip. "Watt Drops a Power Project Threatening Colorado River." *New York Times*, October 30, 1981, A14.

Shaheen, Murad. "Questioning the Water-War Phenomenon in the Jordan Basin." *Middle East Policy* 7, no. 3 (2000): 137–50.

Shannon, Don. "Water Plan Illegal, says Rep. Hosmer." *Los Angeles Times*, March 12, 1955a, 7.

———. "Witnesses Assail Colorado Project." *Los Angeles Times*, March 5, 1955b, 4.

Sharansky, Ira. *Regionalism in American Politics*. New York: Bobbs-Merrill, 1970.

Shen, Dajun. "Planned Water Use System in China." *Water Policy* 14 (2012): 581–93.

———. "River Basin Water Resources Management in China: A Legal and Institutional Assessment." *Water International* 34, no. 4 (2009): 484–96.

Sherman, Gene. "Arizona Fears Federal Politics." *Los Angeles Times*, July 15, 1963, 2.

Shuilibu Huanghe Shuili Weiyuanhui [Yellow River Conservancy Commission]. *Renmin Zhili Huanghe Lishinian* [Sixty years of the people controlling the Yellow River]. Zhengzhou, China: Huanghe Shuili Chubanshe [Yellow River Water Conservancy Press], 2006.

Shuixinxi Wang [Water Information Network]. *Zhou Enlai de zhishui sixiang* [Zhou Enlai's thoughts on water control]. Shuilibu Haihe Shuili Weiyuanhui Zhuban [Hai River Water Conservancy Commission Supervision Office, Ministry of Water Resources], September 28, 2011. Accessed December 16, 2017. http://lishi.zhuixue.net/renwu/zhouenlai/37887.html.

Shumeli, Deborah F. "Water Quality in International River Basins." *Political Geography* 18 (1999): 437–76.

Singh, G. S. *Maharata Geopolitics and the Indian Nation*. Bombay, India: Manaktala and Sons, 1966.

Singh, Mahendra. "A Borderless Internal Federal Space? Reorganization of States in India." *India Review* 6, no. 4 (2007): 233–50.

Singh, Nina. "Inter-State Water Sharing in India." *Man and Development* (December 2008): 31–44.

Singh, Nirvikar Singh, Laveesh Bhandari, Aoyu Chen, and Aarti Khare. "Regional Inequality in India: A Fresh Look." *Economic and Political Weekly* 38, no. 11 (2003): 1069–73.

Singh, Satyajit. *Taming the Waters: The Political Economy of Large Dams in India*. Delhi, India: Oxford University Press, 1997.

Skocpol, Theda. *States and Social Revolutions: A Comparative Historical Analysis of France, Russia, and China*. New York: Cambridge University Press, 1979.

Slocum, Dale. "Apaches Must Protect Future Water Supply." *Fort Apache Scout*, January 31, 1973, 1.

Smith, Andy, and Paul Heywood. *Regional Government in France and Spain*. Constitution Unit, School of Public Policy. 5–6. London: University College London, 2000.

Sneddon, Chris, and Coleen Fox. "Rethinking Transboundary Waters: A Critical Hydropolitics of the Mekong Basin." *Political Geography* 25 (2006): 181–202.

Snider, Annie. "Colorado River: Water Users Scramble as Drought Foretells Scary Future." *Energy and Environment News*, December 24, 2014. Accessed January 2, 2015. http://www.eenews.net/stories/1060011038.

Soldatos, Panayotis. "An Explanatory Framework for the Study of Federated States as Foreign-Policy Actors." In *Federalism and International Relations: The Role of Subnational Units*, edited by Hans J. Michelman and Panayotis Soldatos, 34–53. Oxford: Oxford University Press, 1990.

Sonin, Konstantin. "Provincial Protectionism." *Journal of Comparative Economics* 38, no. 2 (2010): 111–22.

St. Charles, Kean, and E. Householder. "Mohave Legislators Urge Ratification of Colorado River Pact." *Arizona Republican*, November 22, 1922, 1.

Staff Correspondent. "Don't Accept KWDT-II Final Award, Rayaraddi Tells State." *The Hindu*, December 1, 2013. Accessed April 9, 2014. http://www.thehindu.com/news/national/karnataka/dont-accept-kwdtii-final-award-rayaraddi-tells-state/article5409019.ece?ref=relatedNews.

Staff Reporter. "Nehru not for Diverting Krishna Waters Westward." *Times of India*, January 14, 1964, 7.

———. "Sharing Waters of Godavari, Krishna." *Times of India*, September 24, 1960, 8.

Stall, Bill. "Tucson Sips Wisely, Sets Water Example." *Los Angeles Times*, September 29, 1985, 3.

Stanford, Neal. "California Hits 'Bad' Mexico Water Pact." *Christian Science Monitor*, February 6, 1945, 3.

Stanway, David. "In Beautiful China, Local Polluters Still Hold Sway." Reuters, November 22, 2012. Accessed December 5, 2012. http://www.reuters.com/article/2012/11/22/uk-china-congress-environment-idUSLNE8AL00T20121122.

States Reorganization Commission (India). *Report of the States Reorganization Commission*. New Delhi: Government of India Press, 1955.

Stepan, Alfred. "Federalism and Democracy: Beyond the US Model." *Journal of Democracy* 10, no. 4 (1999): 19–34.

Stevens, John. *The Matter of the Colorado River: A Brief by John Stevens, Consulting Engineer for the Colorado River Commission of Arizona*. Denver: Colorado River Commission of Arizona, October 31, 1928.

Stewart, Ian. "More than Just a Line on a Map: The Political Culture of the Nova Scotia–New Brunswick Boundary." *Publius* 20, no. 1 (1990): 99–111.

Stinnett, Douglas, and Jaroslav Tir. "The Institutionalization of River Treaties." *International Negotiation* 14, no. 2 (2009): 229–51.

Susskind, Lawrence, and Shafiqul Islam. "Water Diplomacy: Creating Value and Building Trust in Transboundary Water Negotiations." *Science & Diplomacy* 1, no. 3 (2012), 1–7.

Swain, Ashok. "Odisha Can't Afford a Water War with Chhattisgarh over Mahanadi." *Daily O*, August 12, 2016. Accessed September 1, 2016. http://www.dailyo.in/politics/mahanadi-dispute-odisha-chhattisgarh-transboundary-rivers-modi-sarkar/story/1/12339.html.

Sweet, William. *Report and Supplemental Report of Delph Carpenter, Commissioner for Colorado on the Colorado River Commission*. Vol. 1, *Pamphlets on Water Rights and the Use of Water Power*. Denver: Colorado River Commission, 1923.

Swenden, Wilfried. "Subnational Participation in National Decisions: The Role of Second Chambers." In *Handbook on Multi-Level Governance*, edited by Henrik Enderlein, Sonja Walti, and Michael Zurn, 103–23. Cheltenham, UK: Edward Elgar, 2010.

Swyngedouw, Eric. "Power, Nature, and the City: The Conquest of Water and the Political Ecology of Urbanization in Guyaquil, Ecuador, 1880–1990." *Environment and Planning A* 29, no. 2 (1997): 311–32.

Tarlock, A. Dan. "Federalism Without Preemption: A Case Study in Bioregionalism." *Pacific Law Journal* 27 (1996): 1630–52.

Tarlock, Dan. "One River, Three Sovereigns: Indian and Interstate Water Rights." *Land and Water Law Review* 22, no. 2 (1987): 631–71.

Tarrow, Sidney. *Between Center and Periphery: Grassroots Politicians in Italy and France*. New Haven, CT: Yale University Press, 1977.

———. "National Politics and Collective Action: Recent Theory and Research in Western Europe." *Annual Review of Sociology* 14 (1988): 421–40.

Taylor, Theresa. 1981. "Determining the Parameters of the Navigation Servitude Doctrine." *Vanderbilt Law Review* 34, no. 2 (1981): 461–80.

Teclaff, Ludwik. "Evolution of the River Basin Concept in National and International Water Law." *Natural Resources Journal* 36, no. 359 (1996): 359–91.

Teng, Tse-hui. *Report on the Multi-Purpose Plan for Permanently Controlling the Yellow River and Exploiting Its Water Resources*. Peking, China: Foreign Languages Press, 1955.

The Hindu. "Cauvery: Farmers Threaten Massive Stir." August 28, 2016a. Accessed September 3, 2016. http://www.thehindu.com/news/national/tamil-nadu/farmers-threaten-agitation-in-september/article9040767.ece.

———. "Minister Objects to Water Pact with Telangana." August 26, 2016b. Accessed September 20, 2016. http://www.thehindu.com/news/cities/mumbai/news/minister-objects-to-water-pact-with-telangana/article9032669.ece.

———. "Sharing Without Caring." September 14, 2016c. Accessed September 20, 2016. http://www.thehindu.com/opinion/editorial/editorial-on-cauvery-issue-and-the-dispute/article9104515.ece.

Thelen, Kathleen, and Sven Steinmo. "Historical Institutionalism in Comparative Politics." In *Structuring Politics: Historical Institutionalism in Comparative Analysis*, edited by Sven Steinmo, Kathleen Thelen, and Frank Longstreth, 1–32. Cambridge: Cambridge University Press, 1992.

Thorson, John. *River of Promise, River of Peril: The Politics of Managing the Missouri River*. Lawrence: University Press of Kansas, 1994.

Times News Network. "SC Issues Notices on AP's Krishna Projects." *Times of India*, August 14, 2001, 7.

Times of India. "No Stopping Power at Srisilam: Telangana CM." October 25, 2014. Accessed November 22, 2014. http://timesofindia.indiatimes.com/india/No-stopping-power-at-Srisailam-Telangana-CM-KCR/articleshow/44929080.cms.

———. "30 p.c. of Krishna, Godavari Waters." March 14, 1963a, 5.

———. "Accept Krishna Award: Pant." August 31, 1974, 1.

———. "Centre Against Bengal Plan: Liability in Respect of DVC Projects." February 27, 1958, 9.

———. "Centre Shuns Krishna Dispute." May 17, 1990, 7.

———. "Constitution for Damodar Project Accepted." April 8, 1947, 8.

———. "Distribution of Krishna and Godavari Waters." August 27, 1961a, 11.

———. "Dr. BC Roy Accepted Opposition Plan: State Govt. Control over Affairs of DVC." July 4, 1957, 6.

———. "DVC Dispute." September 13, 1962a, 1.

———. "DVC Hq. Site Issue: Participation in Talks Ruled Out by Bihar." February 21, 1963b, 7.

———. "Fast to Get Krishna Water for Farming." February 20, 1983a, 6.

———. " 'Flagrant Violations' by Maharashtra Alleged." August 5, 1961b, 1.

———. "Future Set-up of DVC." June 25, 1964a, 5.

———. "Gowda Seeks PM's Help: Krishna Plan." February 28, 1995, 6.

———. "Krishna River Project: Two States to Benefit." October 7, 1952, 8.

———. "Krishna Waters for Madras City Suggested." September 15, 1962b, 1.

———. "Krishna Waters for TN: Pact Signed." April 20, 1983b, 1.

———. "Krishna-Goadavari Waters Distribution: State Minister Urges Review of Agreement." November 15, 1960, 1.

———. "Little Benefit to Bihar from DVC: MLA's Charge." December 24, 1959, 6.

———. "M.P.s Urged to Put Pressure on Centre." July 16, 1966a, 1.

———. "Maharashtra Should not Divert Krishna Waters." April 4, 1963c, 5.

———. "Maharasthra's Plan to Divert Krishna Opposed by Mysore, Andhra." August 20, 1964b, 1.

———. "Mysore Wants Equitable Supply of Krishna Waters." June 3, 1961c, 7.

———. "New Board for DVC." February 18, 1954, 11.

———. " 'No Reopening of 1951 Pact on Krishna Waters." July 30, 1961d, 5.

———. "No Award Is Proposed." May 4, 1961e, 9.

———. "Opposition Moves Fail in Assembly." March 26, 1963d, 8.

———. "Proposal from Centre on River Dispute Expected." March 20, 1964c, 11.

———. "Sharing Waters of Krishna, Godavari." April 25, 1963e, 5.

———. "States to Discuss Krishna Award." December 27, 1973, 6.

———. "Upper Krishna Project." April 23, 1966b, 8.

———. "Use of Krishna, Cauvery Rivers." January 11, 1961g, 5.

Times of India News Service. "A.P. Panel for Krishna Waters." *Times of India*, March 26, 1989a, 7.

———. "AP Cannot Violate Krishna Award." *Times of India*, August 24, 1985a, 9.

———. "AP Stake in Krishna Conceeded." *Times of India*, December 11, 1990a, 7.

———. "Abritration Is Preferred: Krishna Waters Dispute." *Times of India*, October 7, 1967a, 1.

———. "Arbitration on Narmada, Krishna Disputes Likely." *Times of India*, April 17, 1968a, 1.

———. "Demand for Completion of Upper Krishna Project." *Times of India*, August 9, 1970, 6.

———. "Deve Gowda Plea on Krishna Waters." *Times of India*, June 21, 1989b, 7.

———. "Diverting Surplus Water of Godavari to Krishna." *Times of India*, November 29, 1963a, 1.

———. "Fears Allayed on Krishna Waters." *Times of India*, July 24, 1985b, 7.

———. "Financial Issues Stall Transfer of DVC System: Parleys in Calcutta." *Times of India*, October 16, 1963b, 3.

———. "Future Set-up of DVC: Talks in Delhi Next Week." *Times of India*, June 22, 1964a, 6.

———. "Headquarters of DVC: Bihar Assembly Demand." *Times of India*, May 11, 1953, 9.

———. "High-Level Inquiry Demanded: Functioning of DVC." *Times of India*, November 20, 1959, 1.

———. "Karnataka Ex-CMs to Meet on Krishna Basin." *Times of India*, August 4, 1992a, 7.

———. "Karnataka Has not Stopped Work on Krishna Project." *Times of India*, March 28, 1998b, 10.

———. "Karnataka Objects to AP's Charges on Funds for Upper Krishna Project." *Times of India*, November 5, 1996, 8.

———. "Karnataka Urges Inclusion of Upper Krishna Project in Ninth Plan." *Times of India*, October 29, 1997, 9.

———. "Koyna Has Made Excess Use of Krishna Waters: Rao." *Times of India*, December 7, 1967b, 13.

———. "Krishna: Mysore Will not Wait Beyond October." *Times of India*, July 22, 1967c, 1.

———. "Krishna Project Will 'Fragment' Forest." *Times of India*, August 10, 1998a, 6.

———. "Krishna Water Used in Excess by Maharashtra." *Times of India*, March 7, 1968b, 10.

———. "Krishna Waters Will Reach Madras on Jan. 1." *Times of India*, December 15, 1995, 7.

———. "Maharashtra May Get More Krishna Waters." *Times of India*, February 7, 1964b, 1.

———. "Maharashtra May Invoke 1956 Act." *Times of India*, August 5, 1963c, 1.

———. "Mysore Demands More of Krishna Waters." *Times of India*, March 30, 1964c, 1.

———. "Mysore Stands Firm on Krishna Waters Dispute." *Times of India*, March 21, 1963d, 7.

———. "Mysore Talks on Krishna Waters Tomorrow." *Times of India*, June 2, 1966, 11.

———. "Mysore Thinking of Legal Steps on Krishna Issue." *Times of India*, July 19, 1967d, 1.

———. "Mysore Threatens to Boycott Talks on Krishna Waters Issue." *Times of India*, September 25, 1967e, 1.

———. "Mysore Will Ask Centre to Settle Krishna Dispute." *Times of India*, July 20, 1967f, 4.

———. "New Scheme for DVC Is Unacceptable to W. Bengal." *Times of India*, August 13, 1965, 10.

———. "Oppn. Muddies Water over Krishna Project." *Times of India*, March 18, 1999, 5.

———. "P.M. Urges Quick Accord on Krishna, Narmada Issues." *Times of India*, April 23, 1968c, 1.

———. "PM Asked to Visit Krishna Projects." *Times of India*, August 19, 1986, 16.

———. "SC Tells States not to Rush to Media on Krishna Water Issue." *Times of India*, January 20, 2000a, 7.

———. "Scrap Damodar Valley Act: Bihar Panel." *Times of India*, September 6, 1992b, 8.

———. "Settling Dispute over Krishna Waters." *Times of India*, December 12, 1962, 5.

———. "State Stakes Claim to Excess Water from River Krishna." *Times of India*, March 25, 2000b, 5.

———. "Storm over Tenughat Dam Issue." *Times of India*, March 15, 1973, 10.

———. "Threat to Break Open Bunds on Krishna." *Times of India*, October 26, 1972, 7.

Tir, Jaroslav, and Douglas Stinnett. "Weathering Climate Change: Can Institutions Mitigate International Water Conflict?" *Journal of Peace Research* 49, no. 1 (2012): 211–25.

Toset, Hans, Nils Gleditsch, and Havard Hegre. "Shared Rivers and Interstate Conflict." *Political Geography* 19, no. 8 (2000): 971–96.

Treisman, Daniel. *The Architecture of Government: Rethinking Political Decentralization.* New York: Cambridge University Press, 2007.

Tsebelis, George. *Veto Players: How Political Institutions Work.* Princeton, NJ: Princeton University Press, 2002.

Tsui, Kai-yuen, and Youqiang Wang. "Between Separate Stoves and a Single Menu: Fiscal Decentralization in China." *China Quarterly* 177 (2004): 75.

Turton, A. R., and L. Ohlsson. *Water Scarcity and Social Stability: Towards a Deeper Understanding of the Key Concepts Needed to Manage Water Scarcity in Developing Countries.* London: Water Issues Study Group, School of Oriental and African Studies, 2000.

Turton, Anthony, Peter Ashton, and Eugene Cloete. "An Introduction to the Hydropolitical Drivers in the Okavango River Basin." In *Transboundary Rivers, Sovereignty and Development: Hydropolitical Drivers in the Okavango River Basin*, edited by Anthony Turton, Peter Ashton, and Eugene Cloete, 9–30. Pretoria, South Africa: African Water Issues Research Unit, 2003.

Tvedt, Terje. "About the Importance of Studying the Modern History of the Countries of the Nile Basin in a Nile Perspective." In *River Nile in the Post-Colonial Age: Conflict and Cooperation Among the Nile Basin Countries*, edited by Terje Tvedt, 1–12. London: I. B. Tauris, 2009.

Uitto, Juha, and Alfred Duda. "Management of Transboundary Water Resources: Lessons from International Cooperation for Conflict Prevention." *Geographical Journal* 168, no. 4 (2002): 365–78.

United Nations Food and Agriculture Organization. "India." Aquastat. 2011. Accessed November 11, 2014. http://www.fao.org/nr/water/aquastat/countries_regions/IND/index.stm.

———. *Sources of International Water Law: Some General Conventions, Declarations, and Decisions Adopted by International Organizations, International Non-Governmental Institutions, International and Arbitral Tribunals, on International Water Resources.* Rome: Food and Agriculture Organization of the United Nations, 1998.

———. FAOLEX Database: Water. Accessed January 8, 2010. http://waterlex.fao.org/waterlex/srv/en/home.

United Nations General Assembly. "United Nations Millennium Declaration A/RES/55/2." September 18, 2000. Accessed January 3, 2010. http://www.un.org/millennium/declaration/ares552e.pdf.

U.S. Army Corps of Engineers. *The US Army Corps of Engineers: A History.* Washington, DC: Corps of Engineers Historical Office, 2007.

U.S. Bureau of Land Management. "California Water Rights Fact Sheet." Western Water Laws. August 15, 2001. Accessed February 17, 2012. http://www.blm.gov/nstc/WaterLaws/pdf/California.pdf.

U.S. Bureau of Reclamation. The Colorado River. December 2008. Accessed February 17, 2012. http://www.usbr.gov/lc/hooverdam/faqs/riverfaq.html.

———. "Upper Colorado River Basin Compact, 1948." October 11, 1948. Accessed July 12, 2012. http://www.usbr.gov/lc/region/pao/pdfiles/ucbsnact.pdf.

U.S. Department of the Interior. *Official Report of Proceedings Before the Department of the Interior in the Matter of: Delaware River Compact Meeting.* Official Report of Proceedings, 4–13. Washington, DC: U.S. Government Printing Office, 1961.

U.S. Department of the Interior. "United States and Mexico Celebrate Partnership for Historic Release of Colorado River Water to Delta, Benefitting Both Nations." Press release, March 27, 2014. Accessed December 16, 2014. http://www.doi.gov/news/pressreleases/united-states-and-mexico-celebrate-partnership-for-historic-release-of-colorado-river-water-to-delta-benefitting-both-nations.cfm.

U.S. General Accounting Office. *River Basin Commissions Have Been Helpful, but Changes Are Needed.* Gaithersburg, MD: U.S. General Accounting Office, 1981.

U.S. House Committee on Irrigation and Reclamation. *Hearings Before the Committee on Irrigation and Reclamation, House of Representatives, Seventieth Congress, First Session, on HR 5773.* Washington, DC: U.S. Government Printing Office, 1928.

U.S. Joint Publications Research Service. *Water Resources of China: Plans and Prospects for Development.* Washington, DC: Clearinghouse for Federal Scientific and Technical Information, 1965.

U.S. Public Works Administration. *Report of the Mississippi Valley Committee of the Public Works Administration.* Federal Emergency Administration of Public Works. Washington, DC: U.S. Government Printing Office, 1934.

U.S. Senate Committee on Interior and Insular Affairs. *Hearings Before the Subcommittee on Irrigation and Reclamation on S.1658 a Bill to Authorize, Construct and Maintain the Central Arizona Project and for Other Purposes.* Senate Committee Report. Washington, DC: U.S. Government Printing Office, 1963.

———. *Hearings Before the Subcommittee on Water and Power Resources of the Committee on Interior and Insular Affairs—Central Arizona Project.* Washington, DC: U.S. Government Printing Office, 1967.

U.S. Senate Committee on Irrigation and Reclamation. *Boulder Canyon Project Adjustment Act. Hearings Before the Committee on Irrigation and Reclamation, United States Senate, Seventy-Sixth Congress, Third Session, on S. 4039.* Washington, DC: U.S. Government Printing Office, 1940.

Vaccaro, Ismael. "Modernizing Mountain Water: State, Industry, and Territory." In *Water, Place, and Equity,* edited by John Whiteley, Helen Ingram, and Richard Perry, 225–48. Cambridge, MA: MIT Press, 2008.

Vaidya, Abhay. "Krishna Valley Irrigation Scheme Makes Rapid Progress." *Times of India.* February 28, 1999, 6.

Vaidyanathan, A. *Water Resource Management Institutions and Irrigation Development in India.* New Delhi, India: Oxford University Press, 1999.

Valsalan, V. M. *Inter-State Water Disputes in India: A New Approach.* New Delhi, India: Central Board of Irrigation and Power, 1997.

Van der Heijden, Hein-Anton. "Political Opportunity Structure and the Institutionalization of the Environmental Movement." *Environmental Politics* 6, no. 4 (1997): 25–50.

Voigt, William. *The Susquehanna Compact.* New Brunswick, NJ: Rutgers University Press, 1972.

Volden, Craig. "Intergovernmental Political Competition in American Federalism." *American Journal of Political Science* 49 (2005): 327–42.

Vorosmarty, Charles, Pamela Green, Joseph Salisbury, and Richard Lammers. "Global Water Resources: Vulnerability from Climate Change and Population Growth." *Science* 289, no. 5477 (2000): 284–88.

Walch, George. *The Engineering Works of the Kistna Delta: A Descriptive and Historical Account*. Vol. 1. Madras, India: Superintendent, Government Press, 1899.

Walton, Brett. "Prompted by Scarcity, Colorado River Basin States Examine Their Lifeline." Circle of Blue. January 24, 2011. Accessed December 16, 2014. http://www.circleofblue.org/waternews/2011/world/prompted-by-scarcity-colorado-river-basin-states-examine-their-lifeline/.

Wang, Weiluo. "Water Management in the People's Republic of China." In *Conflict Management of Water Resources*, edited by Manas Chatterji, Saul Arlosoroff, and Gauri Guha, 88–115. Aldershot, UK: Ashgate, 2002.

Wang, Yahua. "Diagnosis and Recommendations for Transjurisdictional Water Pollution Management in China." *Frontiers of Environmental Science and Engineering in China* 1, no. 4 (2007): 401–17.

———. "Water Governance and State Governance: A Transaction Cost Perspective of China's Unitarianism." In *Workshop in Political Theory and Policy Analysis*. Bloomington, IN: Indiana University, 2009.

New York Times. "Reagan Vetoes a Bill to Settle Arizona Water Rights Dispute." June 2, 1982.

———. "Machine Guns Back Arizona Fight on Dam." November 11, 1934.

Waterbury, John. *Hydropolitics of the Nile Valley*. New York: Syracuse University Press, 1979.

———. *The Nile Basin: National Determinants of Collective Action*. New Haven, CT: Yale University Press, 2002.

Watts, Ronald. *Administration in Federal Systems*. London: Hutchinson Educational, 1970.

———. *Comparing Federal Systems*. Montreal: McGill-Queen's University Press, 2008.

Weaver, R. Kent, and Bert Rockman. "Assessing the Effects of Institutions." In *Do Institutions Matter? Government Capabilities in the United States and Abroad*, edited by R. Kent Weaver and Bert Rockman, 1–41. Washington, DC: Brookings Institution Press, 1996.

Weber, Edward. *Bringing Society Back In: Grassroots Ecosystem Management, Accountability, and Sustainable Communities*. Cambridge, MA: MIT Press, 2003.

Webster, E. J. "Arizona Worried by Water Compact." *New York Times*, January 6, 1935, E6.

Wei, Yehua. *Regional Development in China: States, Globalization and Inequality*. London: Routledge, 2000.

Weingast, Barry. "The Economic Role of Political Institutions: Market-Preserving Federalism and Economic Development." *Journal of Law, Economics, and Organization* 11, no. 1 (1995): 1–31.

Weinthal, Erika. *State-Making and Environmental Cooperation: Linking Domestic and International Politics in Central Asia*. Cambridge, MA: MIT Press, 2002.

Wheare, K. C. *Federal Government*. Oxford: Oxford University Press, 1963.

Wibbels, Erik. "Madison in Baghdad? Decentralization and Federalism in Comparative Politics." *Annual Reviews in Political Science* 9 (2006): 165–88.

Wines, Michael, and Jennifer Medina. "Farmers Try Political Force to Twist Open California's Taps." *New York Times*, December 30, 2015, A1.

Wirsing, Robert. "Hydro-Politics in South Asia: The Domestic Roots of Interstate River Rivalry." *Asian Affairs: An American Review* 34, no. 1 (2007): 3–22.

Wittfogel, Karl. *Oriental Despotism: A Comparative Study of Total Power*. New Haven, CT: Yale University Press, 1957.

Wolf, Aaron. *Hydropolitics Along the Jordan River: Scarce Water and Its Impact on the Arab–Israeli Conflict*. Tokyo: United Nations University Press, 1995.

Wood, John. *The Politics of Water Resource Development in India: The Narmada Dams Controversy*. New Delhi, India: Sage Publications, 2007.

World Bank. *Water Resources Sector Strategy: Strategic Directions for World Bank Engagement*. Washington, DC: World Bank, 2004.

World Water Assessment Programme. *United Nations World Water Development Report 3: Water in a Changing World*. Paris and London: UNESCO/Earthscan Publishers, 2009.

Worster, Donald. *Rivers of Empire: Water, Aridity, and the Growth of the American West*. New York: Oxford University Press, 1985.

Wouters, Patricia, Desheng Hu, Jiebin Zhang, Philip Andrews-Speed, and Dan Tarlock. "The New Development of Water Law in China." *University of Denver Law Review* 7, no. 2 (2004): 243–308.

http://local.xinhuanet.com/dfzf/2006-10/26/content_14237.htm.Huanghe Shuili Weiyuanhui [Yellow River Conservancy Commission]. "'Shuzi Huanghe' gongchengjihua" ['Digital Yellow River' Project Plan]. Huanghe Shuili Weiyuanhui [Yellow River Conservancy Commission]. April 24, 2003. Accessed December 16, 2017. http://www.yellowriver.gov.cn/zwzc/lygh/200612/t20061222_75480.html

———. "Huanghe chunxiu gongcheng quanmian zhankai" [Yellow River spring repair work gets going]. *Renmin Ribao* [People's Daily], March 31, 1951.

———. "Trans-Provincial Compensation Mechanism: A New Path for China's Water Environment Protection." January 30, 2012. Accessed December 16, 2017. http://www.china.org.cn/environment/2012-01/30/content_24507199.htm.

Xinjiang Weiwuer Zizhiqu Dangwei Xuanquanbu Zhuban [Xinjiang Uighur Autonomous Region Party Committee Propaganda Office]. "Xibudakaifa de zhanluexing gongcheng—Daliushu shuili shuniu gongcheng" [Strategic projects of the Great Western Development Plan—the Daliushu Water Conservancy Project]. Tianshan Wang [Tianshan Net]. August 25, 2009. Accessed April 3, 2012. http://www.tianshannet.com/special/content/2007-09/25/content_2199913.htm.

Xu, Gaohong, and Shenglian Guo. "Changjiang liuyu shengji heliuzhong shuishi jiufen chongtu yu duice cuoshi" [Inter-provincial water disputes and conflicts in the Yangtze River basin and potential countermeasures]. *Shuiziyuan Baohu* [Water Resources Protection] 2 (2007): 54–56.

Yang, Dali. *Remaking the Chinese Leviathan: Market Transition and the Politics of Governance in China*. Stanford, CA: Stanford University Press, 2004.

Yang, Hong, and Shaofeng Jia. "Meeting the Basin Closure of the Yellow River in China." *International Journal of Water Resources Development* 24, no. 2 (2008): 265–74.

Yoffe, Shira, and Aaron Wolf. "Water, Conflict and Cooperation: Geographical Perspectives." *Cambridge Review of International Affairs* 12, no. 2 (1999): 197–213.

Yohe, Gary, and Richard S. J. Tol. "Indicators for Social and Economic Coping Capacity—Moving Toward a Working Definition of Adaptive Capacity." *Global Environmental Change* 12, no. 1 (2002): 25–40.

Young, Oran. *The Institutional Dimensions of Environmental Change: Fit, Interplay, and Scale*. Cambridge, MA: MIT Press, 2002.

Zarour, Hisham, and Jad Isaac. "Nature's Apportionment and the Open Market: A Promising Solution to the Arab–Israeli Water Conflict." *Water International* 18, no. 1 (1993): 40–53.

Zawahri, Neda, and Oliver Hensengerth. "Domestic Environmental Activists and the Governance of the Ganges and Mekong Rivers in India and China." *International Environmental Agreements: Politics, Law and Economics* 12, no. 3 (2012): 269–98.

Zeitoun, Mark, and J. A. Allan. "Applying Hegemony and Power Theory to Transboundary Water Analysis." *Water Policy* 10, no. 2 (2008): 3–12.

Zeitoun, Mark, and Naho Mirumachi. "Transboundary Water Interaction I: Reconsidering Conflict and Cooperation." *International Environmental Agreements* 8 (2008): 297–316.

Zetland, David, and Christopher Gasson. "A Global Survey of Urban Water Tariffs: Are They Sustainable, Efficient and Fair?" *International Journal of Water Resources Development* 29, no. 3 (2013): 327–42.

Zhang, Bailin. "Kuaxingzheng quyu shuiwuran jiufen jiejue jizhi yanjiu" [Trans-jurisdictional water pollution dispute resolution mechanism research]. Wuhan Daxue Huanjingfa Yanjiusuo [Wuhan University Environmental Law Research Institute. November 5, 2009. Accessed December 16, 2017. http://aff.whu.edu.cn/riel/article.asp?id=30420

Zhang, Chuncheng. "Huanghe Sanmenxia Daba Gongcheng de Shehui Zeren" [The social responsibility of the Yellow River Sanmenxia High Dam Project]. In *Huanghe yu Henan Luntan* [Forum on Henan and the Yellow River], edited by Yuguo Niu, 265–74. Zhengzhou, China: Huanghe Shuili Chubanshe [Yellow River Conservancy Press], 2008.

Zhang, Jiong, Jia Zhuang, Wanan Lai, and Zhang Qiang. "Zhujiang liuyu kuasheng heliu shuishi maodun ji chuli cuoshi chutan" [Initial research on transboundary water disputes in the Pearl River basin and measures for handling them]. *Renmin Zhujiang* [People's Pearl River] 1 (2010): 7–9.

Zhao, Jilin, and Jincan Wang. "Zunyishi, Xishuixian Cifaju jianli maodun jiufen kuasheng diaochu hezuo jizhi" [Zunyi City and Xizhui County courts establish an inter-provincial cooperation mechanism]. Guizhousheng Renmin Zhengfu Fazhi Bangongshi [Guizhou Provincial People's Government Legal Affairs Office]. April 21, 2009. http://www.gzfzb.gov.cn/html/zunyi/667/.

Zhao, Laijun. *Woguo liuyu kuajie shuiwuran jiufen xiediao jizhi yanjiu: yi Weihe liuyu weili* [Research on coordination mechanisms for transboundary river basin pollution disputes in China: An example from the Wei River basin]. Shanghai, China: Fudan University Press, 2007.

Zheng, Yongnian. *De Facto Federalism in China: Reforms and Dynamics of Central–Local Relations*. Singapore: World Scientific Publishing, 2007.

Zhongguo Jiancai Wang [China Building Materials Network]. "Saima shuini jiang shouyi Daliushu Gabai kaijian" [Saima Cement will benefit from the construction of Daliushu High Dam]. Zhongguo Jiancai Wang [China Cement Network]. February 24, 2011. Accessed December 2, 2012. http://www.c-bm.com/news/2011/2-24/B125952181.shtml.

Zhongguo minzhu tongmeng Guangdongsheng weiyuanhui [China Democratic Alliance Guangdong Provincial Committee]. "Guanyu jiakuai tuijin Zhujiang liuyu kuasheng (qu) shuiziyuan baohu yu shuiwuran fangzhi xiezuo jizhi jianshe de jianyi" [Recommendations on speeding up establishment of an inter-provincial water pollution prevention and cooperation mechanism]. Zhongguo minzhu tongmeng Guangdongsheng weiyuanhui [China Democratic Alliance Guangdong Provincial Committee]. February 4, 2016. http://www.gdmm.org.cn/UserData/DocHtml/2016/2/4/4582059560035.html.

Zhonghua Renmin Gongheguo Zhongyang Renmin Zhengfu [Central People's Government of the PRC]. "Jiedu: Guojia tufa gonggong shijian zongti yingji yuan" [Understanding national critical public incident coordinated emergency planning].

Central People's Government. January 8, 2006. Accessed November 8, 2012. http://www.gov.cn/zwhd/2006-01/08/content_151018.htm.

Zhou, Junqian. "Huanghe shangzhongyou shuitibaochi zongzuo jinzhan huanman" [Progress on Yellow River upper and middle reaches soil conservation work is slow]. *Renmin Ribao* [People's Daily], June 5, 1957.

Zhou, Xuegang. "Unorganized Interests and Collective Action in Contemporary China." *American Sociological Review* 58, no. 1 (1993): 54–73.

Zimmerman, Joseph. *Horizontal Federalism: Interstate Relations*. Albany: State University of New York Press, 2011.

———. *Interstate Water Compacts: Intergovernmental Efforts to Manage America's Water Resources*. Albany: State University of New York Press, 2012.

INDEX

Note: Figures and tables are indicated by an italic *f* and *t* following the page number.